Mechanosensing and Mechanochemical Transduction in Extracellular Matrix

Frederick H. Silver

Mechanosensing and Mechanochemical Transduction in Extracellular Matrix

Biological, Chemical, Engineering, and Physiological Aspects

Foreword by Stephen C. Cowin

 Springer

Dr. Frederick H. Silver
Professor of Lab Medicine and Pathology
UMDNJ-Robert Wood Johnson Medical School
675 Hoes Lane, Piscataway, NJ 08854
fhsilver@hotmail.com

Cover illustration: Diagram illustrating binding of a cell to a collagen fibril through specific integrins attached to the cell membrane. The integrin subunits are α and β.

Library of Congress Control Number: 2005930804

ISBN-10: 0-387-25631-8   e-ISBN 0-387-28176-2
ISBN-13: 978-0387-25631-3

Printed on acid-free paper.

Printed in the United States of America.   (BS/MVY)

9 8 7 6 5 4 3 2 1

springer.com

Foreword

This volume presents a lucid description and timely organization of our present understanding of how the various tissues and organs of the body sense their mechanical environment and pass on this sensory information. In our early education one learns of our five senses (seeing, hearing, smelling, tasting, and touching). Each of these senses is associated with a specialized type of cell that is a transducer of environmental signals and which also transmits this sensory information to the brain. Beyond these five senses there is a larger community of transducer cells in our tissues that communicate their sensory readings to local tissue- or organ-based control systems rather than to the brain. These sensory cells are usually components of multiply connected cellular networks. These networks maintain the health and function of the tissues and organs and serve as local control systems for the tissue or organ's function. Most collagenous tissues, including bone and tendon, are examples of such tissues; organ examples include the kidney and the digestive system. There are many more examples described in this volume. The multiply connected cellular networks in these tissues and organs control the remodeling of the tissues to accommodate changes in their mechanical environment.

The production of the materials for the construction of tissues is one of the better-understood topics concerning tissue formation. Cell and molecular biologists have documented the manufacturing of most of the tissue constituents by cells within or near the tissue. Cells not only manufacture these tissue constituents, but they also have the sensory apparatus for maintaining the tissue and adapting the tissue structure to changed environments, including mechanical load environments. We know that tissues and organs are created in fairly repeatable structural patterns, patterns that are due to both the genetic information and the mechanical environment, but we do not know exactly what percentages of a particular pattern are due to either one of these two factors. We do not know much about the beginning of tissue construction (morphogenesis), but we do know something about the self-assembly methods by which tissues construct themselves and they are described in this volume. When the tissue adapts its structure to accommodate a new mechanical environment, we do not know how the decision is made for the structural adjustment. We do know that tissues and organs grow or reconstruct themselves simultaneously with their active performance of their normal function, but we do not understand the processes that permit tissues and organs to accomplish this. This volume provides a knowledge base for the people who will explore these and other related issues. These are people who think mechanistically about the relations between cell activity and tissue structure. These people generally have a background in engineering or biophysics, but also there is a growing community of biologically educated people who think mechanistically, rather than descriptively, about biological processes.

Stephen C. Cowin
New York Center for Biomedical Engineering Departments of
Biomedical and Mechanical Engineering, the City College, New York

Preface

Many years ago I was told by a sage in the field of collagen that we knew all that we needed to know about the physical nature of collagen. I was surprised at the response and still to this day wonder why anyone would go into research who thought there was nothing new to learn. The purpose of this book is to tell biologists, chemists, physiologists, engineers, and anyone who listens, that just the opposite exists. There is a rich world of new information that is waiting to be explored on mechanotransduction. To hope to achieve an understanding of the processes involved we need scientists of all backgrounds and experiences. I hope this book acts as a stimulus to attract any and all that can contribute to this field. I also urge that those who are interested not let all the terminology scare them from contributing to this field. We need more self-motivated people to answer the fundamental physical questions about mechanotransduction. This only will be achieved by promoting better interaction among individuals in life sciences, chemistry, physiology, and engineering. Although there has been at lot of discussion of interdisciplinary education and research we need more emphasis on problem solution rather than adding to the unbounded data that exist in the literature. Much of the information on mechanotransduction cannot be compared because the model systems used are different in many studies. We need to stop emphasizing the funding of interesting observations and pay more attention to the development of more generalized studies that emphasize how systems work. This is essential to prolongation of life and the amelioration of diseases. We have reached a limit in successfully applying trial-and-error solutions to complex healthcare problems.

We spend our whole lives working against a gravitational field that contributes to the failure of our host systems. However, we really don't understand how the gravitational field affects biological systems. The purpose of the approach used is to attempt to illustrate how basic chemistry and physics can be applied to try to understand some basic phenomena that are observed. At no point in the book do I mean to imply that the subject is simple. The simple approach used only is meant to be a starting point for more complex analyses. It is my hope that this text will stimulate new thinking about how man has evolved over a period of several million years and the role of gravitational forces in the development and maturation of the human host.

Acknowledgments

I would like to take this opportunity to thank all my students and colleagues who have provided inspiration and guidance to pursue this project. I had the opportunity to be part of the Harvard and MIT educational systems during the 1970s and early 1980s at a time when much of the emphasis was on learning and thinking. Much of the philosophy of that period has influenced my continued interest in the interface between the biochemical and physical aspects of life. I am particularly thankful to my last group of graduate students at Rutgers, Pat Snowhill, Paul Seehra, Istvan Horvath, Gino Bradica, and Joseph Freeman, for helping me explore the final pieces to the puzzle that has fascinated me for two decades. I would also like to thank them for helping me prepare many of the diagrams on mechanochemical transduction and to thank Dr. David Christiansen for preparing many of the other diagrams used in this text. They have inspired me to complete this book under conditions that would challenge any academic. I hope that at least one reader will be as stimulated by this material to continue the pursuit of knowledge in this field as I have been.

Contents

1
Introduction to Mechanochemical Transduction in Tissues

1.1 Background to Tissue Structure and Mechanochemical Transduction

Life as we know it on earth reflects the influence of gravity on our development, maturation, and aging. Although the molecular biological revolution has attracted an enormous amount of attention in the press and in the scientific literature, with the hope that gene manipulation will cure many diseases, we have only recently learned of the influence of gravitational and other external forces on expression of genes and cellular and tissue function. As we show later in this book, the cellular control mechanisms by which environmental influences affect cell behavior are similar for mechanical forces as well as other external influences. This suggests that altering the external forces acting on tissues may modify gene expression. We indirectly manipulate gene expression every day through exercise and space flight, yet we fail to consider in depth the potential power of this approach in modulating growth and development, wound healing, and aging. When I was a young scientist, just a few years ago, one of my colleagues told me that we knew everything about fibrous collagen that was important. This was long before we had a suspicion that fibrous collagen is one of the key players in transmitting forces between the external environments and cells. The goal of this text is to introduce to the young scientist and engineer the concepts necessary to begin to understand how external and internal forces modulate cellular processes in tissues.

Mechanical forces play a role in the development and evolution of tissues found in the human body. Gravitational forces acting on mammalian tissues increase the net muscle forces required for movement of vertebrates. As body mass increases during development, musculoskeletal and other tissues are able to adapt their size to meet the increased mechanical requirements. However, the control mechanisms that allow for rapid growth in tissue size during development are altered during maturation and aging. The role of mechanical forces in controlling tissue growth, development, aging, and disease processes is an important subject in the design and development of implants that are needed to restore the function of damaged tissues. Without an understanding of the influence of these forces on tissue and cellular metabolism, the design of synthetic implants becomes an art as opposed to a science.

From the point of view of biomedical engineering, the human body is composed of a structural framework to which the various organs and tissues are attached. The functional units include the heart, lungs, gastrointestinal (GI) system, brain, and other organs and the structural framework includes the musculoskeleton and connective tissue matrix that holds the organs together and allows for locomotion. The organs must work in concert to ensure the proper metabolite levels and blood flow are maintained whereas the musculoskeleton allows for packaging these organ units into a system that can move itself from place to place. Without the organ units acting in a cooperative fashion, life as we know it could not be sustained; in the absence of a movable framework we would be dependent on our environment for transportation to other locations. The structural framework is necessary to perform a number of functions including: (1) supporting the organ systems against destruction, (2) organizing individual units of cells into tissues and organs, (3) forming connections and conduits between different tissues and organs, (4) storing elastic energy produced during locomotion, and (5) transducing external and internal mechanical loads into changes in gene expression and protein synthesis. The structural framework is termed connective tissue or extracellular matrix and is found in either soft (nonmineralized) or hard (mineralized) forms. One purpose of this text is to study the structure of the extracellular matrix that is found in soft and hard connective tissue, and to examine the mechanisms by which mechanical forces are transduced into changes in cellular function. Understanding the relationship between external and internal mechanical loading and the resulting turning on and off of genes during development, aging, and disease is key to improving healthcare and prolonging life. We show later that all cells appear to be under the influence of both internal and external forces and that the balance of these forces, the net force, may be important in dictating whether tissue is formed or destroyed.

Extracellular matrix (ECM), composed of collagen and elastic fibers, proteoglycans, is the ubiquitous substrate for cell adhesion found throughout the human body. It serves as the stimulus for cell growth and differentiation, and it adapts to changes in external mechanical loading to provide mechanical support to tissue. It is well known that connective tissue cells adapt their ECM to changes in externally applied mechanical loads during wound healing and development. For this response to occur, a feedback mechanism must exist by which cells that sense mechanical stress via their substrate respond by altering patterns of protein expression, thus remodeling their ECM to meet changing mechanical requirements. In addition to their ability to adapt to externally applied loads, cells have the ability to generate their own internal forces through the production of cytoskeletal tension. These externally applied forces and internal cytoskeletal forces appear to be integrated with other environmental signals, which are then transduced into a biochemical response in the cell cytoplasm and nucleus. Thus both developmental biology and human homeostasis involve constant integration of external and internal mechanical signals into biochemical processes that take place in cells. It is the balance between these mechanical stimuli and biochemical processes that is maintained during normal homeostasis.

1.1.1 Significance

Development of new healthcare technologies as well as the rapid diagnosis of disease requires an integrated understanding of chemistry, biology, medicine, and engineering, because the various biochemical pathways involved in homeostasis are linked to tissue structure and function. Mechanical and biochemical feedback loops exist that integrate mechanical and chemical events that are required for development and homeostasis. These mechanical and chemical feedback loops are altered during aging and maturation and along with preprogrammed cell death lead to initiation of disease processes. For example, using echocardiography, the reflection of ultrasonic waves at interfaces between the inner and outer walls of blood vessels or the heart is used to diagnose aortic dilatation, progression of atherosclerotic lesions, cardiac hypertrophy, and the potential need for drug or device intervention. Mechanical forces play a major role in progressive changes seen in blood vessels that are involved in the disease process. Mechanical forces play an important role in tissue homeostasis of musculoskeletal tissues as evidenced by the bone resorption and muscle atrophy experienced by astronauts, as well as in disease processes such as the deposition of lipids and the progression of atherosclerosis. Therefore, it is not only important to be able to measure the structure and normal function of tissues, but it is important to be able to understand the relationship between external mechanical loading to tissues and the resultant changes to gene expression and protein synthesis.

Almost every phase of biomedical engineering involves the measurement, replacement, or modification of tissues and organs. For this reason it is essential that workers in biology, medicine, and biomedical engineering understand the relationship among biochemical pathways, mechanical loading, tissue structure, and normal function. The purpose of this text is to integrate this material in order to provide a conceptual framework needed to understand how external forces affect tissue metabolism.

It is now recognized that epithelial, endothelial, and a variety of other "parenchymal" cell types respond to external mechanical loading by changing expression of certain genes and regulating synthesis of several types of macromolecules. Although many biological products are affected by mechanical loading we limit our discussion to the effects of external mechanical loading on changes in cell cytoplasmic structure and changes to the extracellular matrix surrounding these cells. This requires that we have an in-depth understanding of cell and ECM macromolecular structure.

1.1.2 Background Definitions

There are a number of terms that must be defined in order to proceed with introducing the concepts explored in this book. The human body is composed of a variety of tissues that includes organs such as the lungs, heart, liver, and spleen which provide both biochemical and biophysical functions such as detoxifying chemicals found in the blood (liver) and filtering expired cells from the blood (spleen), providing exchange of CO_2 and O_2 (lungs), and circulating blood through the cardiovascular system (heart). Although each of these organs contains different cell types with different organizations, the extracellular matrix that is composed of cells, collagen, and elastic fibers, proteoglycans and cell attachment factors form the structural support. Thus, the human body is composed of extracellular matrix (ECM) containing macromolecules that form a continuous interface with the surrounding cells. The ECM can be highly aligned with large collagen fibers such as those found in tendon; in this case it is termed dense regular connective tissue. It can also be loosely woven as is found in the top layer of skin and termed loose connective tissue. In organs, the ECM forms what is termed the parenchyma, the scaffold of the tissue, and the cells form the functional units. ECM is also found in conduits such as blood and lymphatic vessels, planar sheets such as the pleural and peritoneal membranes that separate different anatomical structures, and in dense and loose connective tissue found in musculoskeletal tissue and in surface lining structures (epithelial and mucosal tissue line the inner and outer surfaces of body).

Fifty years ago, ECM was considered to be a passive scaffold for cells; today we are aware that ECM–cellular communication occurs through mechanical force transduction that affects gene expression and ECM synthesis and that mechanochemical transduction is a key element in modern human physiology and molecular biology.

Connective tissue and components of the ECM provide more than structural support to cells that form organs and other tissues of the body. The interactions between these tissues that occur during development and growth can only be understood after we analyze the structure and properties of each of the components. There are numerous types of cells found in the ECM that continuously form and remodel the material found in their extracellular matrix. Material outside the cell, termed extracellular matrix, consists of collagen fibrils and fibers that form a continuous network (see Table 1.1). Collagen fibers contain other nonfibrous materials that not only form bridges between collagen fibers but also form attachments to the cell membrane. As diagrammed in Figure 1.1, collagen fibrils bind directly to cells via specific attachment molecules termed integrins. In addition, cells have binding sites for other macromolecules such as fibronectin that is

TABLE 1.1. Location and composition of connective tissue components

Component	Composition	Building blocks (repeat unit)	Location
Collagen	Protein	Amino acids	Found in banded extracellular fibrils, basement membranes and thin fibrils extracellularly
Elastin	Protein	Amino acids	Structureless component of elastic fibers
Microfibrillar protein	Protein	Amino acids	Fibrous core of elastic fibers
Proteoglycans	Protein core & large side chains (glycosaminoglycans)	Amino acids and disaccharides	Associated with collagen fibers
Hyaluronan	Polysaccharide	Disaccharides	Found between collagen fibers
Fibronectin	Glycoprotein	Amino acids and monosaccharides	Attachment factor between collagen fibers and fibroblasts
Laminin	Glycoprotein	Amino acids and disaccharides	Attachment factor, component of basement membrane

found in the ECM loosely associated with collagen fibers. Different cell types manufacture a number of integrins; each type of integrin is specific to a particular macromolecule found in ECM.

The fibrous components of tissues are analogous to nylon fibers used to construct fabrics in the clothing industry or the steel wire that is an integral part of the belt that prevents blowouts in automobile tires. In ECM the fibers are composed of collagen, a protein containing three polypeptide chains, elastic tissue containing an amorphous polypeptide termed elastin, and a fibrillar component termed microfibrillar protein or fibrillins. Collagen fibers in ECM prevent overexpansion and failure as does the steel wire in tires; however, in addition, collagen fibers transmit stresses in tissues to cells that are transduced into chemical signals that help maintain the homeostasis of these tissues. Elastic fibers are found primarily in tissues including vessel wall and skin where they contribute to tissue shape recovery after tissue unloading.

ECMs contain a variety of cell types (see Table 1.2) such as fibroblasts in the skin, chondrocytes in cartilage, and osteoblasts in bone. The role of these cells is to synthesize and deposit the ECM surrounding the cell and to change the amount and location of the ECM in response to trauma and changes in external loading. For instance, collagen synthesis in these cells is typically up-regulated by the application tensile forces and is down-

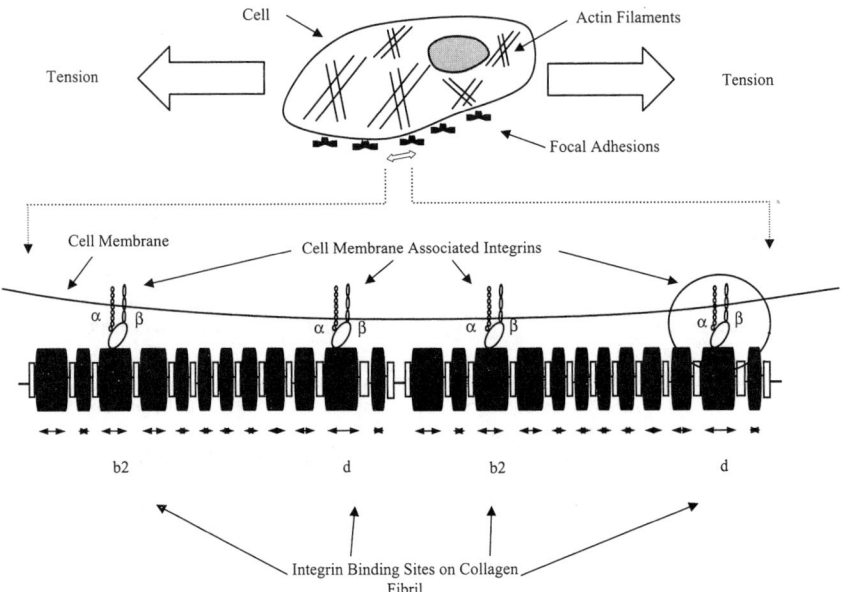

FIGURE 1.1. Diagram illustrating binding of a cell to a collagen fibril through specific integrins attached to the cell membrane. The integrin subunits are α and β.

regulated by the application of compression. These cells form attachments to collagen and elastic fibers as well as to cell attachment factors such as fibronectin and hyaluronan. External forces borne by collagen and elastic fibers found in ECM are transferred to cells via cell surface integrins; some of the external loads are dissipated via connections among the cell surface and fibronectin and hyaluronan located between collagen and elastic fibers. Fibronectin is a glycoprotein that is involved in mediating the attachment to collagen fibers; cells can directly bind to collagen via integrin receptors. Laminin is a component of basement membranes and facilitates binding of epithelial cells to type IV collagen networks that are also present in

TABLE 1.2. Cells found in connective tissue

Cell	Location
Chondrocytes	Cartilage
Epithelial	Cover exposed body surface; skin
Endothelial	Line inside of blood vessels
Smooth muscle	Found in vessel walls
Fibroblasts	Found in skin, tendon and cardiovascular tissue

basement membranes. Basement membranes form the interface between epithelial and endothelial cells and the surrounding ECM. To complicate matters even more, there are about twenty types of collagen found in ECM and several types of proteoglycans. The fiber-forming collagens consist of types I, II, and III and form the structural support for most tissues (see Table 1.3). Other classifications of collagens include the fibril-associated collagen with interrupted triple helices (FACIT) and the nonfibril-forming collagens. The specific roles of these collagens are the subject of much speculation.

Historically the material between collagen fibers in tissues has been referred to as interfibrillar matrix. The major components of the inter-fibrillar matrix include proteoglycans, hyaluronan, and water (see Table 1.1) and serve to bind the fibers together and prevent interfiber friction. In skin, the collagen fibers undergo alignment and slippage during mechanical loading that allows skin to bend over joints and absorb impact loads. The proteoglycans that are found between the collagen fibers are important in supporting elastic energy storage as well as in dissipating applied loads. Proteoglycan degradation in the ECM is the first change seen that is associated with the development of osteoarthritis and loss of articular cartilage. This loss precedes wearing away of the cartilage surface that leads to limited locomotion due to excessive pain.

Throughout this text we refer to the relationship between external mechanical loading and cellular response that affects the physiology of ECM. Inasmuch as mechanical terms are used repeatedly, some of these terms are listed in Table 1.4. These terms refer to the ability of tissue to change its shape and to deform. Strain refers to the changes in tissue dimensions with respect to the initial dimensions, and extensibility refers to the strain a tissue can undergo before failure. The force per unit cross-sectional area of a tissue is referred to as stress; the stress at which a tissue tears is

TABLE 1.3. Location and tissue source of some of the different collagens

Collagen type	Tissue or organ	Location
I	Tendon, skin, bone and fascia	Thick extracellular fibrils and fibers
II	Cartilage	Thin fibrils around cartilage cells
III	Cardiovascular tissue	Intermediate size extracellular fibrils
IV	Basement membranes	Network forming component
V	Tendon, skin and cardiovascular tissue	Pericellular matrix around cells
VI	Cardiovascular tissue, placenta, uterus, liver, kidney, skin, ligament, and cornea	Extracellular matrix
VII	Skin	Anchoring fibrils
VIII	Cardiovascular tissue	Around endothelial cells
IX	Cartilage	Extracellular matrix
X	Cartilage	Extracellular matrix
XI	Cartilage	Extracellular matrix

TABLE 1.4. Definitions of mechanical terminology used in text

Terms	Definitions
Strain	change in length/initial length
Extensibility	strain at failure
Stress	force/cross-sectional area
Stiffness	change in stress/change in strain
Toughness	strain energy absorption: area beneath stress-strain curve

referred to as the ultimate strength. Stiffness is a measure of the ratio of change in stress divided by the change in strain and is the resistance a material offers when a strain is applied.

1.2 Cell and ECM Macromolecular Structure

All cells found in ECM are complex organizations of macromolecules. They are arranged in such a manner as to allow small molecules and ions to diffuse into the cell and for these components to be synthesized into macromolecules used in forming structural materials within and outside the cell. Macromolecules are made up of long chains of repeat units that are connected end to end in the form of proteins, nucleic acids, lipids, and polysaccharides.

1.2.1 Cellular Components

Cells are composed of proteins, nucleic acids, lipids, ions, and water. These materials are discussed further in Section 2.1 and are organized into functional compartments that include the cell and nuclear membranes, cytoplasm, nucleus, and organelles (Figure 1.2). The organelles that are important in this book include endoplasmic reticulum (ER), mitochondria, ribosomes, Golgi apparatus, and lysosomes (Table 1.5). Cellular changes associated with mechanical loading or due to the presence of an implant are studied by evaluation of the cell structure before and after loading or after contact with a surface. Changes in cellular components such as changes in the staining pattern or swelling are the first indication that cell injury is occurring and that cell death may follow.

The normal cell has an outer layer with a chemical structure similar to fat that has been esterified. It is found in the form of a cell membrane that separates the fluid and tissue outside a cell from the organelles inside the cell. The cell or plasma membrane is a lipid bilayer that contains proteins on the inner and outer surfaces (Figure 1.3). The function of the cell membrane is to: (1) maintain ionic and chemical concentration gradients; (2) carry specific surface markers and receptors such the human leukocyte antigens, growth factors, integrins, and hormone receptors; (3) participate in intracellular communication; (4) regulate cell growth and differentiation; and (5) participate in the transduction of external and internal mechanical signals.

Figure 1.3 shows a diagram of the fluid mosaic model of the cell membrane that incorporates a discontinuous lipid bilayer composed of phospholipids and cholesterol. Proteins and glycoproteins are embedded or attached to the membrane in this model. Peripheral proteins, such as the glycoprotein fibronectin, are found attached to the cell membrane whereas integral proteins including the proteoglycan syndecan are found at least partially embedded in the cell membrane. Peripheral proteins are involved in maintaining cellular shape and transducing mechanical forces across the cell membrane and integral proteins are involved in cell recognition and transport of molecules across the cell membrane. Transplantation of tissues from one unrelated host to another involves matching the cell surface antigens in order not to stimulate immune responses that lead to rejection. The cell surface antigens are major histocompatibility markers found attached to the cell surface.

The cell membrane is involved in interaction between cells allowing for the flow of ions and electrical impulses between neighboring cells. Cell-to-cell attachment via specific types of specialized junctions allows for the exchange of proteins and ions. One example is that of epithelial cells that exhibit gap junctions, tight junctions, and desmosomes. Some of these junctions can be mechanically active and contract after a mechanical loading.

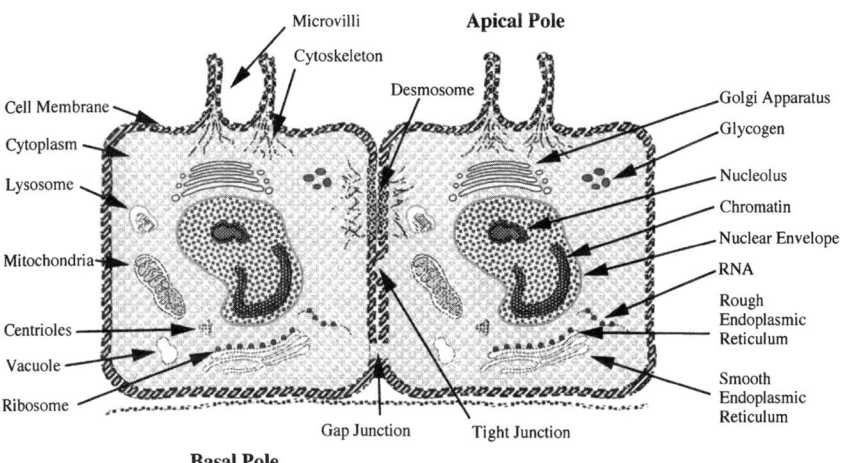

FIGURE 1.2. Diagram of cell components. Generalized diagram showing components of cell including ER, mitochondria, ribosomes, Golgi apparatus, lysosomes, and other components.

TABLE 1.5. Components of the normal cell and their functions

Cell component	Composition	Function
Cell membrane	Lipid bilayer, containing surface proteins (peripheral proteins), proteins totally embedded in the membrane (intregal proteins), and glycoproteins partially embedded in the membrane	Maintains ionic and chemical concentration gradients, cell-specific markers, intercellular communication, regulates cell growth and proliferation
Cytosol or cytoplasm	Water, ions, soluble proteins	Contains enzymes and structures for generation of energy (ATP[a]) in the absence of oxygen (TCA[b] cycle), activation of amino acids, carrying out specialized cell functions
Endoplasmic reticulum	Membrane-enclosed channels	Involved in transport of proteins for extracellular secretion and modification or detoxification of chemicals
Mitochondria	Contains membrane-lined channels to which enzymes are attached that generate ATP from glucose	Involved in TCA cycle, respiratory chain, and oxidative phosphorylation
Ribosomes	Small and large subunits composed of ribosomal RNA; strands of messenger RNA form a complex with large subunits	Involved in synthesis of enzymes and structural components and proteins for extracellular release
Golgi apparatus	Membrane-lined tubular system that forms stacks	Packaging of proteins in vesicles for extracellular release
Lysosomes	Membrane-lined vescicles containing hydolytic enzymes	Involved in breakdown to intracellular and extracellular material
Nucleus	Membrane-limited area of cell containing nucleolus and chromatin	Site of synthesis of RNA and chromatin, involved in cell division

[a] ATP = adenosine triphosphate.
[b] TCA = tricarboxylic acid cycle.

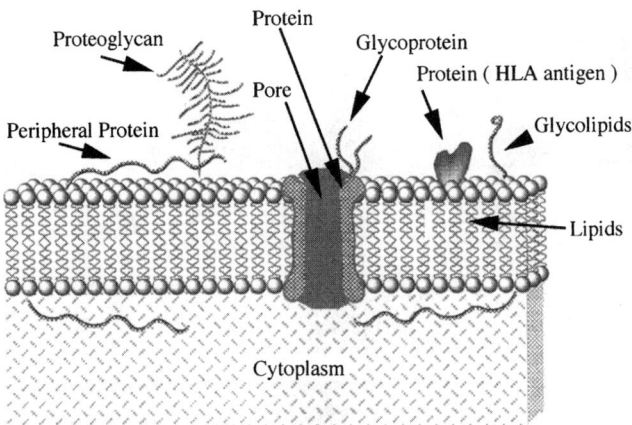

FIGURE 1.3. Diagram of cell membrane. The plasma membrane is a lipid bilayer containing peripheral proteins, including HLA antigens, proteoglycans, and integral proteins, such as pore-forming proteins that transverse the cell membrane.

The contacts between two adjoining cell membranes (see the gap junctions in Figure 1.2) are stabilized by specific cell adhesion molecules (CAMs), which include the Ca^{+2} dependent cadherins in adherence junctions, and by connexins in gap junctions. The connexins are components of the connexon channels between cells. Spontaneous contraction of neighboring cells leads to cellular tension and mechanical transmission of tensile forces through adherens junctions. Results of recent studies suggest that N-cadherins in fibroblasts transmit mechanical forces applied to adherens junctions by activating stretch-sensitive calcium-permeable channels, leading to an increased actin polymerization. Analogously, connexin 43 is present in gap junctions between cells including chondrocytes of knee cartilage, in growth plate, and between fibrocartilagelike cells at tendon and ligament insertions. Mechanical forces exerted at cell junctions can affect intercellular communications.

Typically, the integrity of the cell membrane is evaluated by identifying the percentage of cells taking up a vital dye termed trypan blue. Cells that take up trypan blue have defective cell membranes and therefore external loading or contact with materials that cause cells to have faulty membranes are likely to cause cell injury and death. Cell cytotoxicity is evaluated by contacting a surface of a material or the extract of a material or implant with cells in culture. If more than five percent of the cell population stains with dye after contact with the material or an extract then it is considered cytotoxic.

Cells normally exhibit polarity; that is, they have a top, bottom, left, and right sides. Most cells such as the epithelium present an apical pole that is characterized by many ruffles in the cell membrane termed microvilli and a basal pole that is in contact with a basement membrane. This polarity is very important because normal cell function can only be expressed if the cell has the correct orientation.

The material within the cell membrane is gellike and is termed the cytoplasm or cytosol. It is composed of ions, water, soluble proteins, and enzymes that are involved in generation of energy in the form of ATP by a process termed the tricarboxylic acid cycle (TCA) or Krebs cycle in the absence of oxygen (Figure 1.4). It is also involved in activation of amino acids for protein synthesis (Table 1.5 and Figure 1.4).

In addition to organelles, the cell cytoplasm contains actin filaments that make up the cellular cytoskeleton that controls shape. Myosin and α-actinin are also found in the cytoplasm and are believed to be involved in cell contraction. Other filaments including intermediate filaments, tubulin, calmodulin, and spectrin form networks within the cytoplasm that modify cell and organelle mobility and shape.

Endoplasmic reticulum (ER) is a branching system of membrane-limited channels that are found within the cytoplasm (Table 1.5 and Figure 1.2). These channels are 40 to 70nm wide and are enclosed by a membrane similar to the plasma membrane. In some cases the channels are covered

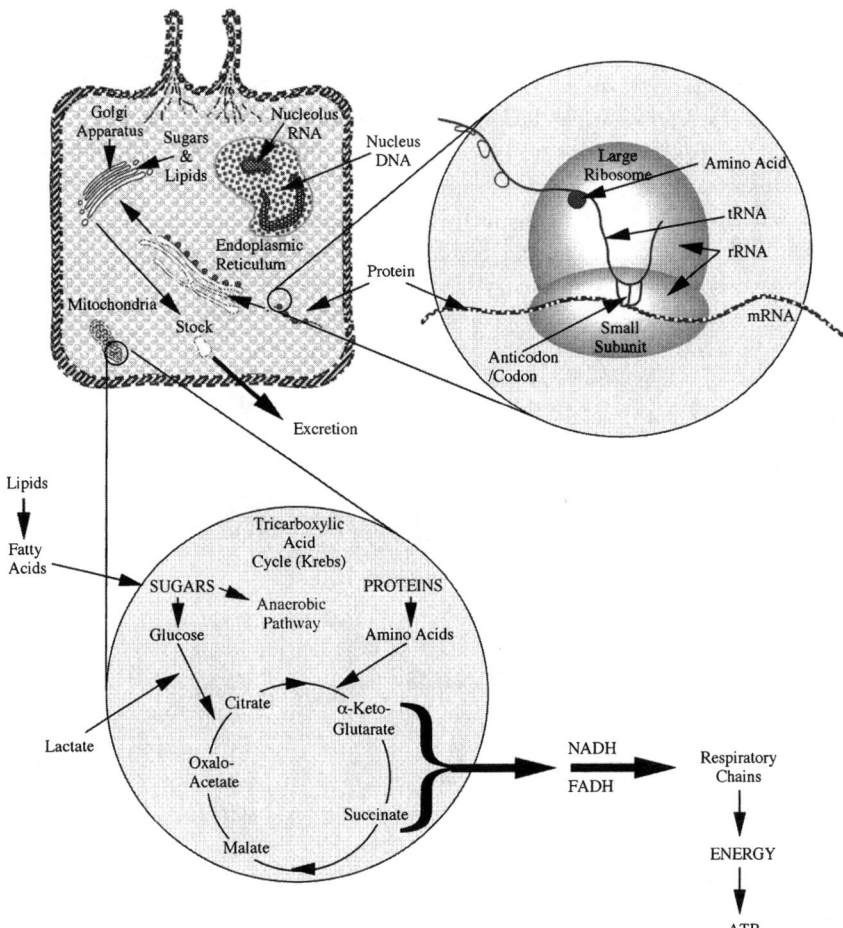

FIGURE 1.4. Krebs cycle and protein synthesis. This diagram illustrates the generation of ATP through the Krebs cycle and the protein synthesis that occurs within the cell cytoplasm.

with spheres 20 nm in diameter; this is termed rough ER. The spheres are ribosomes and are the site of the synthesis of proteins. The ER serves as a conduit for export extracellularly for the proteins synthesized within the cell cytoplasm. In the absence of ribosomes, the ER is termed smooth (SER). Proteins such as collagen are synthesized on the ribosomes, pass into the center of these channels, and are transported into the Golgi apparatus where they are packaged into vesicles for release from the cell. Membranes that line the SER are involved in modification and detoxification of low molecular weight materials that are released by implants. Implants or biomaterials that stimulate proliferation of the SER result in its proliferation leading to an increase in cell size. Increased cell size in the presence of an implant is an indication that the implant may cause irreversible cell changes. Active proliferation of the ER suggests that a cell is responding to increased mechanical loading.

Mitochondria are cigar-shaped organelles (Figure 1.2) that are separated from material found in the cell cytoplasm by a double membrane. Both inner and outer mitochondrial membranes are similar to plasma membranes. The inner membrane is connected to a series of folded channels (cristae) upon which the enzymatic reactions of the acid cycle (TCA), respiratory chain, and oxidative phosphorylation occur (Figure 1.4). These reactions are required to generate ATP from glucose in the presence of oxygen. Mitochondria are prevalent in cells that are actively secreting proteins such as collagen in response to increased external mechanical loading and in cells that are synthesizing enzymes required for modification or detoxification of chemicals.

Proteins are synthesized on ribosomes, which are made of small and large subunits containing nucleic acids. The small subunit (see Figure 1.4) is shaped like a donut that has been cut in half. The large subunit is spherical and contains a notched groove on the top surface. A strand of messenger ribonucleic acid (mRNA) is found between the notched groove of the large subunit and the hole in the center of the small subunit; mRNA in association with the large and small ribosomal subunits acts as a template for protein synthesis. Synthesis of enzymes and structural proteins used within the cell occurs on free ribosomes found within the cytoplasm. Proteins that are synthesized for release from the cell are synthesized on ribosomes attached to endoplasmic reticulum.

The Golgi apparatus (Figure 1.2) is a membrane-bound system of tubes that is connected to the endoplasmic reticulum. The tubular system is in the form of individual tubules that make up a winding system of cisternae that form stacks termed a dictyosome. Proteins synthesized on ribosomes attached to the inner membrane of the endoplasmic reticulum are transported into the Golgi apparatus where they are packaged into vesicles. Vesicles 40 to 80nm in diameter at the outside of the cisternae are used to release these proteins extracellularly during wound healing and during bone formation. Extracellular matrix cells synthesize large quantities of collagen and other proteins and therefore the Golgi apparatus is prominent. Coincidental with synthesis of new proteins is the removal of old proteins via a process termed phagocytosis.

Phagocytosis is a process by which foreign or old autologous proteins are ingested by the cell and then removed by fusion with lysosomes; these structures are vesicles containing hydrolytic enzymes. Lysosomes (Figure 1.5) are membrane-bound vesicles that are used to break down proteins, nucleic acids, sugar polymers, and other materials that are either extracellular or

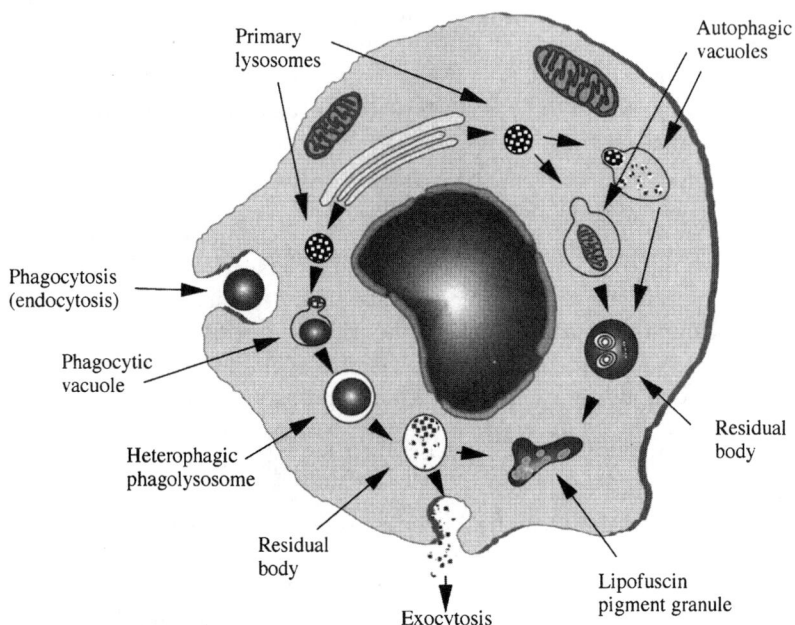

FIGURE 1.5. Lysosomes and phagocytosis. Dead and foreign material are removed via phagocytosis and degradation of macromolecular material using lysosomal enzymes.

intracellular. Lysosomes are 0.1 to 0.8 μm in diameter and contain a variety of hydrolytic enzymes. Cells rich in lysosomes include polymorphonuclear leukocytes (neutrophils), monocytes, and macrophages. These white cells are involved in the inflammatory process to remove dead cellular debris and damaged extracellular matrix. Excessive external mechanical loading or implantation of a medical device leads to trauma and cell death that stimulates migration of inflammatory cells and phagocytosis. Inflammatory cells clean up the debris by attempting to eat the material including the damaged collagen or an implant. When tissues are unloaded some of the cellular and noncellular materials are removed by phagocytosis.

The nucleus of the cell (Figure 1.2) is composed of a porous nuclear membrane, the nucleolus, and soluble materials. The nucleolus contains ribonucleic acids (RNA) and genetic materials also termed chromatin that code for the proteins synthesized upon the ribosomes in the cell cytoplasm. The nuclear membrane is continuous with the outer membrane of the endoplasmic reticulum. Messenger RNA synthesized in the nucleus is transported across the nuclear membrane and is involved in protein synthesis. It fits into the groove between the large and small rRNA subunits (Figure 1.2) where it acts as a template for tRNA (transfer RNA) to add the appropriate amino acids to the growing protein chain. mRNA is synthesized from a DNA template that is in the form of a double helix (Figure 1.6). DNA in turn is associated with proteins termed histones and the complex makes up the chromosomes or genetic material. There are 23 sets of chromosomes containing all the genetic material (DNA) required to synthesize all the proteins found within the human body. Chromosomes within the

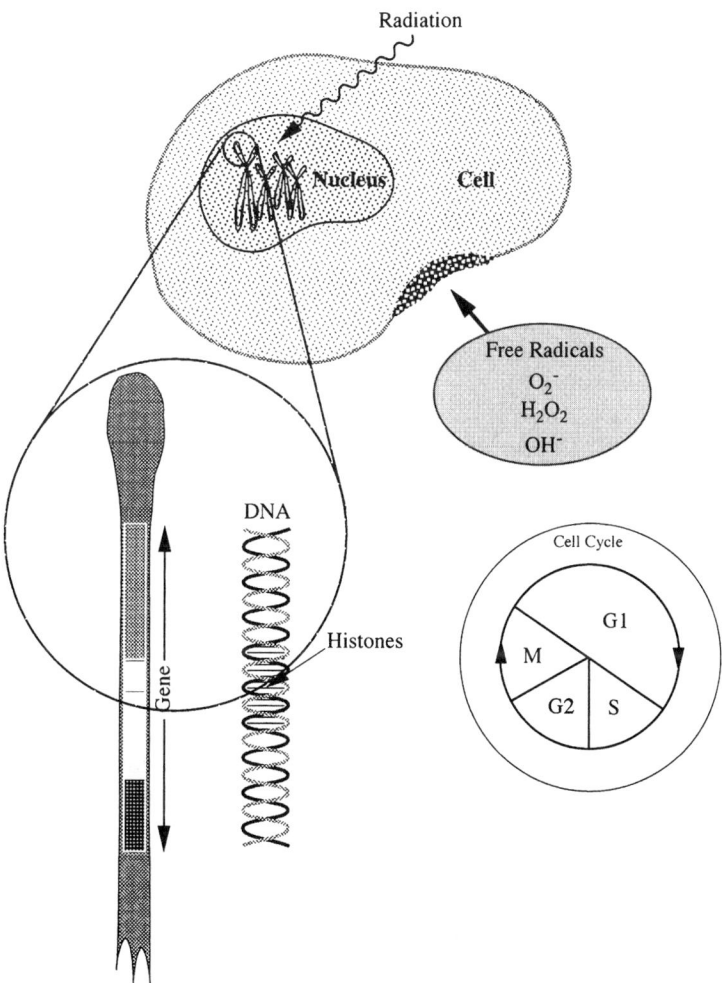

FIGURE 1.6. DNA, histones, and cell division. Shown are (1) the structure of DNA, histones and genetic material; (2) changes in genetic material via radiation or free radical induced damage; and (3) the cell cycle, synthesis of DNA and mitosis.

cell are characterized by the size and shape of their long and short legs. They all look like the letter x with varying lengths and geometries. External mechanical forces are transduced into changes in expression of genes found on chromosomes located in the nucleus and lead to a variation in the expression of mRNA that is synthesized into proteins in the cytoplasm.

Ribosomal RNA is synthesized and packaged within the nucleolus of the cell. The nucleic acids that are synthesized within the nucleus pass through pores in the nuclear membrane into the cytoplasm. The nucleolus appears as a dense spot in the nucleus under the light microscope after staining with special dyes. The staining characteristic of the chromatin, which contains the genetic material of the cell, is somewhat different. Chromatin is acidic and therefore stains darkly with basic dyes. Loosely coiled chromatin (euchromatin) stains lightly with basic dyes whereas tightly coiled DNA (heterochromatin) stains darkly with basic dyes. Increases in external mechanical loading result in increased amounts of heterochromatin. Resting cells exhibit increased amounts of euchromatin. The nuclear staining characteristics can be useful in determining the activity of cells that are exposed to changes in mechanical loading. Division of cells within the capsule surrounding an implant indicates that the capsule thickness is likely to increase and that the implant is causing a reaction. Contraction of the tissue that surrounds an implant appears to be stimulated by mechanochemical transduction by fibroblasts found in the capsule.

During cell division (see M phase in Figure 1.6) heterochromatin is separated and arranged into distinct clumps of genetic material, the chromosomes. The karyotype is a fingerprint of the genetic material and is obtained by preventing termination of cell division. This allows for observation of the size and shape of the genetic material found in any cell that is useful to determine if contact with an implant or external mechanical loading causes changes in the genetic material termed a mutation. Cellular replication can be associated with an implant in both positive and negative senses. In a positive sense wound healing is associated with cell replication and synthesis of nucleic acids and proteins. In a negative sense cell replication in wounds after remodeling has occurred is reflective of an abnormal scarring process. Therefore, the analysis of cell replication around an implant must be done in conjunction with other information concerning the cell to be correctly interpreted.

1.2.2 Macromoleular Structure

The principal building blocks of tissues are large molecules referred to as macromolecules or polymers. Without these large molecules life as we know it would not be possible because these moieties are responsible for the completion of most biological processes. Biological macromolecules can be broken down into four classes of large molecules, namely, proteins, polysaccharides (sugar polymers), nucleic acids, and lipids.

The differentiation among these classes of large molecules is a result of differences in the repeat unit, that is, the chemical structure that is repeated over and over again to make a large chain. Another way of stating this fact is that the properties of long chains of repeat units linked together are very dependent on the chemistry of the chain. The physical properties of long chained molecules also depend on the rotational freedom about the backbone as diagrammed in Figure 1.7. It is very interesting to note that regardless of the exact chemistry of the backbone of a macromolecule the physical behavior is fixed. What this boils down to is that the modulus (stiffness) or resistance of a polymer to deformation is independent of the backbone chemistry. What is dependent on the backbone chemistry is the temperature at which a particular behavior is observed; that is, all polymers behave at some temperature like a rubber band (easily reversibly stretched). The temperature at which a polymer behaves like a rubbery material is termed the glass transition temperature. The glass transition temperature is affected by the chemistry of the repeat unit. The glass transition temperature is dependent on the chemistry of the repeat unit by virtue of how it influences the backbone flexibility. The relationship between the chemistry of the backbone of a polymer and its rubberiness is a bit more complex than just analysis of the backbone rotational freedom; however, that discussion is beyond the scope of this book.

Now if we consider a chain of carbon atoms similar to that observed in lipids and poly(ethylene) we can next ask the question about how this chain exists in three dimensions. We know from general chemistry that a chain of

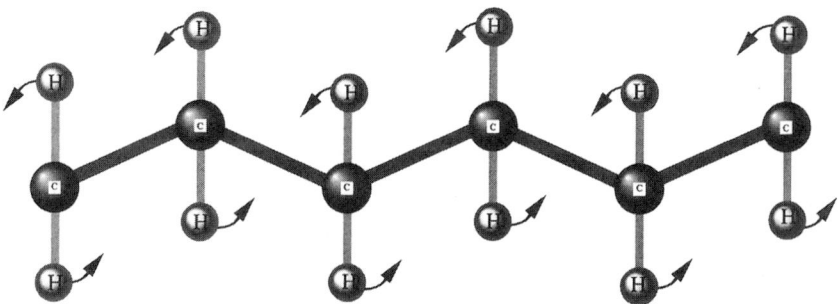

FIGURE 1.7. Mobility of hydrocarbon chain. The diagram shows mobility of a polymer chain composed of carbon atoms attached by single bonds with hydrogen side chains. Rotational freedom of the single carbon-to-carbon bonds allows hydrogen atoms to rotate freely about the backbone.

TABLE 1.6. Minimum interatomic contact distances[a]

Contact atoms	Contact distances normal, (minimum distance), Å
Carbon to Carbon	3.20 (3.00)
Carbon to Nitrogen	2.90 (2.80)
Nitrogen to Nitrogen	2.70 (2.60)
Carbon to Oxygen	2.80 (2.70)
Nitrogen to Oxygen	2.70 (2.60)
Oxygen to Oxygen	2.70 (2.60)
Carbon to Hydrogen	2.40 (2.20)
Hydrogen to Nitrogen	2.40 (2.20)
Hydrogen to Oxygen	2.40 (2.20)
Hydrogen to Hydrogen	2.00 (1.90)

[a] Distances and coordinates given in Å (10^{-10}m).

atoms that are covalently bonded together must have a fixed bond length and bond angle. Therefore our carbon chain must adhere to these principles and all carbon to carbon bonds must have a bond length of about 1.5 Å and a bond angle of 108° (please check a general chemistry book for those of you who do not follow this discussion). Now let us consider our carbon chain with fixed bond angles; it turns out that the carbon chain because of the bond angle easily forms a planar zigzag in 3-D (see Figure 1.7). When the hydrogen atoms are attached to the backbone we must next consider how these are arranged in space.

It turns out that the allowable conformations of a polymer chain can be determined by considering atomic bond angles and bond lengths and the fact that the distance between two nonbonded atoms must not be closer than the sum of the van der Waals radii (see Table 1.6). Each conformation, which is a set of dihedral angles within the backbone, can be assessed by making a computer model knowing the atomic coordinates of all atoms in the repeat unit and then translating and rotating a repeat unit through the available combinations of dihedral angles. This can be done using vector mathematics and the resulting conformational plot gives the allowable values of the backbone angles for a polymer. This plot is known as Ramachandran plot (see Figure 1.8) for proteins and is modified by interactions between side chains and between polymer molecules. The conformational freedom of a macromolecule is related to the flexibility and the mechanical properties through the relationship between the number of available conformations and Boltzmann's constant k for chains that deform without interchain and intrachain interactions. In this manner chain flexibility and stiffness under these conditions is related to the number of available conformations of a macromolecule.

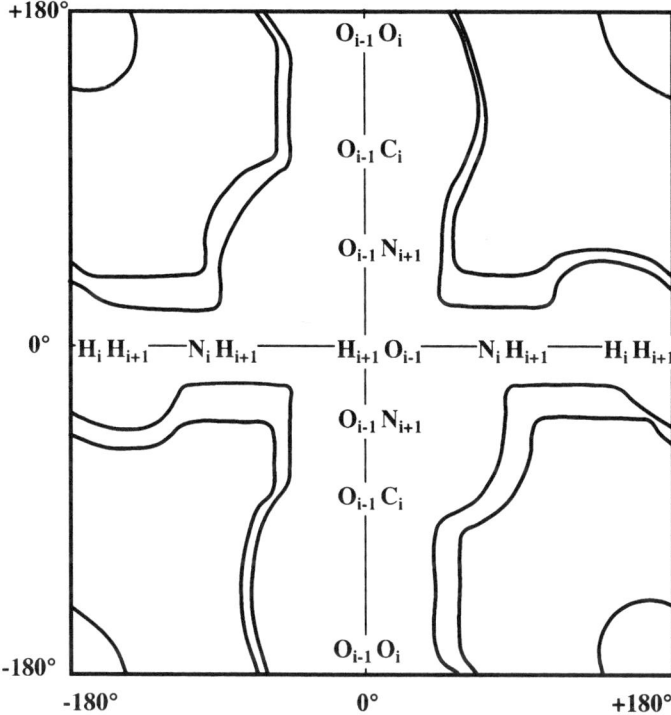

FIGURE 1.8. Allowed conformations for a dipetide of glycine. Plot of fully (outer solid lines) and partially (inner solid lines) allowed combinations of ψ (vertical axis) and ϕ calculated using a hard sphere model with normal (outer solid lines) and minimum (inner solid lines) atomic contact distances. Dihedral angles (ϕ, ψ) in center of plot (0°, 0°, 0°, 180° and 180°, 180°) are unallowed because of contacts between backbone atoms in the neighboring (ith, ith + 1 and ith − 1) peptide units. The contacts between nitrogen (N), hydrogen (H), oxygen (O), and carbonyl groups (CO_2H) prevent allowed conformations in the center of this diagram. This diagram can be constructed using coordinates of the atoms within the peptide unit and inter-atomic distances (see Tables 2.2 and 2.3). To construct this diagram a standard dipeptide unit (see Figure 2.8) is formed by translating the first peptide unit along the line between the α carbons and the flipping the second unit into the *trans* configuration (Figure 2.9). The values of ϕ and ψ were varied using matrix multiplication, and the interatomic distances were checked for all nonbonded atoms. Pairs of ϕ and ψ that have allowable interatomic distances are found between sets of solid lines shown as the allowable conformations.

1.3 Mineralized Versus Nonmineralized Tissues

Although part of the human body is made up of cells and soft collagenous tissues, much of the musculoskeletal tissue is made up of collagen fibers that are impregnated with calcium, phosphate, hydroxyl, and carbonate groups in the form of hydroxyapatite. Although it is not clear how mineralization is turned on and off in human tissues, it is clear that mechanical loading and mechanochemical transduction are important stimuli for mineralization. In the case of mineralized tissues there are a number of noncollagenous proteins found in addition to collagen. They include osteopontin, bone sialoprotein, osteocalcin, osteonectin, and phosphoryn and their role in bone mineralization is not known. What is known is that the major building blocks for bone are similar to other tissues in that collagen fibers, cells, cell attachment factors, water, and ions are the major components.

It is intriguing to try to analyze if differences in mechanochemical transduction mechanisms could explain why some tissues mineralize normally and others do not. This would make it possible to attempt to analyze why heart valves and vessel walls mineralize abnormally in diseases leading to valvular failure or cardiac insufficiency.

1.4 Cell Cytoskeleton, Extracellular Matrix, and Mechanochemical Tranduction

It has been hypothesized that forces exerted by the extracellular matrix on cells may be in equilibrium with forces exerted by cells on the extracellular matrix. Ingber proposed that forces are transmitted to and from cells through the extracellular matrix with changes in mechanical forces and cell shape acting as biological regulators (see Silver et al., 2003 for a review). He further hypothesized that cells use a tension-dependent form of architecture, termed tensegrity, to organize and stabilize their cytoskeleton. Mechanical interactions between cells and their extracellular matrix appear to play a critical role in cell regulation by switching cells between different gene products.

Extracellular matrix components are linked to cytoskeletal elements via cell surface receptors termed integrins. Integrins have been implicated in mediating signal transduction through the cell membrane in both directions. Integrin adhesion receptors are heterodimers of two different subunits termed α and β (Figure 1.1). They contain a large extracellular matrix domain responsible for binding to substrates, a single transmembrane domain, and a cytoplasmic domain that in most cases consists of 20 to 70 amino acid residues. They mediate signal transduction through the cell membrane in both directions: binding of ligands to integrins transmits signals into the cell and results in cytoskeletal reorganization, gene expression, and cellular differentiation (outside-in signaling); on the other side, signals within the cell can also propagate through integrins and regulate integrin–ligand binding affinity and cell adhesion (inside-out signaling).

Eukaryotic cells attach directly to extracellular matrix collagen fibers via integrin subunits $\alpha 1\beta 1$ and $\alpha 1\beta 2$ through a six-residue sequence (glycine-phenylalanine-hydroxyproline-glycine-glutamic acid-arginine) that is present in the b2 band of the collagen positive staining band (Figure 1.1). Integrins are transmembrane molecules that associate via their cytoplasmic domains with a number of cytoplasmic proteins including vinculin, paxillin, tensin, and others, which are all involved in the dynamic association with actin filaments. In cultured cells, integrin-based molecular complexes form small (0.5 to 1 μm) or point contacts known as focal adhesion complexes (see Figure 1.1) and elongated streaklike structures (3 to 10 μm long). The elongated structures are associated with actin and myosin containing filament bundles (stress fibers) known as focal contacts or focal adhesions. Results of recent studies suggest that integrin-containing focal complexes behave as mechanosensors exhibiting directional assembly in response to local force. It has been reported that collagen-binding integrins are involved in down-regulating collagen synthesis and up-regulating enzymes that degrade collagen when fibroblasts are grown in a relaxed collagen gel.

Effects of mechanical forces have been studied on a variety of cell types including isolated fibroblasts and fibroblasts cultured in a collagen matrix. Fibroblasts cultured on flexible-bottom surfaces coated with components of ECM including fibronectin, laminin, or elastin were observed to align perpendicular to the force vector. Mechanically loaded cells grown on laminin or elastin or other substrates expressed higher levels of procollagen mRNA and incorporated more labeled proline into protein than did unstressed cells. Fibroblasts in cell culture that are not aligned with the force direction show a severalfold increase in collagen-degrading enzymes suggesting that cells that are unable to align with the direction of the applied load remodel their matrix more rapidly than oriented cells.

Fibroblasts grown in a three-dimensional collagen lattice have been shown to align themselves with the direction of principal strain and adopt a synthetic fibroblast phenotype characterized by induction of connective tissue synthesis and inhibition of matrix degradation. Fibroblasts grown in collagen lattices can generate a force of approximately 10 N as a result of a change in cell shape and attachment. Fibroblasts have the ability to maintain a tensional homeostasis of approximately 40–60×10^{-5} N per million cells. Cell contraction of 3-D collagen matrices was observed to be opposite to the direction of applied loads. Increased external loading is followed immediately by a reduction in cell-mediated contraction.

It is clear from these studies that external mechanical loading not only affects the cellular response through changes in genetic expression by cells, but it also sets into motion a cascade of events that balances cellular biochemical and biomechanical responses with changes in the environmental mechanical loading. Whether this occurs such that human development can meet the evolving physical demands of the environment or this occurs to minimize body mass if the environmental demands are met, is unclear. It is clear that tissue and organ size may in part be a consequence of the environmental mechanical demand on the organism.

1.5 Internal Stresses in Tissues

Extracellular matrices (ECMs) found in musculoskeletal, cardiovascular, and dermal tissue are all under tension under normal physiologic conditions even in the absence of external loading. This tension not only fulfills cosmetic functions (e.g. smooth skin is much more appealing than wrinkled skin), it also sets up a state of dynamic mechanical loading at the collagen fibril–cell interface that stimulates mechanochemical transduction. Mechanochemical transduction as defined in this book is the effect of stresses and strains on the expression of genes and the regulation of cellular protein synthesis that result from changes in mechanical loading. By definition, at equilibrium all external forces acting on tissues and organs and collagen fibrils must sum to zero. In addition, increases in external loading result in increases in internal stresses acting on collagen fibrils and at the collagen fibril–cell interface. Beyond this effect, the observation that cells grown in collagen lattices exert a contractile force suggests that under normal physiologic conditions cells apply tension to the attached ECM. This tension not only leads to dynamic active stresses that are applied to the collagen network, but also to incorporation of passive tension in the collagen fibrils during development and maturation of tissue scaffolds.

The forces that operate in skin have been studied extensively for over 100 years and serve as an example of the existence of passive stresses existing in the collagen fibrils found in skin ECM. It is well established that the collagen fibrils are laid down approximately parallel to Langer's lines, which characterize the direction of principal tension in skin. In 1862, Langer found that circular holes punched in the skin of cadavers became elliptical, with the major axis of the ellipse directed along the direction of maximum tension (see Silver et al., 2003 for a review). In the absence of external loading, cadaver skin in which the cells are not viable is under a biaxial tension; the magnitude of the tension varies from location to location, with the tension highest in the skin from the limbs. Beyond the passive tension that exists in the skin of cadavers, in living skin there are both passive and active tensional forces. A circular punch defect made in living skin increases in size (a 2.0 mm biopsy site increases to at least 3.0 mm depending on the location) and remains approximately circular after the skin plug is removed.

The circular nature of the skin defect in living skin, as opposed to the elliptical defect in cadaver skin, underscores the ability of skin fibroblasts to maintain approximately equal tension in both directions within the plane of the skin. This suggests that the active tension exerted by skin cells occurs over and beyond the passive tension in the collagen network within skin, thus supporting the conclusion that mechanochemical transduction at the collagen fibril–cell interface must regulate tissue tension.

The concept that both active and passive stresses exist internally within ECM is important because any external stress that is applied affects the balance between active cellular-induced stresses and passive stresses transferred through the collagen fibrils of the matrix. At equilibrium any increase in external loading transmitted through the collagen fibrils found in the ECM must lead to an increase in the active stress produced by the attached cells. This increased active stress, or the resulting increase in cell strain, may cause conformational changes in membrane components or in cell–cell junctions that activate cellular pathways that modify gene expression and protein synthesis. Of particular interest is how mechanical loading may up-regulate expression of genes required for production of collagen and the resulting synthesis of new ECM and the down-regulation of genes that control production of collagenolytic enzymes that degrade the ECM.

Other examples of tissues exhibiting active and passive tension include cornea, cardiovascular tissue, and cartilage. When a corneal transplant is trephined out from a cadaver eye, the corneal material to be transplanted shrinks from about 8.5 mm to about 8.0 mm as a result of unloading of the passive and active tensions that exist. In the cardiovascular system, passive and active stresses along the longitudinal and transverse directions of the vessel wall provide in situ strains that are as high as 50% in the carotid artery.

The basis of the active tension exerted by cells has been studied by growing a variety of cell types in collagen matrices. When isolated fibroblasts are grown in a reconstituted collagen matrix, they contract the matrix as a result of active tension exerted by the cells on the matrix. In addition, the cells respond differently when the matrix is stressed as opposed to unstressed. Fibroblasts cultured in collagen matrices not only actively contract the matrix; they also remodel it. These examples underscore the importance of passive and active stresses in mechanobiology of a variety of extracellular matrices and suggest that mechanical loading is somehow intrinsically related to genetic expression of the resident cells.

1.5.1 Internal Stresses Acting Within Cartilage

Several reports suggest that the curling or distortion of intact cartilage occurs when it is removed from the underlying bone because it is under tension. Removal of the superficial layer of rib cartilage leads to straightening of the curled cartilage suggesting that it is the superficial zone that is under tension. Removal of a circular defect from articular cartilage results in differential retraction of the edges depending on the depth of the defect; the edges retract more in the superficial zone as compared to the deeper zones after a circular defect is removed with a punch. This observation underscores the presence of active and passive tensile forces in articular cartilage. Normal human cartilage, with an intact superficial zone, curls similarly to rib cartilage when removed from the underlying bone. Osteoarthritic human cartilage, however, which lacks the superficial zone, doesn't exhibit curling when cartilage is removed from the underlying bone, indicating that the loss of active and passive tension is observed in tissues affected by this disease.

Interpretation of this curling phenomenon is complicated by observations of the swelling behavior of articular cartilage. Fry and Robertson suggested that this swelling is a potential mechanism for the existence of residual stresses and strains in cartilaginous tissues (see Silver et al., 2003 for a review). In one study, the swelling-induced residual strain at physiologic ionic strength was estimated to be tensile and vary from 3 to 15%. Increasing the salt concentration from physiologic levels to 2.0 M decreased the extent of curling, suggesting that breakage of electrostatic interactions decreases the curling phenomenon; these workers concluded that the surface zone greatly limits swelling of the entire cartilage.

Isolated chondrocytes grown in type I and type II collagen–glycosaminoglycan matrices contract the matrix and immunochemically stain for alpha-smooth muscle actin. These results suggest that chondrocytes can also generate active tension and may be responsible for maintaining the tension in articular cartilage.

1.6 Mechanical Properties of Tissues

Because ECMs store, transmit, and dissipate energy as part of their physiologic function, it is important to understand and characterize their mechanical properties. The mechanical properties of developing and adult tissues have been studied extensively. It is important to relate changes in mechanical properties of ECMs and mechanical loading to structural changes that are observed at the microscopic and gross levels.

For example, the mechanical properties of developing tendons rapidly change just prior to the onset of locomotion. The ultimate tensile strength (UTS) of developing chick extensor tendons increases from about 2 MPa at 14 days of development to 60 MPa 2 days after birth. This rapid increase in UTS is not associated with changes in fibril diameter, but is associated with increases in collagen fibril length, which can be related to the mechanical properties of tendons. Therefore changes in mechanical loading that are associated with the onset of locomotion not only affect mechanochemical transduction but also affect the resulting mechanical properties of the affected tissues.

1.7 Mechanochemical Transduction Processes

During cell adhesion, the initial binding of integrins to their ECM ligands leads to their activation and clustering, and to assembly of focal adhesion complexes that serve as "assembly lines" for signaling pathways. The signaling pathways include a large number of affector molecules that are influenced by mechanical loading and may directly trigger mitogen-activated protein kinases (MAPK) pathways that lead to changes in gene expression and protein synthesis (Figure 1.9).

G proteins are another family of membrane proteins believed to modulate mechanochemical transduction pathways. Mechanical stimulation changes the conformation of G protein that leads to growth-factorlike changes that initiate secondary messenger cascades leading to cell growth. It has been reported that cyclic strain of smooth muscle cells significantly decreased steady-state levels of G protein and adenylate cyclase activity. Muscular stimulation also appears to be coupled with G protein activation in small arteries.

In addition to activation of signaling pathways, mechanical stress triggers activation of stretch-activated ion channels. These ion channels have been identified in a number of cells and have been studied extensively in muscle cells. Stretch-activated channels are reported to permit passage of cations including Ca^{+2}, K^+, and Na^+, although a few anion channels are reported to be sensitive to mechanical stimulation. In muscle cells Ca^{+2} influx through voltage-gated channels induces a transient elevation in intracellular Ca^{+2} levels. Ca^{+2} influx is activated by mechanical stimulation and leads to membrane depolarization and cell contraction. The presence of extracellular Ca^{+2} appears to be a requirement for its influx as a result of stretch-induced cell contraction in muscle cells.

1.8 Scope of Book

One objective of this book is to relate the structure and mechanical properties of ECMs that make up mineralized and nonmineralized tissues to mechanochemical transduction processes that are important in normal development and tissue homeostasis. Although we know a lot about how tissue is formed based on tissue morphology and gene expression, very little is known about how mechanical forces exerted on human tissues lead to increased size and cell content of tissues. The lifting of weights by an individual during weight training not only increases muscle size but also leads to an increased amount of skin that covers the muscles.

In subsequent chapters we analyze how the macromolecular components that are found in tissue give rise to the complex biological structures that make up the intracellular and extracellular materials. To do this we need to consider how stereochemistry of the units contained in macromolecules

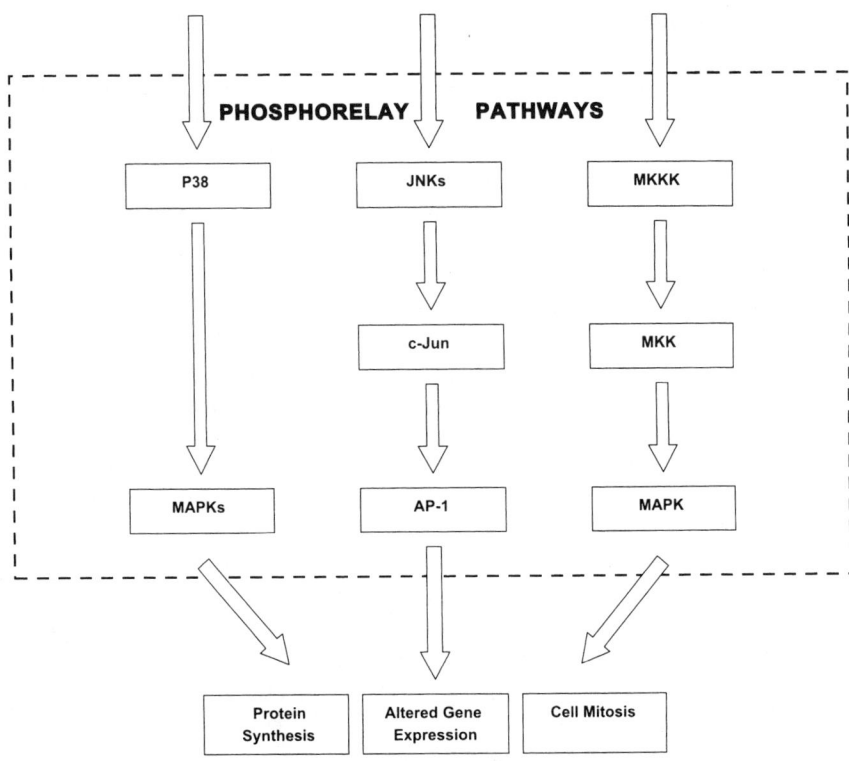

FIGURE 1.9. Diagram of activation of the phosphorelay system by physical forces. Mitogen-activated protein kinases (MAPKs) are part of a phosphorelay system, which are activated through cell membrane and cytoskeletal stretching. Tension applied to the extracellular matrix (ECM) occurs as a result of mechanical, osmotic, hydrostatic, fluid, and electromechanical effects, leading to stretching of the cell membrane. Cell mitosis, gene expression, and protein synthesis occur by generation of secondary messengers that activate the pathways shown. These pathways of the phosphorelay system include the extracellular signal-regulated kinase (ERK-1/2) pathway, which is part of the MAPK kinase kinase (MKKK) pathway (*right*), the c-Jun kinase (JNK) (JNK1, JNK2 and JNK3) pathway (*middle*), and the p38 pathway (*left*). The ERK-1/2 pathway is stimulated by growth factors, cytokines, and G protein and can lead to increased cell mitosis. The JNK pathway is stimulated by growth factors and environmental stresses, altering gene expression and controlling programmed cell death. The p38 pathway regulates protein expression of many cytokines such as interleukin-1, which has been implicated in modulating the response to mechanical loading in a number of tissues such as cartilage.

leads to the formation of long chained molecules with the characteristic properties. Beyond this we need to relate the macromolecular chain structure to the desired macroscopic properties and note how these properties reflect macromolecular organization at the supramolecular level of structure.

Although the mechanical behavior of tissues has been studied extensively, we know that interpretation of this behavior is made more complex by the time-dependence, also termed viscoelasticity. Therefore, we examine how the time-dependence can be corrected for in terms of the structural components. The goal of the mechanics sections is to relate mechanical behavior to the composition, arrangement, and environmental conditions that affect each tissue type. Finally, we examine how external forces balance with internal forces to modulate mechanochemical transduction.

Suggested Reading

Ruoslahti E, Perschbacker MD. New perspectives in cell adhesion: RGD and integrins, *Science*. 1987;238:491.

Silver FH, Christiansen DL. *Biomaterials Science and Biocompatibility*. New York: Springer-Verlag; 1999: Chapters 1, 8.

Silver FH, DeVore D, Siperko LM. Role of mechanophysiology in aging of ECM: Effects of changes in mechanochemical transduction, *J Appl. Physiol.* 2003;95:2134.

2
Macromolecular Structures
in Tissues

2.1 Introduction

The human body is a miraculous dynamic structure. When we are young it grows in size and changes shape depending on genetic inheritance and environmental influences such as mechanical loading. Perhaps the most obvious change is the increase in height and weight that is experienced during our adolescent years. These changes are influenced by gravity because the size of our musculoskeleton increases in proportion to the effort we exert against gravity. In addition, the muscle mass we gain reflects the environmental influences such as how much weight we must lift or how fast we must be able to run if we are hunters. Evolutionwise, hunters who couldn't run or lift heavy animals didn't survive to evolve into modern man. If we lived on the moon man would not have needed to have these attributes and we could exist in a more compact evolutionary form. The structural materials in tissues have evolved based on the required mechanical demands on cells and tissues. In humans, the primary structural material is fibrous collagen. This protein is not only stiff, but it can store elastic energy during locomotion and in this manner allow for fast efficient running. In species that do not locomote or fly, other structural materials have evolved including chitosan and other sugar-based polymers.

Materials of construction of the human body include proteins, polysaccharides, lipids, and nucleic acids. These macromolecules not only compose the protein networks that form the scaffold of both mineralized and nonmineralized tissues, but they also make up the structural materials within the cell including the organelles that power and control gene expression and protein synthesis. Proteins form the structural materials of extracellular matrix in the form of collagen and elastin; in addition they make up enzymes and cell surface markers to name a few of their functions. In this text we are most concerned with the role of proteins as structural materials. A few examples of protein-containing molecules of significance include collagens, myosin, actin, tubulins, integrins, and proteoglycans. The basic repeat unit in proteins, also termed polypeptides, is the peptide unit that is

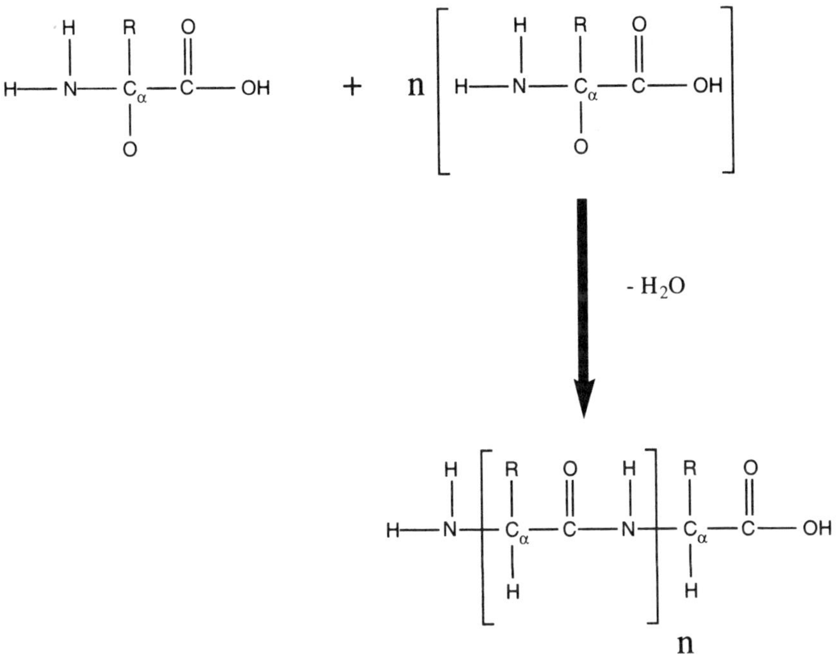

FIGURE 2.1. Formation of a polypeptide. Polypeptides are formed when a condensation reaction occurs between amino acids, releasing water when the amino and acid end groups react.

formed as two amino acids condense to form a dipeptide and water (see Figures 2.1 and 2.2). Proteins are synthesized for transport extracellularly on the endoplasmic reticulum and then transported through the Golgi apparatus for release extracellularly. Proteins are synthesized on ribosomes

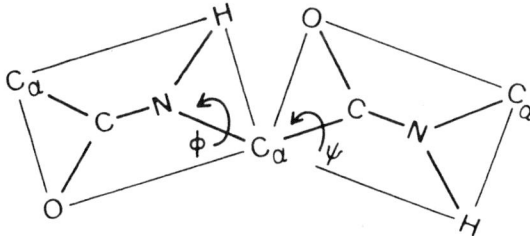

FIGURE 2.2. Diagram of a peptide unit: the peptide unit formed by condensation polymerization is enclosed in a box that is within the plane of the paper. All atoms shown are to a first approximation found within the plane of the paper. The unit begins at the first alpha carbon, C_α, and ends at the second alpha carbon, C_α. Two angles (ϕ, ψ) are sites of free rotation along the backbone of the chain and exist between adjacent peptide units. Both ϕ and ψ are defined as positive for counterclockwise rotation looking from the nitrogen and carbonyl carbon positions towards the alpha carbon between these atoms.

that consist of large and small subunits that translate the genetic code of the cell into a sequence of repeat units.

Polysaccharides are found as sugar polymers that are components of the extracellular matrix (hyaluronan), as carbohydrates (starch), and as energy stores in humans (glycogen). There are numerous repeat units found in polysaccharides but many are derivatives of the simple sugars, glucose and galactose, which are six-membered ring structures (see Figure 2.3). The repeat unit of polysaccharides consists of a six-membered sugar ring linked via an oxygen molecule to the next six-membered ring. Precursors of polysaccharide molecules are synthesized in the cell cytosol and polysaccharides are assembled at the cell membrane or inside the cell depending on whether the polysaccharide is to be released extracellularly or used intracellularly.

Lipids are a heterogeneous group of long hydrocarbon chains that can have one of the following: a free carboxyl group (fatty acids), an ester group

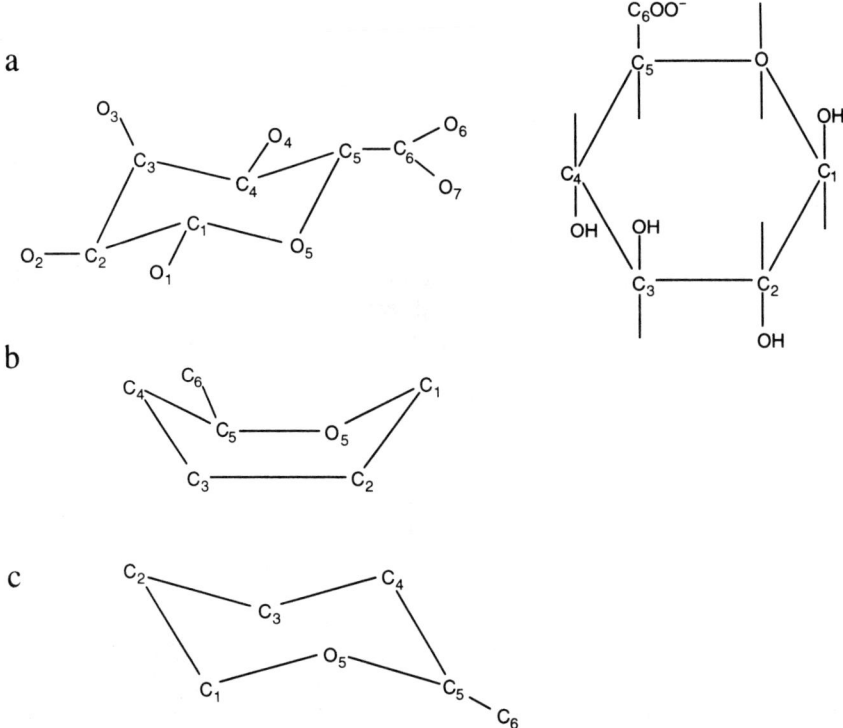

FIGURE 2.3. Boat and chair forms of β-D-glucuronic acid. (**a**) C_1 chair conformation (left) and planar projection (right); (**b**) Boat conformation; and (**c**) ^1C chair conformation of β-D-glucuronic acid. The latter two conformations are energetically less stable than the C_1 chair conformation.

(neutral fats), or derivatives of fatty acids such as membrane phospholipids (see Figure 1.3). Derivatives of phosphatidic acid (phosphatidyl choline, phosphatidyl serine, phosphatidyl inositol) are membrane components that are modified hydrocarbon chains. Rotation around carbon-to-carbon bonds in the backbone gives these molecules great flexibility. The membrane of mammalian cells contains a wide array of polymer structures including lipids, polysaccharides, and proteins. The function of these polymers depends on the size, shape, and flexibility. Polymer molecules of importance in the cell membrane are phospholipids, hyaluronan, cell surface proteo-glycans, class I and II histocompatibility markers, and integrins.

The final class of macromolecules to be covered in this book is the nucleic acids. These polymers are composed of a nitrogenous base (either a purine or a pyrimidine), a five-membered sugar ring (a pentose), and phosphoric acid. Deoxyribonucleic acids (DNA) contain the purines, adenine and guanine, the pyrimidines, cytosine and thymine, 2-deoxyribose, and phos-phoric acid. DNA is found in the cell nucleus in double-stranded form in the chromosomes. Ribonucleic acid (RNA) contains purines (adenine and guanine), pyrimidines (cytosine and uracil), ribose, and phosphoric acid. RNA is found in the small and large subunits of ribososmes (rRNA), as a copy of the genetic material (messenger (m)RNA) and for adding amino acids to a growing protein chain (transfer (t)RNA). The repeat units and structures of DNA and RNA are shown in Figure 2.4.

2.2 Protein Structure

Proteins make up the bulk of the structural materials found in vertebrate tissues. They comprise the soft and hard ECMs that allow for locomotion, protection from environmental contamination, and for the hair and upper layer of skin that is on the outer surfaces of the body. In addition, they make up the bulk of the structural materials found in the cell cytoskeleton and provide a scaffold on which cells reside in the organs and tissues of the body. For this reason it is important to understand structure–property relation-ships for proteins.

2.2.1 Stereochemistry of Polypeptides

In the case of proteins the building blocks that are synthesized into a long polymer chain are amino acids. When amino acids are added together they form peptide units that are more easily visualized because they are con-tained within a single plane (see Figures 2.5 to 2.7) except for the side chain or R group. The sequence of amino acids is important in dictating the manner in which polymer chains behave because the sequence dictates whether a chain can fold into a specific three-dimensional structure or whether the polymer chain does not fold. Specific examples of these prin-

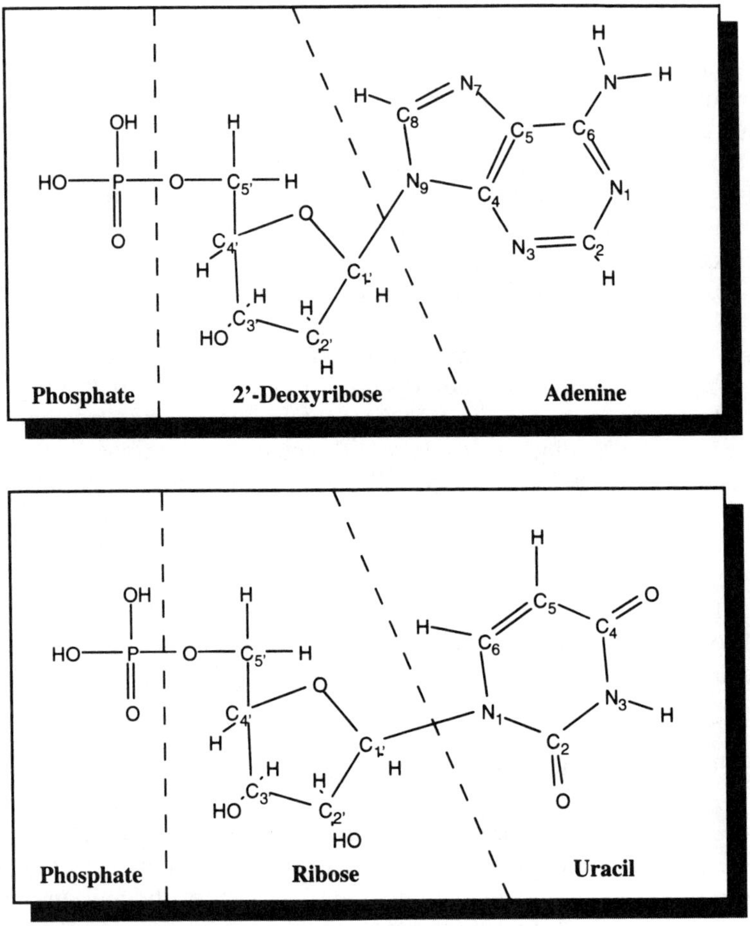

FIGURE 2.4. Structure of nucleic acids. The diagram shows the linkage between phosphate, sugars, and a purine or pyrimidine for DNA (top) and RNA (bottom).

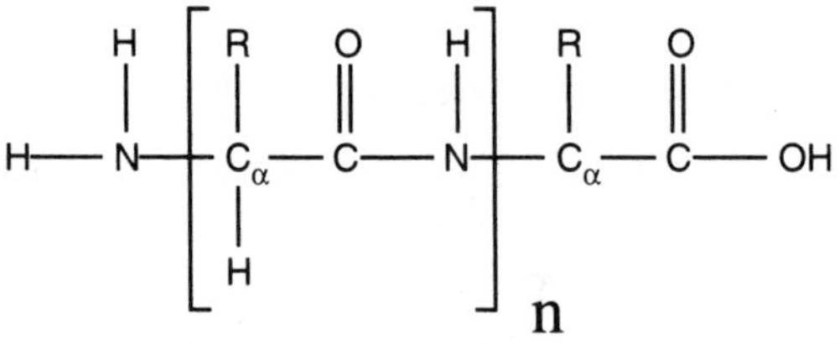

FIGURE 2.5. Chain structure of a polypeptide. The repeat unit of a polypeptide chain consists of amino acid residues, each characterized by a specific amino acid residue designated by R. The size of the polypeptide chain is dictated by n, the degree of polymerization.

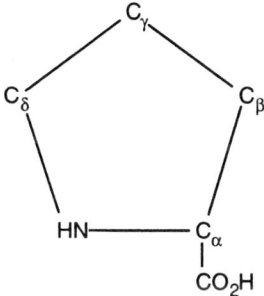

FIGURE 2.6. Chemical structure of proline. The structure of proline differs from that of other amino acids because a ring is part of the polypeptide backbone. The chain backbone includes the amide nitrogen (N), Cα, and the carbonyl carbon (CO₂H).

ciples include keratin that folds into an α helix because of its primary sequence of amino acids and elastin that contains segments with random chain conformations. α helices in keratin pack together to form twisted ropes with very high modulus and tensile strength; these high modulus values are necessary to protect skin from mechanical damage. Elastin forms short segments with random chain structure that result in a material with a much lower modulus and ultimate tensile strength compared to keratin. A relationship between protein structure and mechanical properties exists. Amorphous polymers such as elastin behave as rubberlike materials whereas force-transmitting rigid proteins such as collagen are in extended conformations (i.e., the chain is stretched out in space) and behave as do stiff ropes.

Proteins contain 20 different R groups; actually there are 19 different amino acid side chains or R groups and the 20th, which is proline, is a ring

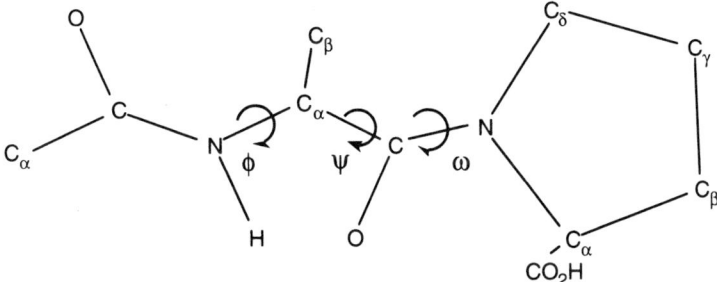

FIGURE 2.7. Polypeptide chain structure containing proline. Chains containing proline lack the flexibility of other peptides because the proline ring has only one available angle for backbone rotation. Rotation occurs around the angles φ, ψ, and ω.

that does not have two angles of free rotation. Therefore, the general representation of amino acids in the common side chains found in proteins is shown in Figure 2.5. Proline is the only amino acid found in proteins that is an exception to this rule, inasmuch as the amino acid side chain is part of the backbone as shown in Figures 2.6 and 2.7. The presence of proline and a hydroxylated form of proline, hydroxyproline, are characteristic of collagen and collagenlike polypeptides, which form a triple-helical structure. The primary consequence of the incorporation of proline into the backbone is that it stiffens the three-dimensional structure. Below we explain why this happens.

There are 20 "standard" amino acids that are commonly found in proteins as illustrated by Table 2.1. The sequence of these amino acids dictates the shape of the resulting polypeptide chain, and the presence of large R groups limits mobility of the chain backbone by preventing rotation as diagrammed in Figure 2.7. The structure of a polypeptide and its mobility is very dependent on the location of the alpha carbon (Cα; see Figures 2.8 to 2.10). Also interfering with rotation about the backbone is the possibility

TABLE 2.1. Standard amino acids

Amino acid	Abbreviation	R group
Alanine	Ala	$-CH_3$
Arginine	Arg	$-CH_2-CH_2-CH_2-NH-CH \begin{smallmatrix} NH_2 \\ NH_2 \end{smallmatrix}$
Asparagine	Asn	$-CH_2-\overset{\overset{\displaystyle O}{\|\|}}{C}-NH_2$
Aspartic acid	Asp	$-CH_2-\overset{\overset{\displaystyle O}{\|\|}}{C}-O^-$
Cysteine	Cys	$-CH_2-SH$
Glutamic acid	Glu	$-CH_2-CH_2-\overset{\overset{\displaystyle O}{\|\|}}{C}-O^-$
Glutamine	Gln	$-CH_2-CH_2-\overset{\overset{\displaystyle O}{\|\|}}{C}-NH_2$
Glycine	Gly	$-H$

TABLE 2.1. *Continued*

Amino acid	Abbreviation	R group
Histidine	His	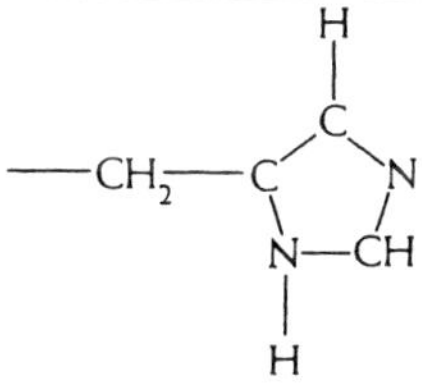
Isoleucine	Ile	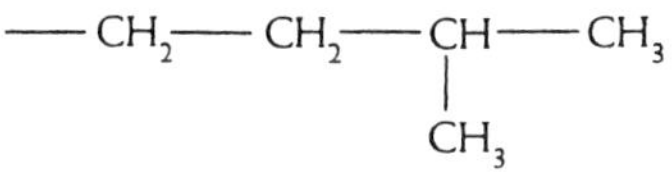
Leucine	Leu	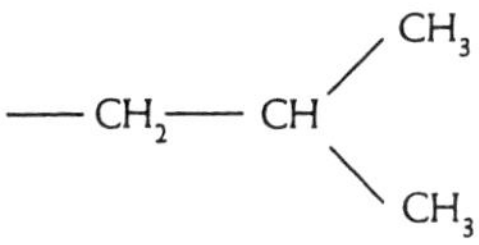
Lysine	Lys	$—CH_2—CH_2—CH_2—CH_2—NH_3+$
Methionine	Met	$—CH_2—CH_2—S—CH_3$
Phenylalanine	Phe	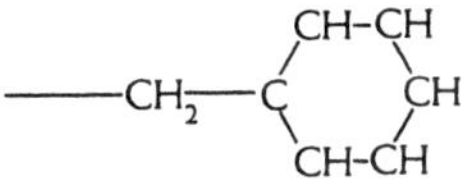
Proline	Pro	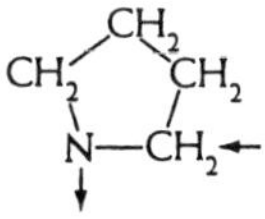
Serine	Ser	$—CH_2OH$
Threonine	Thr	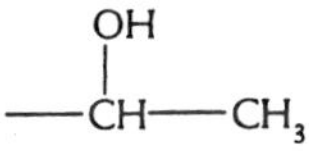
Trytophan	Trp	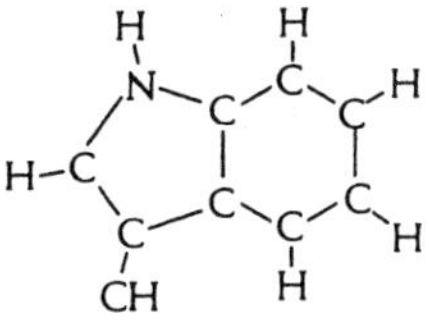
Tyrosine	Tyr	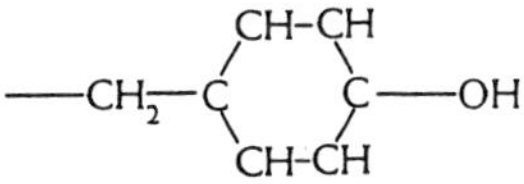
Valine	Val	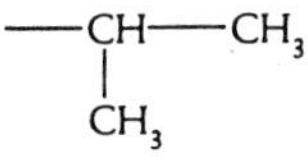

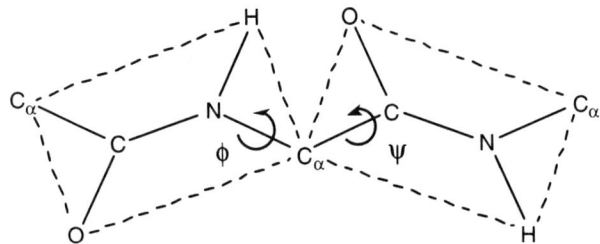

FIGURE 2.8. Diagram of a dipeptide. The diagram shows a dipeptide of glycine in the standard conformation 180°, 180°. To a first approximation, all atoms seen in the dipeptide backbone are found within the plane of the paper. For a dipeptide of glycine, the only atoms not within the plane are the side chain hydrogens. The unit begins at the first Cα and goes to the second Cα. Both φ and ψ are defined as positive for clockwise rotation when viewed from the nitrogen and carbonyl carbon positions toward the Cα.

that each amino acid can be in either the D or L form because the carbon to which the R group is attached, the alpha carbon in chemical terms, is asymmetric, that is, has four different chemical groups attached to it. Therefore there are two different chemical forms of amino acids that rotate a plane of polarized light differently (the L form rotates the plane to the left and the D form to the right).

To make a long story short, the structure and properties of the resulting protein synthesized with all D or L amino acids differ. The L form is predominantly found in proteins of higher life species and can be deciphered from the D form by the position of the R group with respect to the backbone. Inasmuch as the alpha carbon (Cα) is attached to three different groups of atoms, the L form cannot be simply rotated around the carbonyl carbon (carbon to which oxygen is attached)–alpha carbon bond. The L form is defined by moving along the peptide chain beginning at the end that contains the free carboxyl acid group (free COOH end of chain) and moving toward the free amino end with the alpha carbon as the point of reference. The amino acid side chain (R group) is on the left as shown in Figure 2.10 (bottom). When the side chain is on the right the D form is

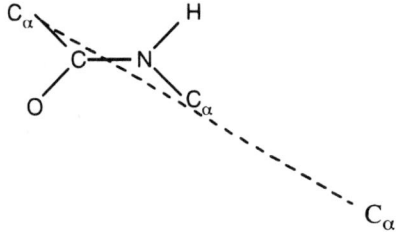

FIGURE 2.9. Location of the second peptide unit. Dipeptides are constructed by knowing the coordinates of atoms in the first peptide unit (using X-ray diffraction) and then translating along the line between the first and second Cα by a distance of 0.37 nm.

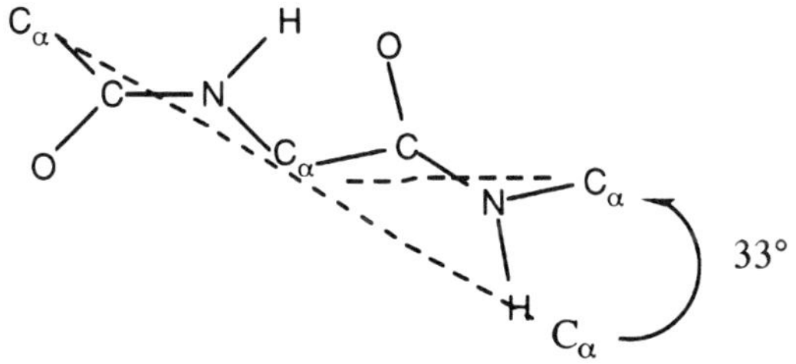

FIGURE 2.10. (Top) Graphical construction of a dipeptide. Once the second peptide unit is located (see Figure 2.12), it is rotated through an angle of 33° clockwise to generate a dipeptide in the standard conformation, with $\phi = 180°$ and $\psi = 180°$.

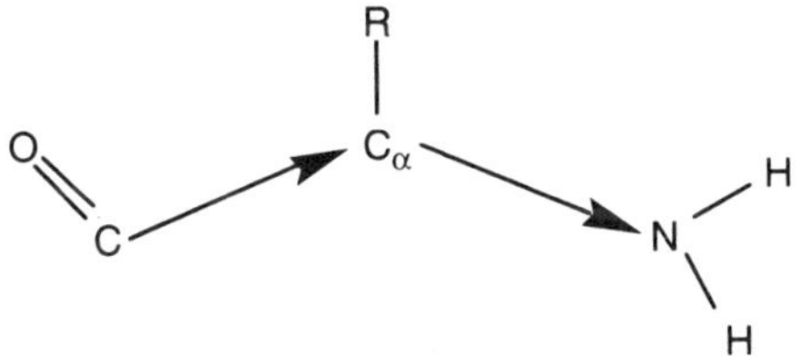

FIGURE 2.10. (Bottom) Difference between L and D amino acids. The two naturally occurring forms of amino acids differ in the position of the R group with respect to the backbone. An L-amino acid has the R group on the left if viewed along the chain from the free carbonyl end to the free amino end as shown. If the R group is on the right, it is defined as the D form. The predominant form found in proteins is the L form, although some amino acids in the D form are present in proteins.

present. This is important because the position of the side chain dictates how a chain can fold in three dimensions.

2.2.2 Primary and Secondary Structure

From a structural viewpoint, a polypeptide is composed of planar peptide units as shown in Figure 2.8. The usefulness of considering the peptide unit as opposed to the amino acid is that the peptide unit is almost planar as opposed to the amino acid, which has atoms that are in more than one plane. To illustrate this point, the coordinates of atoms in the peptide unit are given in Table 2.2 and nonbonded atoms cannot be closer than the sum of the minimum atomic distances (Table 2.3). Note that all the atoms from the first alpha carbon ($C\alpha$) to the second alpha carbon do not have a z-coordinate. These coordinates come from X-ray diffraction studies on proteins and represent the average coordinates found among many pro-

TABLE 2.2. Coordinates of atoms in peptide unit

Atom	Coordinates (Å) x, y, z
C_α (first amino acid)	0, 0, 0
Carbonyl carbon	1.42, 0.58, 0
Oxygen	1.61, 1.79, 0
Nitrogen	2.37, −0.33, 0
Hydrogen	2.19, −1.31, 0
Cα (second amino acid)	3.70, 0, 0
C_β (first carbon of side chain)	0.51, 0.72, 1.25

teins. The 3-D structure of a dipeptide is shown in Figure 2.8 in the standard conformation.

The standard conformation is obtained by taking the first peptide unit and rotating it about the line between C1 α and C2 α by 180 degrees, translating the rotated peptide unit along the line between C1 α and C2 α by the distance between C1 α and C2 α and then rotation counterclockwise along the extension of the line C1 α to C2 α by 33 degrees as shown in Figures 2.9 and 2.10. If we define the angle of rotation of the bond containing the atoms N–C α as phi (ϕ) and the angle of rotation of bond containing C α and C (carbonyl) as psi (φ), the conformation shown is arbitrarily defined as phi = 180 and psi = 180 degrees.

Using a computer and matrix multiplication techniques the first and second peptide units can be rotated through all possible combinations of values of phi and psi. For each possible set of these dihedral angles, the distances between all pairs of nonbonded atoms can be compared using a set of minimum interatomic contact distances (Table 2.3); if the distance between any set of atoms is smaller than the minimum contact distance then that conformation is not allowed. This boils down to the fact that two atoms cannot be closer than the sum of the van der Waals radii of each atom or electron repulsion occurs. Therefore, by determining the values of phi and

TABLE 2.3. Minimum interatomic distances for nonbonded atoms

Nonbonded atom pairs	Contact distances normal (minimum), Å
Carbon to carbon	3.20 (3.00)
Carbon to nitrogen	2.90 (2.80)
Nitrogen to nitrogen	2.70 (2.60)
Carbon to oxygen	2.80 (2.70)
Nitrogen to oxygen	2.70 (2.60)
Oxygen to oxygen	2.70 (2.60)
Carbon to hydrogen	2.40 (2.20)
Hydrogen to nitrogen	2.40 (2.20)
Hydrogen to oxygen	2.40 (2.20)
Hydrogen to hydrogen	2.00 (1.90)

psi that result in conformations that are not prevented by electron repulsion, we can determine the number of available conformations (combinations of phi and psi that are allowed).

Conformational maps show values of phi and psi that fall into two regions: the first region contains points that are always allowable and the second region contains points that are sometimes allowed. As shown in Figure 2.11, all points within the inner lines are always allowed and those within the outer solid lines are sometimes allowed. Hence the entropy (or rotational freedom of a peptide) is obtained from the area surrounded by the solid lines on a conformational map (Figure 2.11) and is proportional to the number of allowable conformations of a dipeptide unit. Please note a flexible chain with a lot of rotational freedom is easier to stretch than a rigid chain. We define the flexibility of a polypeptide as the natural logarithm of the number of allowable conformations, #, times Boltzmann's constant (Equation (2.1)).

$$S = \text{entropy or chain flexibility} = k \ \ln(\#) \qquad (2.1)$$

What Figure 2.11 tells us is that a conformational map for a dipeptide of glycine (the side chain in glycine is very small, just a hydrogen) has mostly allowed or partially allowed conformations and therefore polyglycine is flexible. One question that you might ask is how do we know that the conformational plot for a polypeptide is the same as for a dipeptide? The answer is that because the side chain points away from the backbone for most conformations the atoms in the side chains are separated by more than the sum of the van der Waals radii. However, below we discuss several highly observed conformations of proteins in which the conformational map is an overestimation of the flexibility because of interactions between atoms more than two peptide units apart in space.

A chain of carbon atoms bonded together is ideally flexible in the same manner that polyglycine is flexible. In fact a more practical conclusion of viewing the conformational plot of polyglycine is that poly (ethylene), and lipids are composed of carbon chains bonded together and therefore are ideally flexible macromolecules. Most polypeptides are made up of sequences of amino acids that are not identical and have side chains bigger than glycine. Therefore, protein structure in general is more complicated than that of polyglycine as is discussed next.

Polypeptides, however, are composed of amino acids with side chains that are longer and therefore the area of allowed conformations is reduced when an alanine (Figure 2.12), aspartic acid (Figure 2.13), or a proline (Figure 2.14) is added to the second peptide unit. Finally, the conformational map for a dipeptide of proline–hydroxyproline is dramatically reduced. Rings in the backbone of any polymer reduce the ability of the polymer backbone to adopt numerous conformations and thereby stiffen the structure.

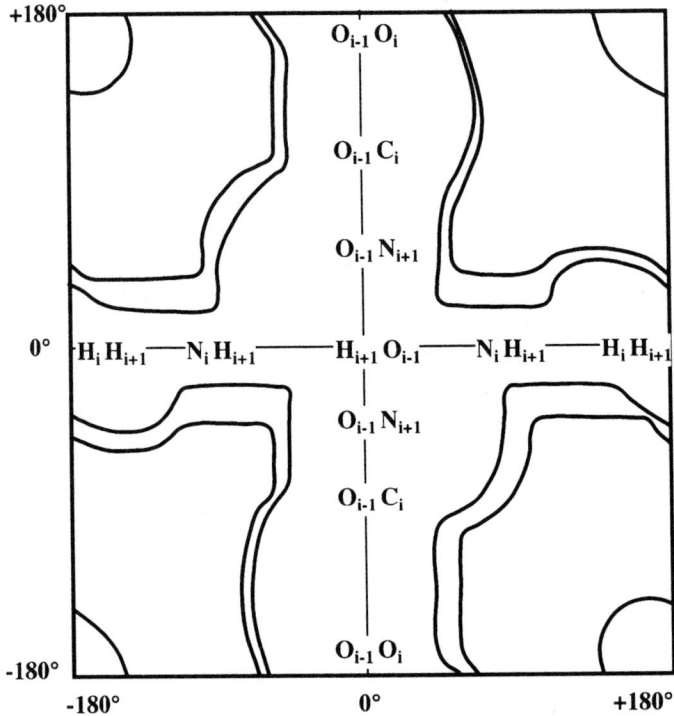

FIGURE 2.11. Allowed conformations for a dipeptide of glycine. Plot of fully (outer solid lines) and partially (inner solid lines) allowed combinations of ψ (vertical axis) and ϕ calculated using a hard sphere model with normal (outer solid lines) and minimum (inner solid lines) atomic contact distances. Dihedral angles (ϕ, ψ) in center of plot ($0°$, $0°$; $0°$, $180°$ and $180°$, $180°$) are unallowed because of contacts between backbone atoms in the neighboring (ith, ith + 1 and ith − 1) peptide units. The contacts between nitrogen (N), hydrogen (H), oxygen (O), and carbonyl groups (CO_2H) prevent allowed conformations in the center of this diagram. This diagram can be constructed using coordinates of the atoms within the peptide unit and inter-atomic distances (see Tables 2.2 and 2.3). To construct this diagram, a standard dipeptide unit (see Figure 2.8) is formed by translating the first peptide unit along the line between the α carbons and then flipping the second unit into the *trans* configuration. The values of ϕ and ψ were varied using matrix multiplication, and the interatomic distances were checked for all nonbonded atoms. Pairs of ϕ and ψ that have allowable interatomic distances are found within the inner and outer solid lines shown as the allowable conformations.

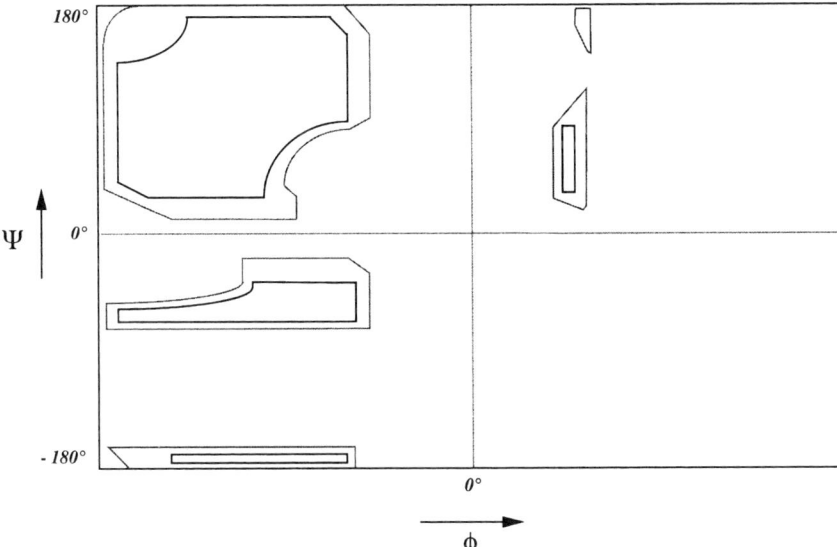

FIGURE 2.12. Conformational plot for glycine–alanine. Plot of allowable angles for peptides containing a repeat unit of glycine and alanine showing totally (outer solid lines) and partially (inner solid lines) allowed conformations determined from normal and minimum interatomic distances.

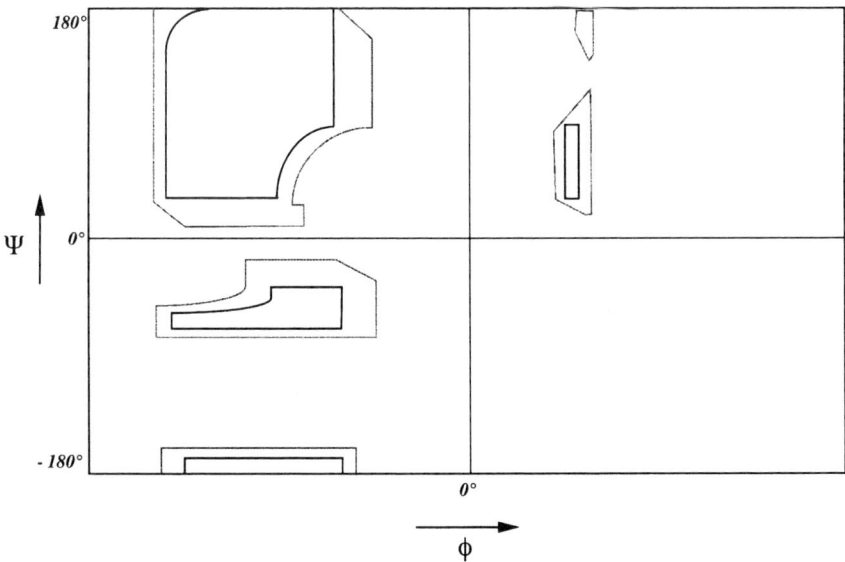

FIGURE 2.13. Conformational plot for glycine–aspartic acid. The allowable conformations are only slightly reduced compared to the plot in Figure 2.12 because the side chain length is increased in going from alanine to aspartic acid.

For polypeptides the addition of a methyl or slightly longer side reduces the number of available conformations to the point where two major regions of allowable conformations stand out (see Figure 2.15). It turns out that when the conformations that are most frequently observed in proteins are tabulated, they fall into these two regions. These conformations include the α helix (α in Figure 2.15; not to be confused with the α chain helix in collagen), β sheet (β in Figure 2.15), and the collagen triple helix (C in Figure 2.15). The regions that these conformations fall into are given by phi between −45 and −120 degrees and psi between 150 and −90 degrees. We learn later that at least the β sheet and the collagen triple helix are highly extended structures and therefore are very rigid and stable; the α helix can be extended into a β sheet and therefore is more deformable.

The mechanical properties of polymer chains that do not exhibit interactions between the side chains and the backbone, or one part of the backbone and another part of the backbone, are related to the number of available conformations and hence the chain entropy. As we discuss later, the stiffness of a polymer chain that does not exhibit bonding with other parts of the chain is related to the change in the number of available conformations. It turns out this refers to random chain polymers of which elastin, poly(ethylene) at high temperatures, and natural rubber are discussed in this text. As we stretch a polymeric chain we reduce the number

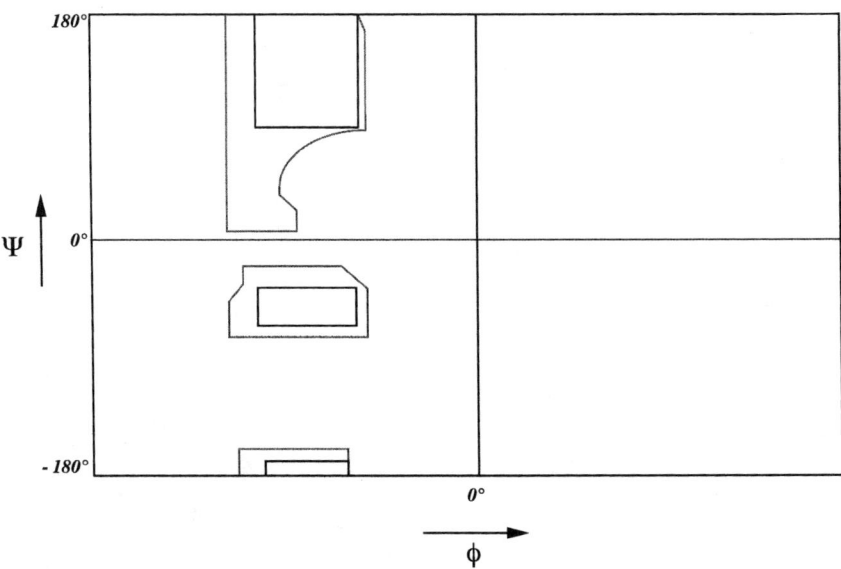

FIGURE 2.14. Conformational plot for glycine-proline. Addition of proline to a dipeptide further reduces the number of allowable conformations when compared to Figures 2.12 and 2.13.

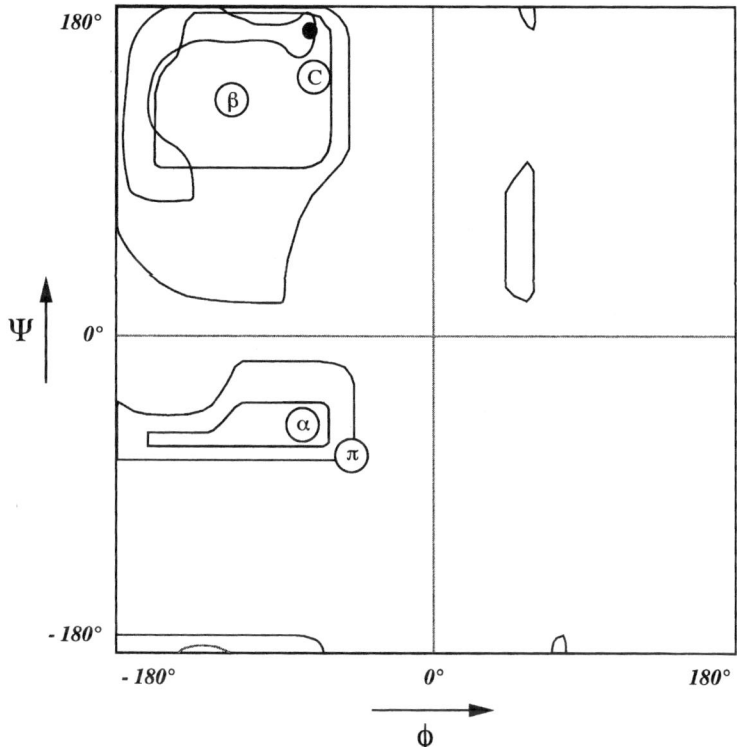

FIGURE 2.15. Conformational plot showing location of α helix, β sheet, and collagen triple helix. The plot shows the localization of the predominant chain structures found in proteins, including the α helix (α), β sheet (β), and collagen triple helix (C). The π stands for a helix that does not occur in nature.

of conformations and the resulting flexibility, thereby increasing the stiffness and resistance to deformation. When a random chain polymer is unloaded, the mobility and flexing of the backbone result in a return of the chain to its original range of rotational motions.

Unfortunately, in general the behavior of proteins is more complex because the α helix, β sheet, and collagen conformations are found repeated in different structures in mammals. As we saw above, a large polymer molecule that has small side chains and no rings in the backbone has the potential to constantly rearrange itself like an eel moving through water. A single conformation is a single set of dihedral angles that characterize the rotational state of each peptide unit that makes up the backbone of the molecule. As a molecule moves by diffusion, it changes its conformation by changing the set of dihedral angles that characterize each dipeptide unit. Folding of a polypeptide chain into the most commonly observed conformations, the α helix, β sheet, and collagen triple helix, occurs because of the inherent flexibility of peptide chains; however, once it folds into a particular structure it is held in place by secondary and other forces.

An interesting perspective is to look at the free energy change associated with transition from a flexible chain into a folded conformation. From thermodynamics we know that the free energy of the transition must be negative for it to occur spontaneously. Let us review a few definitions at this point to refresh your memory. The Gibbs free energy, G, is a measure of the total energy that a system of macromolecules has as well as the entropy (flexibility). Thus for any process to occur spontaneously (within our lifetime) the change in free energy ΔG must be equal to the change in enthalpy ΔH, minus the temperature times the change in entropy $T \Delta S$ (see Equation (2.2)). The enthalpy ΔH of a macromolecule is related to the number of covalent Pc, dispersive Pd, and electrostatic bonds Pe, that are formed as well as the translational Et and rotational kinetic energy Er of the chains (see Equation (2.3)).

$$\Delta G = \Delta H - T \Delta S \tag{2.2}$$

$$H = -(Pc + Pd + Pe) + Et + Er \tag{2.3}$$

Therefore the change in enthalpy associated with a transition from a flexible chain to a folded chain involves a change in the number of bonds (the P term) and a change in the flexibility of the chain (S term). We can calculate the change in the S term by a change in the area of allowable conformations (ideally this is k times the natural logarithm of the area under Figure 2.15 divided by one because there is one final conformation). For the process to be spontaneous the change in P must be positive (this means we must form bonds). Formation of a covalent bond lowers the enthalpy by 100 kcal/moles whereas hydrogen and electrostatic bonds lower it by between 1 and 5 kcal/mol. Finally, van der Waals or dispersive forces lower it by 0.01 to 0.2 kcal/mol. Electrostatic and hydrogen bonds occur between atoms with partial charges and can be quantitatively assessed using Coulomb's law, Equation (2.4), where $q1$ and $q2$ are the partial charges on the atoms involved and E is the permutivity of the medium between the charges and rij is the separation distance between charges. Dispersive energy Pd is calculated using the Lennard–Jones 6–12 potential energy function where A and B are two constants specific to the two atoms involved (see Equation (2.5)).

$$Pe = q1q2/(4\pi E\ rij) \tag{2.4}$$

$$Pd = (A/rij^{12}) - (B/rij^{6}) \tag{2.5}$$

In the sections to follow we show that hydrogen bonds are the primary type of bonds that stabilize helical and extended polypeptide conformations.

These structures differ in the different hydrogen bond patterns that occur. Therefore, it is the hydrogen bond pattern that stabilizes folding of polypeptide chains.

2.2.3 Supramolecular Structure

Although protein primary structure (sequence of amino acids) determines how polypeptides fold, proteins that form isolated helices do not possess much mechanical stability. Another way of looking at this is that structural stability of proteins involves transfer of stress between structural units. For this reason the polypeptides that make up our skin, hair, and tendons are assembled into structures with larger diameters and lengths. This implies that tensile-bearing tissues require a higher level of structural hierarchy than do nonload-bearing tissues. Because all structural tissues in vertebrates bear at least tensile loads this suggests that these tissues have levels of structural hierarchy above the primary folding of helices.

Stress transfer is achieved functionally by connecting folded polypeptide chains into continuous networks through electrostatic, hydrogen, hydrophobic, and covalent bonds. An example of the necessity for covalent bonds to stabilize protein structure comes from studying ECMs from animals that have been feed crosslink inhibitors such as beta amino proprionitrile that inhibit collagen crosslinking. The skin and tendons of animals fed this inhibitor tear easily and are rich in collagen molecules that can be removed by immersion in aqueous solvents. For this reason stress transfer and mechanochemical transduction by proteins in the ECM require protein stabilization through chemical crosslinking.

2.2.3.1 Primary and Secondary Structures of Proteins

There are really four classes of protein structures that make up part or all of a biological macromolecule. It should be pointed out that some molecules contain several different structural types connected by sequences that have other structures. The four classes that we discuss include α helix, β sheet, collagen triple helix, and random chain structure. The combinations of backbone angles that are given on the conformational plot shown in Figure 2.15 define these four classifications of macromolecular structures found in proteins. The one exception is the random chain structure that is characteristic of polypeptide sequences such as those found in elastin. In this case the conformational angles are not fixed but allow almost free mobility of polymer chains. However, even in elastin the freely rotating segments are connected via α helical regions and regions with β turns.

2.2.3.2 α Helix

The most commonly occurring helical structure observed in proteins and the first that was worked out is the α helix. The work of Linus Pauling,

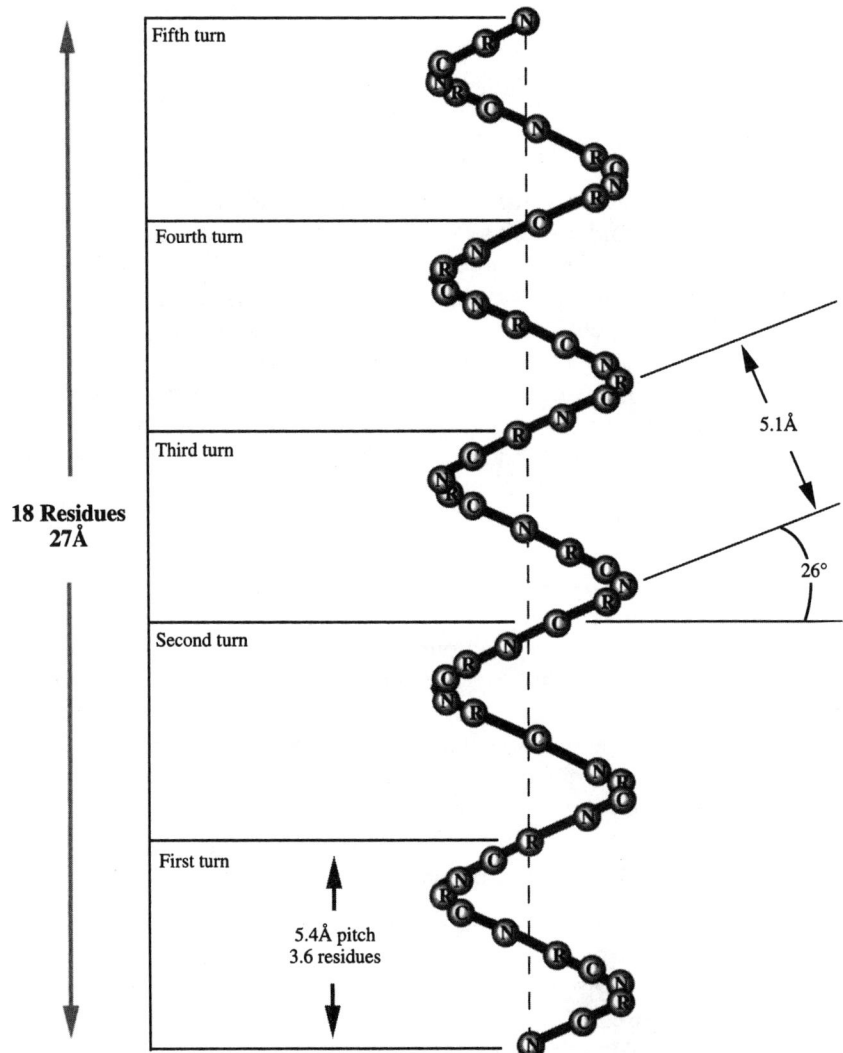

FIGURE 2.16. Structure of a helix. The structure of the α helix is characterized by 3.6 amino acid residues (R groups) per turn of the helix over an axial distance of 0.54 nm. This is consistent with 18 residues per five full turns of the helix.

before he became interested in vitamin C, was instrumental in showing that this helix has 3.6 amino acids per turn, an exact repeat of the helix every 5.4 Å and an axial rise per residue of 1.5 Å as is illustrated by Figure 2.16 (Pauling and Corey, 1951). In the α helix, the amino acid side chains are directed radially away from the axis of the helix. The helix is stabilized by formation of hydrogen bonds between the carbonyl oxygen of one amino acid residue, which has a slight negative charge and the hydrogen of the amino group of a residue four amino acids farther down the chain, which has a slight positive charge. Of course this requires that the sequence of the amino acids can accommodate a hydrogen bond pattern that is almost perpendicular to the axis of the molecule; protein sequences containing either proline or hydroxyproline cannot form an α helix. α helices are either right- or left-handed (right-handed chains run clockwise as you look from one end to the other whereas left-handed chains are counterclockwise); right-handed forms are most commonly observed. The α helix is very stable because of the numbers of hydrogen bonds formed as well as the linear nature of the bond.

Macromolecules of importance to the biomaterials scientist having some portion composed of α helical structure include hemoglobin, myosin, actin, fibrinogen, and keratin. The α helix is a rather condensed structure because the rise per residue is 1.5 Å and as such is quite different from that of the collagen triple helix and the β structure of silk. The rise per residue in the two latter structures is about twice that found in the α helical structure. For this reason the extensibility of the α helix is greater than that of the collagen triple helix and the β structure and in the case of keratin, tensile deformation of the α helix leads to formation of a β structure.

Although α helices are abundant in proteins, the average length is fairly short, that is, 17 Å long containing about 11 amino acids or three turns. Therefore, α helices are typically found in short domains within proteins and not as continuous stretches. Keratins that make up hair and the most superficial layer of skin contain a central domain with an α helical component of 310 amino acids or about 46.5 nm in length. Keratin is an example of a protein with a fairly long α helix. The amino acid composition that favors α helix formation is fairly broad with the exception of proline and serine.

2.2.3.3 β Sheet

In a similar manner to α helices, extended structures can be held together by hydrogen bonds with the hydrogen bonds running perpendicular to the chain axis. Silk is an example of a protein that is found in the β structure. The amino acid composition of silk is rich in glycine (44.5%), alanine (29.3%), and serine (12.1%) amino acids with small hydrocarbon side chains that form sets of antiparallel hydrogen bonds between molecular chains. Models of polypeptide chains with sequences of poly(gly-ala) and poly(ala-gly-ala-gly-ser-gly) show that the most probable structure contains all the gly residues on one side of the chain and all the ala residues on the other side of the chain, and therefore by packing the chains in

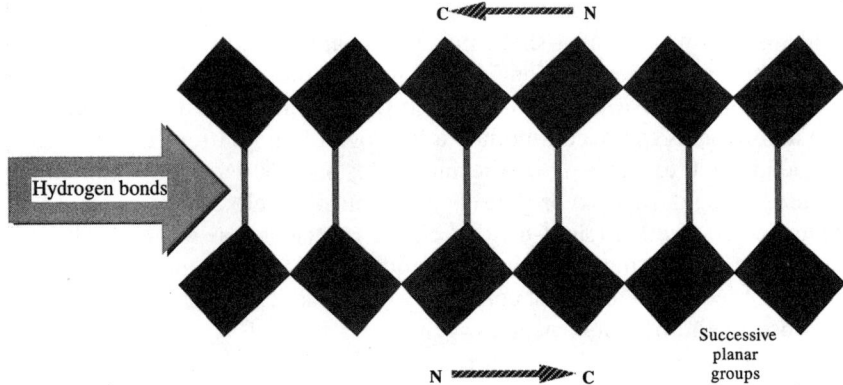

FIGURE 2.17. Hydrogen bonding in antiparallel β sheet. Antiparallel hydrogen bonding between carbonyl and amide groups within the peptide unit stabilizes the β extended conformation.

antiparallel fashion the side chains fit neatly into the empty spaces between the chains (see Figure 2.17). The rise per residue in the β structure is about 3.5 Å.

2.2.3.4 Collagens

The collagens are a family of structural proteins that contain stretches of triple helix that are interrupted by nonhelical regions. Although the fibril-forming collagens found in the ECM are molecules with helical regions about 300 nm long and form fibrils and large fibers seen in ECMs, there are other collagen types that do not form fibrils and have short triple-helical segments. The fibril-forming collagens include types I, II, III, V, and XI; they self-assemble into cross-striated fibrils with the characteristic 67 nm repeat. Fibril-associated collagens, FACIT collagens, are found on the surface of collagen fibrils and appear to connect fibrillar collagens to other components of the ECM. In this text we focus on the fibril-forming collagens that are the structural elements found in vertebrate ECMs and have a characteristic repeat pattern that is observed in the electron microscope (Figures 2.18 and 2.19).

Types I, II, and III collagen form the fibrous network that prevents premature mechanical failure of most tissues and acts to transmit stress to and from cells. The molecular sequences of these collagens are known and they are composed of approximately 1000 amino acids in the form of Gly-X-Y with small nonhelical ends before and after these sequences. All of these collagen types form continuous triple-helical structures that pack laterally

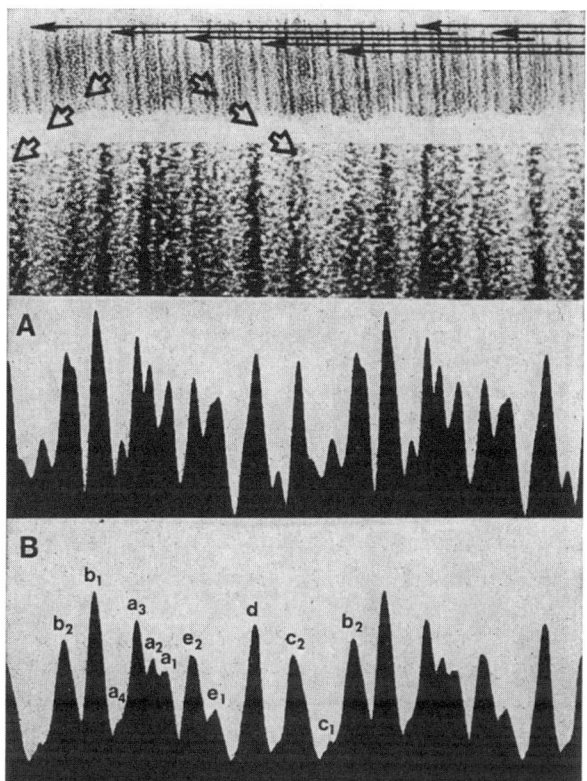

FIGURE 2.18. Positive staining pattern of collagen fibrils: transmission electron micrograph of collagen fibril from rat tail tendon stained with uranyl acetate (top). Uranyl ions electrostatically bind to both negatively and positively charged amino acid residues creating a "positive staining pattern" which represents the location of charged residues along the axis of the triple helix. Arrows indicate that the banding pattern at top is magnified to reveal 12 separate dark lines. Passing a beam of light along the axis of the photographic negative containing this banding pattern results in the peaks shown in (A). The scan shown in (B) was theoretically synthesized using the molecular packing pattern shown in Figure 2.19 and the axial position of charged residues in type I collagen. Band numbers b_2 to c_1 are used to identify the 12 different bands within the repeat period, D, in type I collagen fibrils.

into a quarter-stagger structure in tissues to form characteristic D-periodic fibrils as shown in Figure 2.19. These fibrils range in diameter from about 20 nm in cornea to over 100 nm in tendon. In tendon, collagen fibrils are packed into fibril bundles that are aligned along the tendon axis. In skin, type I and III collagen fibrils form a nonwoven network of collagen fibrils that aligns with the direction of force. In cartilage, type II collagen fibrils

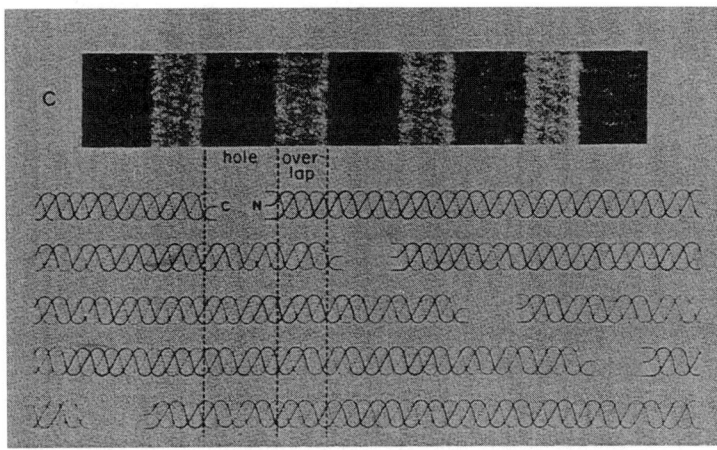

FIGURE 2.19. Packing diagram showing lateral arrangement of collagen molecules in fibrils: collagen molecules observed by electron microscopy in native fibrils are characterized after staining with heavy metals by a repeat pattern consisting of a light region followed by a dark region (top). This repeat pattern occurs in the presence of metal stains such as phosphotungstic acid used for transmission electron microscopy. If short staining times are used, the stain penetrates the holes within collagen fibrils, resulting in a "negative staining pattern." As diagrammed below the staining pattern, this arises because neighboring collagen molecules are staggered laterally by about 22% of the molecular length with a hole of about 13.5% longitudinally between neighboring molecules. This is termed the "quarter stagger" model and is an accurate one-dimensional projection of a type I collagen fibril.

form oriented networks that are parallel to the surface (top layer) whereas in the deeper zones they are more randomly oriented with respect to the surface. The ability of collagen fibers to store and transmit energy is related to the staggered crosslinked structure of the molecules in collagen fibrils.

2.2.3.4.1 Collagen Triple Helix

The other protein structure that we are concerned with is the collagen triple helix. Our knowledge of the structure of collagen comes as a result of early studies on the amino acid composition, the structure of peptides derived from collagen, and the analysis of the X-ray diffraction pattern of collagen fibers. Early compositional studies were important in establishing that collagen was characterized by a high content of glycine, proline, and hydroxyproline. This did not fit in with the established amino acid profile for proteins that form α helices or β sheet structures and therefore a new structure was proposed for this sequence. However, it was clear that the presence of proline and hydroxyproline would result in an extended structure based on the conformational plot. After cleavage of collagen with acid it was determined that glycine accounted for about 33% of the amino acid residues, and proline and hydroxyproline together accounted for another 25% of the amino acid residues. It was demonstrated by the early 1950s, that the dipeptides that made up collagen were mostly Gly-Pro and Hyp-Gly. This observation led to the hypothesis that every third residue in the collagen structure was probably glycine and the proposed structure must accommodate the rigid proline and hydroxyproline residues.

The other evidence that proved important in unraveling the structure of collagen came from understanding the X-ray diffraction pattern. Prior to 1940 biophysicists recognized that when an X-ray beam passes through a tendon or a tissue containing oriented collagen fibers, spots appear on a photographic plate positioned behind the fiber. Near the meridian (the vertical axis) of the exposed photographic plate, arcs appeared at a position that was equivalent to a spacing of 2.86 Å. These arcs were found in oriented samples of different types of connective tissue, ranging from mammoth tusk to sheep intestine. The repeat of 2.86 Å was thought to be the displacement per amino acid along the axis of the molecule. In 1954, Ramachandran and Kartha (1955) proposed a model consisting of three parallel chains linked to form a cylindrical rod, with the rods being packed into a hexagonal array. A year later Ramachandran and Kartha (1955) modified this model by adding an additional right-hand twist of 36° every three residues in a single chain (see Figure 2.20).

Further refinement of the structure continued for the next 40 years and has led to the understanding that the hydroxyproline has been shown to stabilize the molecule by forming a hydrogen bond between the polypeptide chains via a water molecule as well as a direct hydrogen between the carbonyl group on one chain and amide hydrogen on another chain within groups of three amino acids (Figure 2.21). In addition, it is believed that the molecule can be broken up into rigid and partially flexible regions associated with the gap (four molecules in cross-section) and overlap regions (regions with five molecules in cross-section) within the fibril structure and that crosslinks hold the fibril together (see Figure 2.22). The collagen molecule is found inside the cell in a precursor form termed procollagen (see Figure 2.23). Once the ends of the procollagen molecule are removed within the collagen fibril it has a length of 300 nm and width of about 1.5 nm.

Recently it has been demonstrated, by analysis of the flexibility from conformation maps of dipeptide sequences, that the collagen triple helix contains rigid domains separated by domains with increased flexibility (see Figure 2.23) (see Silver et 2003, for a review). Autocorrelation of the peptide sequences in collagen, using Fourier analysis, demonstrates that there is a period of the flexible domains in both the molecule and

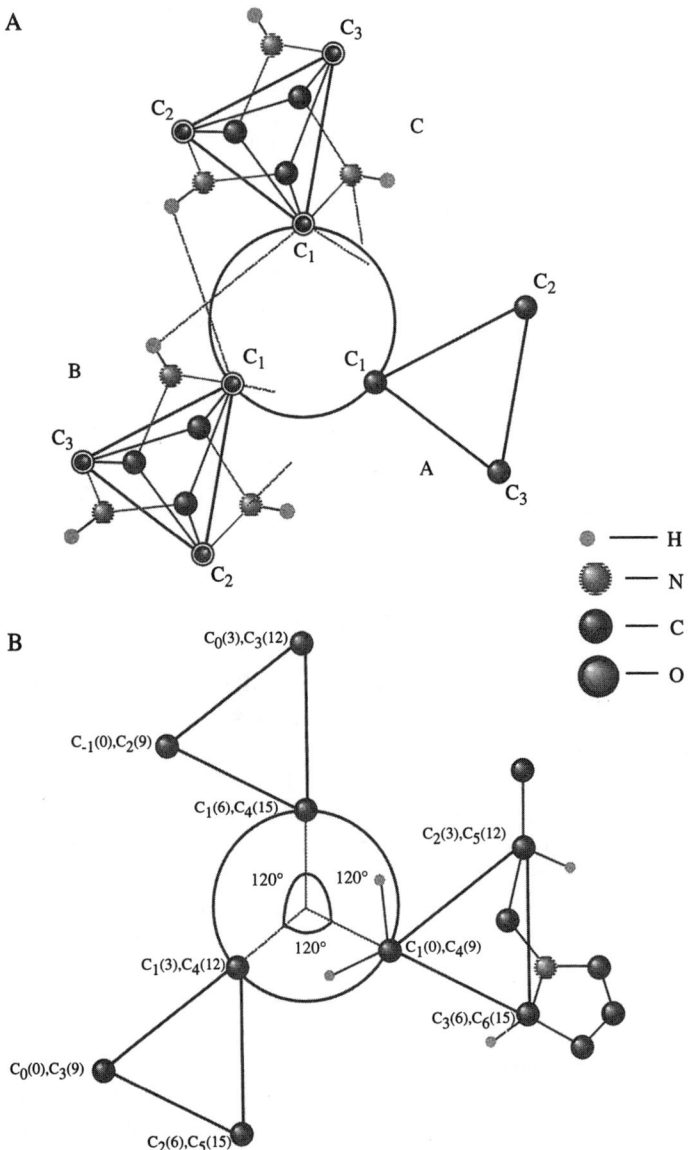

FIGURE 2.20. Models of collagen structure. (**A**) Model of three parallel left-handed helixes of collagen showing the location of Cα (C) for chains A, B, and C. Note all glycines are found in C-1 position because this is the only amino acid residue that can be accommodated at the center of the triple helix. Later studies by Ramachandran and co-workers indicated that the three chains are wrapped around each other (**B**) in a right-handed superhelix. The axial rise per residue is 0.29 nm, and the axial displacement of different Cα atoms is shown in parentheses in angstroms.

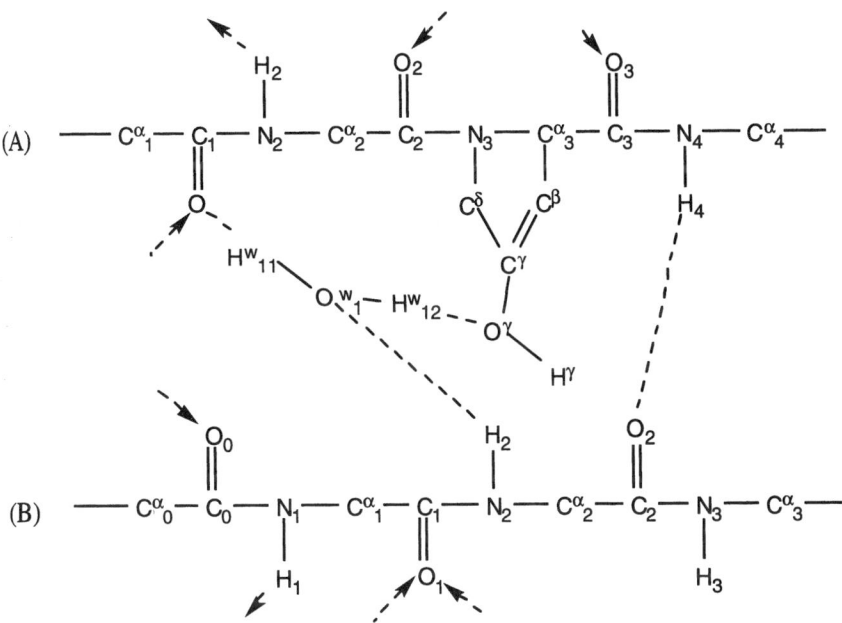

FIGURE 2.21. Stabilization of collagen triple helix. The diagram shows hydrogen bonding between the amide hydrogen in position 4 on chain A and the carbonyl oxygen in position 2 on chain B. A second water-mediated hydrogen bond occurs when hydroxyproline is present in position 3 on chain A between the carbonyl oxygen in position 1 on chain A and the amide hydrogen in position 2 on chain B.

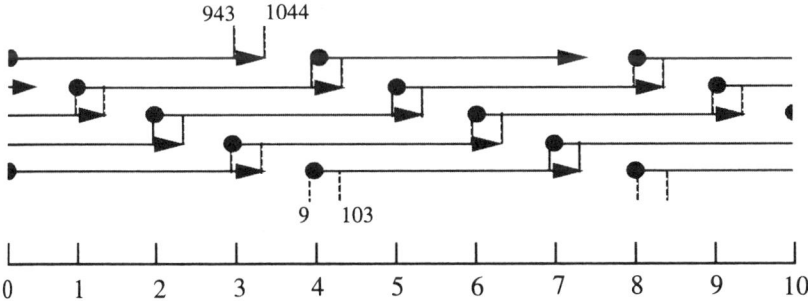

FIGURE 2.22. Cross-linking of collagen molecules in quarter-stagger packing pattern of collagen in fibrils. Each molecule is 4.4-D long (where D is 67 nm) and is staggered by D with respect to its nearest neighbors. A hole region of 0.6 D occurs between the head (circles) of one molecule and the tail of the preceding molecule (arrowheads).

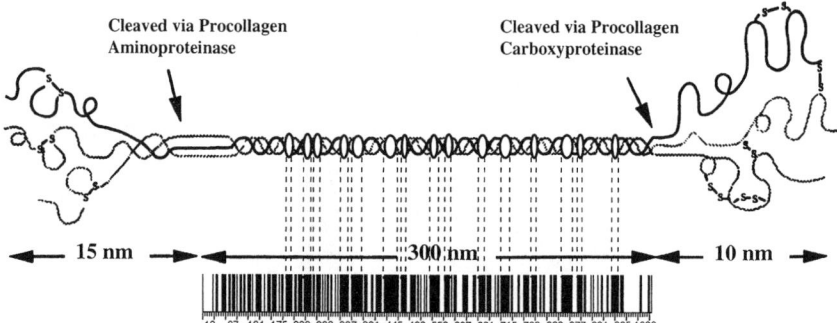

FIGURE 2.23. Diagram illustrating the structure of procollagen, the precursor form of collagen. Procollagen is cleaved to collagen enzymatically during collagen fibril formation. The collagen triple helix is composed of alternating flexible domains (circle) that alternate with rigid domains. The alternation of these domains can be seen as a series of dark (flexible) and light (rigid) regions at the bottom. The collagen triple helix is about 300 nm long and 0.15 nm wide.

collagen fibril. These flexible sequences in collagen are believed to be important in energy storage during mechanical deformation and in attachment to the cell surface through integrin molecules.

Although helical structures dominate the structural hierarchies found in proteins, random chain structures similar to those found in natural rubber also are found in tissues. The most-studied random chain polymer found in vertebrate tissues is elastin.

2.2.3.4.2 Random Chain Coils

Although true randomly coiled chains don't exist in vertebrate tissues, because these structures are susceptible to rapid hydrolysis, sequences that are found in proteins such as elastin are believed to form rubberlike regions that store energy entropically by changing conformational angles during mechanical deformation.

Elastic fibers form the network in skin and cardiovascular tissue (elastic arteries) that is associated with elastic recovery. Historically the recovery of skin and vessel wall on removal of mechanical loads at low strains has been attributed to elastic fibers. Elastic fibers are composed of a core of elastin surrounded by microfibrils 10 to 15 nm in diameter composed of a family of glycoproteins recently termed fibrillins. Fibrillins are a family of extracellular matrix glycoproteins (MW about 350,000) containing a large number of cysteine residues (cysteine residues form disulfide crosslinks). Several members of the family have been described. The common molecular features include: N and C terminal ends with 47 tandemly repeated epidermal growth-factorlike modules separated by a second repeat consisting of eight cysteine residues and other structural elements. Several possible structures for fibrillin have been postulated including unstaggered parallel arrangements and staggered parallel arrangements (Figure 2.24).

Elastin is typically considered as an amorphous protein consisting of random chain sequences connected by α helical regions. The elastin content varies in elastic fibers such as those found in skin. Elastic fibers are termed oxytalan fibers in the upper dermal layer of skin and they are termed elaunin fibers in the deeper dermis where their elastin content is higher. In vessel wall elastic fibers have recently been differentiated based on histological staining patterns suggesting that differences in mechanical properties of different vessel walls may in part be due to differences in elastin

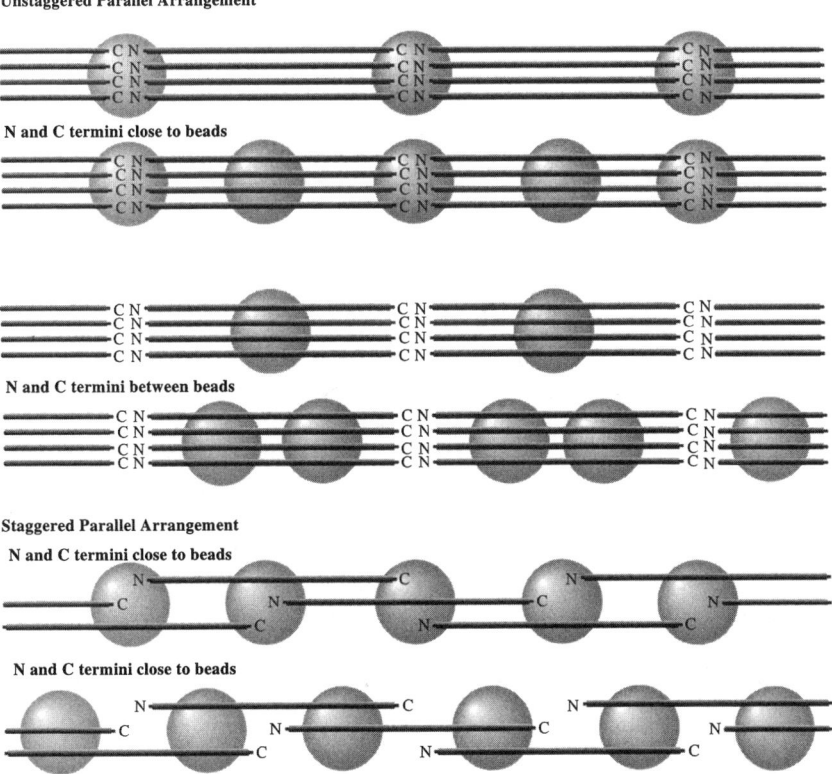

FIGURE 2.24. Structural models for fibrillin in beaded microfibrils. Fibrillin molecules are modeled as being arranged head to tail in parallel arrangements. In some of the models, the molecules are staggered with respect to their neighbors. The circles represent areas where the beads are observed.

concentrations. Mechanically, elastic fibers are in parallel with collagen fibers in skin and vessel wall.

Elastin is a macromolecule synthesized as a 70,000 single peptide chain, termed tropoelastin and secreted into the extracellular matrix where it is rapidly crosslinked to form mature elastin. The carboxy-terminal end of elastin is highly conserved with the sequence Gly-Gly-Ala-Cys-Leu-Gly-Leu-Ala-Cys-Gly-Arg-Lys-Arg-Lys. The two Cys residues that form disulfide crosslinks are found in this region as well as a positively charged pocket of residues that is believed to be the site of interaction with microfibrillar protein residues. Hydrophobic alanine-rich sequences are known to form α helices in elastin; these sequences are found near lysine residues that form crosslinks between two or more chains. Alanine residues not adjacent to lysine residues found near proline and other bulky hydrophobic amino acids inhibit α helix formation. Additional evidence exists for β structures and β turns within elastin thereby giving an overall model of the molecule that contains helical stiff segments connected by flexible segments.

2.2.4 Examples of Other Proteins

Most proteins are a combination of the four structures described above arranged into three-dimensional structures. This results in proteins that have a variety of functional units that have very precise three-dimensional structures. Below we take a look at some of the proteins that are involved in mechanobiology of vertebrate tissues.

2.2.4.1 Keratins

Keratins are found in the superficial layer of skin as well as in the intermediate filaments that support the cell cytoskeleton. Intermediate filaments (IF) are proteins 80 to 120 Å in diameter found in cytoskeletal fibers that reinforce cells. They have molecular weights ranging from 40,000 to 210,000 and are composed of a central domain of 310 to 350 amino acids. There are six types of intermediate filaments; types I and II are composed of keratins that are the largest group of IFs. There are about 30 different protein chains that make up 20 epithelial keratins and 10 hair keratins. Epidermal keratinocytes synthesize two major pairs of keratin polypeptides: K5/K14 of the basal layer and K1/K10 of the cells in the suprabasal layer. Epithelial keratins are expressed in pairs with type I (K between 10 and 20) having acidic groups and type II (K 1 to 9) having neutral and basic groups. It is believed that keratins are expressed as pairs with one acidic and one neutral/basic chain in each molecule.

The basic structural feature of each double stranded (type I/type II hybrid) molecule is the presence of four interrupted α helical sequences termed 1A, 1B, 2A, and 2B which are interrupted by three nonhelical sequences termed L1, L1–2, and L2 (see Figure 2.25). In addition, there is

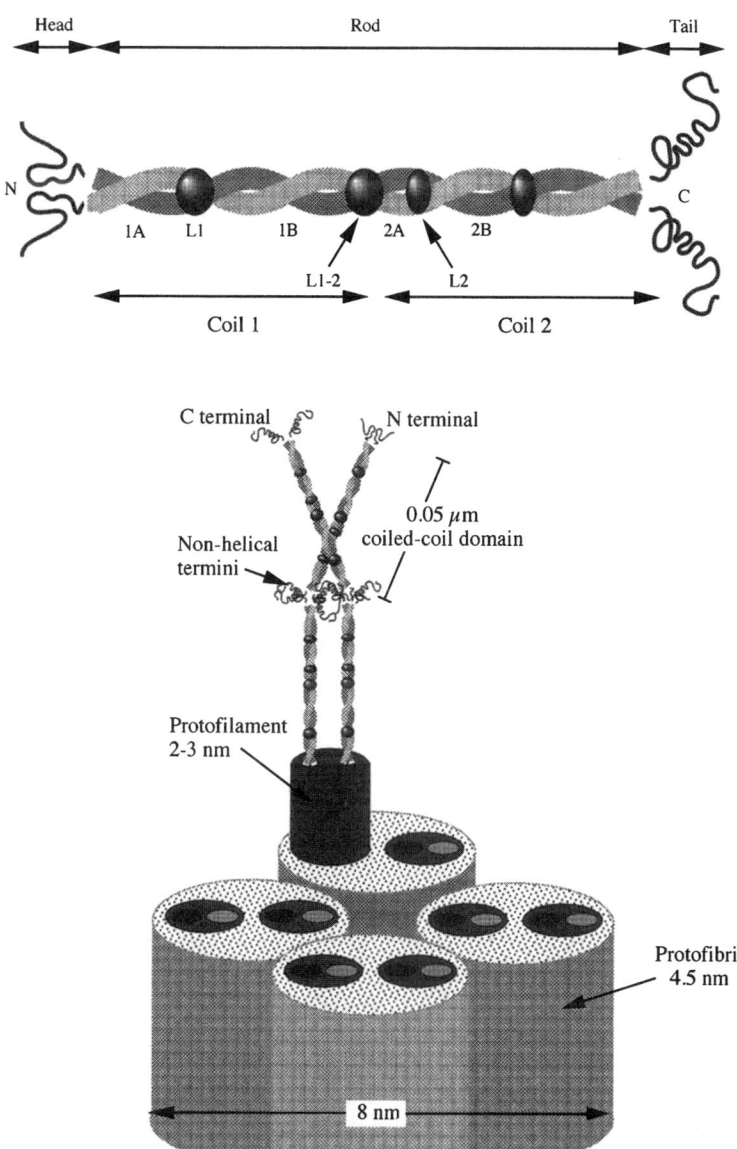

FIGURE 2.25. Structure of keratin protofibrils. The diagram illustrates the structure of keratin in intermediate filaments containing α helical sequences. Two coils are wound around each other and then packed into protofilaments. Eight protofilaments are packed into a filament.

a head and tail domain added to the interrupted α helical domains. The head and tail regions are different for type I and type II keratins. Filament assembly requires at least one type I and one type II keratin that associate parallel and in register to form a coiled-coil. Two coiled-coil dimers then form a ropelike structure in an antiparallel unstaggered or nearly half-staggered array termed a tetramer or protofilament (diameter 2 to 3 nm). Two tetramers form a protofibril (4 to 5 nm) and four protofibrils form filaments 8 to 10 nm in diameter (Figure 2.25).

2.2.4.2 Actin and Myosin

Actin is a protein that is found in cells in the form of filaments and in conjunction with tropomyosin and troponin, forms thin filaments in muscle. Actin exists in two states, a monomeric globular state (overall 3-D sphere-like structure), termed G-actin, and an assembled state, termed filamentous or F-actin. G-actin in the presence of actin binding proteins (ABP) and CA^{+2} and Mg^{+2} assembles into actin filaments in the cell cytoplasm. Actin filaments act as structural supports within the cell cytoskeleton. The 3-D structure of G-actin has been determined from X-ray diffraction studies on complexes containing Ca-ATP-G-actin-DNase I containing small amounts of Mg^{+2} ions (DNase I is added to inhibit F-actin formation). The structure of G-actin at a 2.8 Å resolution consists of small and large domains each divided into two subdomains. ATP is bound in a cleft between the two domains. The actin molecule is composed of helical domains connected by domains containing β extended structures. Actin molecules polymerize into filaments that are coiled-coils described by a left-handed helix with a pitch of 5.9 nm and a right-handed helix with a pitch of 72 nm (Figure 2.26).

Myosin is an enzyme that catalyzes hydrolysis of ATP and converts the energy released into movement through muscle contraction. During muscle contraction, an array of thick filaments containing myosin actively slides by an array of thin filaments containing actin. Myosin has a molecular weight of about 500,000 and consists of six polypeptide chains: two heavy chains (molecular weight 400,000) and two sets of light chains, with each light chain having a molecular weight of 20,000. The molecule consists of a long tail connected to two globular heads as illustrated in Figure 2.27. Each globular head is composed of about 850 amino acids residues contributed by one

FIGURE 2.26. Structure of F-actin. The diagram illustrates assembly of G-actin into double helical segment of F-actin.

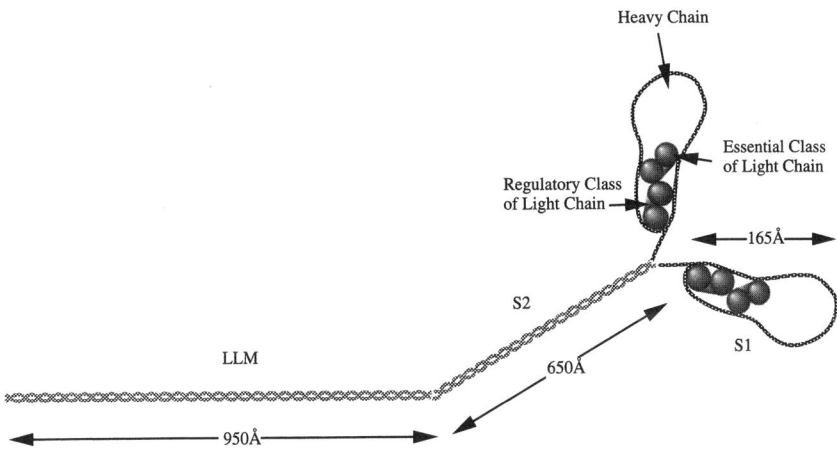

FIGURE 2.27. Structure of myosin. The diagram shows the structure of myosin, which is composed of a helical segment with a flexible bend that is attached to two head units.

of the two heavy chains and two of the light chains. The remaining portions of the heavy chains form an extended coiled-coil that is about 1500 Å in length. The globular head contains the ATP binding sites and the actin-binding region; and the long rodlike portion of myosin forms the backbone of the thick filament. The myosin head has a length of over 165 Å and is about 65 Å wide and 40 Å deep at its thickest end. The secondary structure is dominated by many long α helices and the myosin head is characterized by several prominent clefts and grooves that allow myosin to bind both ATP and actin.

2.2.5 Cell Attachment Factors

The fibronectins are one class of high molecular weight multifunctional gly-coproteins that are present in soluble form in plasma (0.3 g/l) and other bodily fluids, and in fibrillar form in extracellular matrix. They bind to cell surfaces and other macromolecules including collagen and gelatin, the unfolded form of collagen, as well as fibrinogen and DNA. Fibronectin mediates cell adhesion, embryonic cell migration and wound healing. It is composed of two chains, α and β with molecular weights of about 260,000 that are covalently linked via two disulfide bonds near the carboxy termini. This macromolecule may contain some β structure and contains a heparin-binding domain, a cell-binding domain, and a collagen/gelatin-binding domain. The molecule can adopt an extended conformation at high ionic strength or a compact one at physiologic ionic strength suggesting that the molecule is flexible (Figure 2.28). Other types of cell adhesion molecules

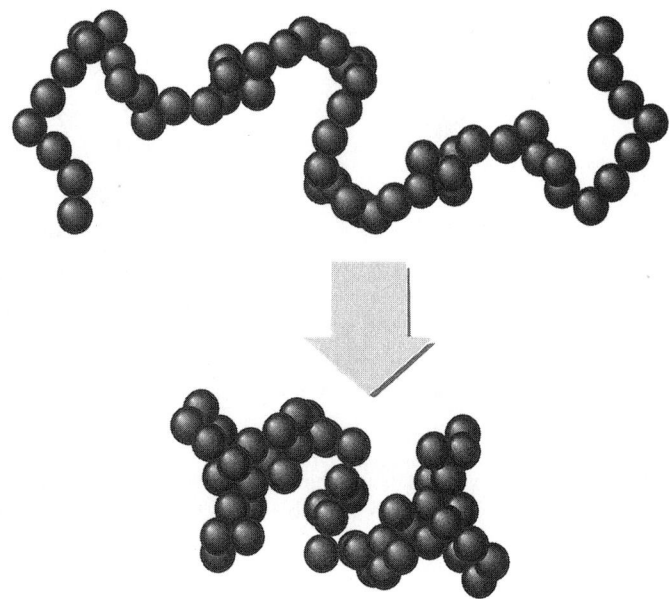

FIGURE 2.28. Structural changes in fibronectin. Structure of fibronectin at high (top) and low (bottom) ionic strength. Note collapse of molecule at low ionic strength.

include laminin, chondronectin, osteonectin, and a variety of other glyco-proteins that moderate adhesion between cells and their ECMs.

Laminins are a family of extracellular matrix proteins that are found in basement membrane and have binding sites for cell surface integrins and other extracellular matrix components. They consist of α, β, and γ chains with molecular weights between 140,000 and 400,000. These chains associate through a large triple helical coiled-coil domain near the C-terminal end of each chain (see Figure 2.29). Eight different laminin chains have been identified, α1, α2, α3, β1, β2, β3, γ1, and γ2. The most extensively character-ized of the seven forms of laminin is laminin-1 (α1β1γ1), which assembles in the presence of calcium to form higher-ordered structures in basement membranes with type IV collagen (Figure 2.29).

2.2.6 Integrins

Integrins are a family of membrane glycoproteins consisting of α and β sub-units (see Figure 1.1). The binding site appears to contain sequences from both subunits, and their cytoplasmic domains form connections with the cytoskeleton. In this manner integrins form a connection between the cytoskeleton and extracellular matrix. In accomplishing this there are 11 α subunits and 6 β subunits forming at least 16 integrins including glycoprotein IIb/IIa expressed by megakaryocytes (cells in bone marrow from which platelets are derived) and platelets, LFA-1, Mac-1, and p150/95 are expressed by leukocytes. Many integrins bind to components of the extracellular matrix including collagen, fibrinogen, fibrin, laminin, and other proteins. Specific examples include the interaction of endothelial cells with core protein of basement membrane perlecan; the expression of $\alpha v\beta 3$ integrin by blood vessels in wound granulation tissue during angiogenesis (formation of new blood vessels); adhesion of platelets and other cell types to collagen involving integrin $\alpha 2\beta 1$; adhesion of natural killer cells (T cells) to fibronectin via VLA-4 and VLA-5; and adhesion of neutrophils to collagen via $\alpha 2$ integrins. Other integrins bind to cell membrane proteins, mediating cell–cell adhesion.

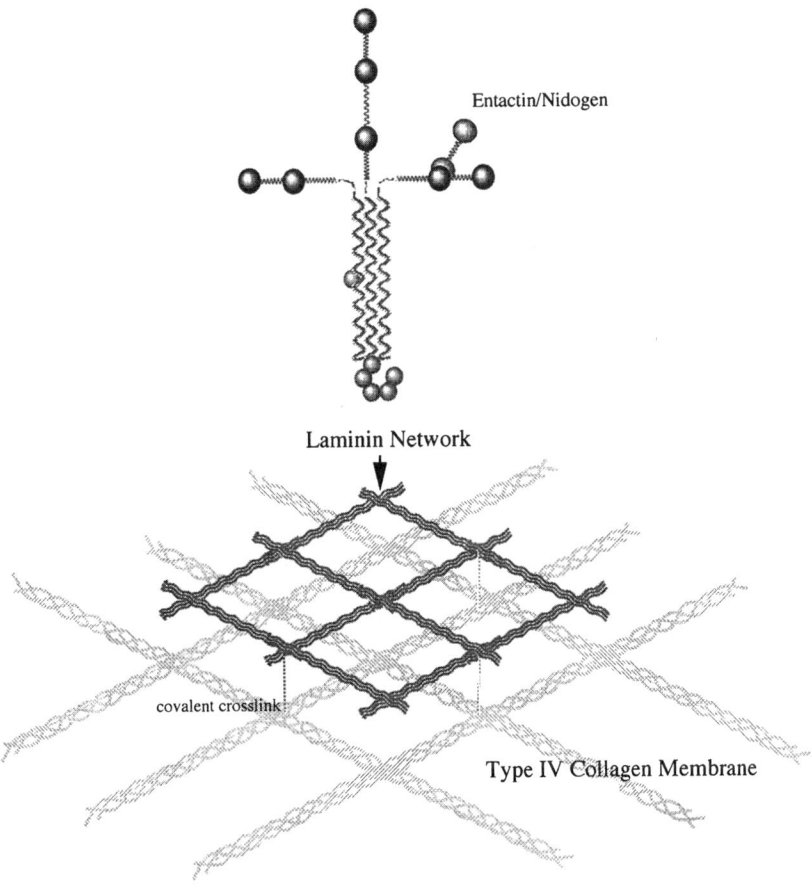

FIGURE 2.29. Network structure of laminin in basement membranes. Basement membranes contain a laminin network that is covalently cross-linked to a type IV collagen network.

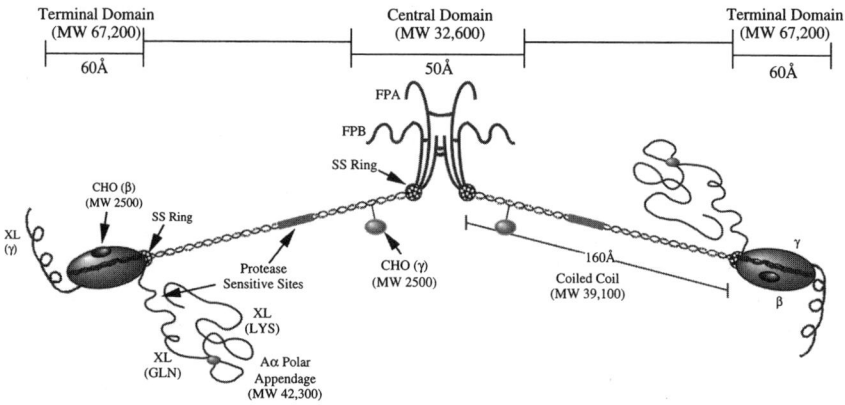

FIGURE 2.30. Domain structure of fibrinogen. The fibrinogen molecule is composed of globular end regions separated from a central domain by helical threads. The central domain has cross-links that hold all the polypeptide chains together. Fibrinopeptides A and B (FPA and FPB) are found in the central domain.

2.2.7 Fibrinogen

Fibrinogen is a plasma protein, formed in the liver that is the basis for the formation of a blood clot. The blood concentration of this protein is about 3% and in the presence of other blood proteins pieces of the fibrinogen molecule, fibrinopeptides A and B, are cleaved and fibrinogen is polymerized to form a fibrin clot (Figure 2.30).

2.2.8 Tubulin

Tubulin is a self-assembling protein that forms microtubules (MTs) within the cell. MTs are involved in a variety of cell functions including vesicle movement, chromosome segregation, and cell motility. MTs are assemblies of heterodimeric proteins, α/β-tubulins. The protein consists of two subunits, a modified α-tubulin and a modified β-tubulin. The two monomers have about 40% of the same amino acid sequences and their structures are similar except for a few differences in the loops. Each monomer contains a pair of central β sheets surrounded by α helices. The tubulin monomers self-assemble into hollow structures, termed microtubules.

2.3 Polysaccharide Structure

Although sugar polymers are less abundant on a mass basis throughout vertebrate tissues they still are important components of the cell surface and extracellular materials. Unlike proteins that are made up of four basic structural forms, polysaccharides are found in a variety of conformational forms. These forms are a consequence of the flexibility of polysaccharides.

2.3.1 Stereochemistry of Sugars

The stereochemistry of polysaccharides is found using procedures similar to those used with polypeptides. The only real difference is that the repeat unit is a sugar and not a peptide unit. Polysaccharides are found as free molecules such as glycogen, starch, hyaluronan, and as side chains on molecules such as proteoglycans. There are three possible conformations of the basic glucose repeat unit: ^{4}C1 or C1 chair conformation, ^{1}C4 chair conformation, or the boat conformation (see Figure 2.3). The most observed conformation of glucose and its derivatives is the C1 chair conformation, also known as the ^{4}C1 conformation. Macromolecular repeat units of glucose can be linked through oxygen atoms that are either up, α linkage, or down, β linkage. D and L sugar isomers occur based on the position of the OH groups on the sugar ring.

2.3.2 Stereochemistry of Polysaccharides

Polysaccharides are large molecules formed when sugar rings are polymerized. In this book we are interested in hyaluronan and the sugar side chains of proteoglycans because these macromolecules make up the interfibrillar matrix that surrounds collagen and elastic fibers and cells. These molecules are highly flexible and allow tissues such as cartilage to compress during joint loading and also assist in collagen fibril rearrangement during loading by allowing interfibrillar slippage.

Hyaluronan (HA) is a component of every tissue or tissue fluid in higher animals. The highest concentrations are found in cartilage, vitreous humor, and umbilical cord and even blood contains some HA. At physiologic pH, the molecule has been shown to adopt a helical structure and therefore the molecule can be modeled as a series of helical segments that are connected with flexible segments. However, based on results of solution studies, the molecular structure of HA appears to be more complicated.

Stereochemical calculations have been conducted on polysaccharides similar to those done on polypeptides to determine the allowable conformations for different repeat units. Figure 2.31 is a diagram showing a repeat unit containing β-D-glucuronic acid and β-D-N-acetyl glucosamine (hyaluronan). Both units are derivatives of glucose and are connected via oxygen linkages where the position of the linkage is either oriented equatorially (i.e., the bonds of the side chains are not perpendicular to the chain backbone), which is termed a β linkage, or they can be perpendicular to the backbone, which is termed an α linkage. If the linkage involves the carbon at the first and third position we refer to it as either an α or β 1–3 linkage

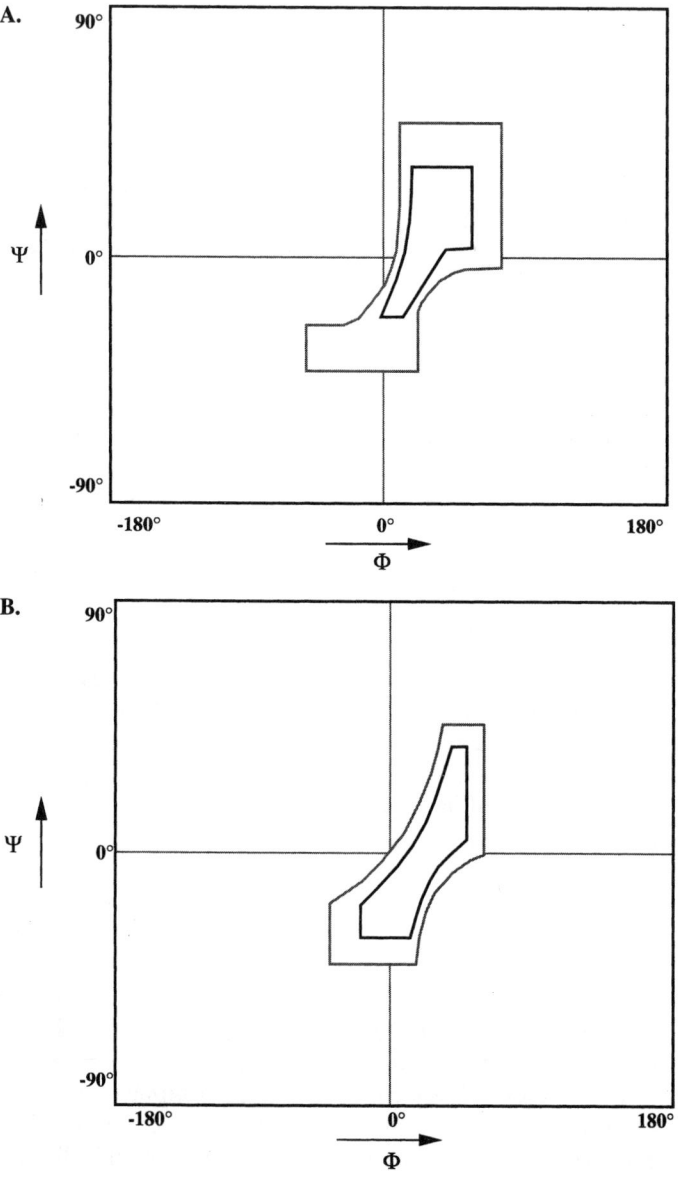

FIGURE 2.31. Conformational plots for β-(1–3) and β-(1–4) linkages in hyaluronan. Plots of fully allowed (inner solid lines) and partially allowed (outer solid lines) conformations of φ and ψ for (**A**) D-glucuronic acid, which is β-(1–3)-linked to N-acetyl glucosamine, and (**B**) N-acetyl glucosamine, which is β-(1–4)-linked to D-glucuronic acid. Note that the allowed conformations center around 0°,0° and show that hyaluronan has flexibility.

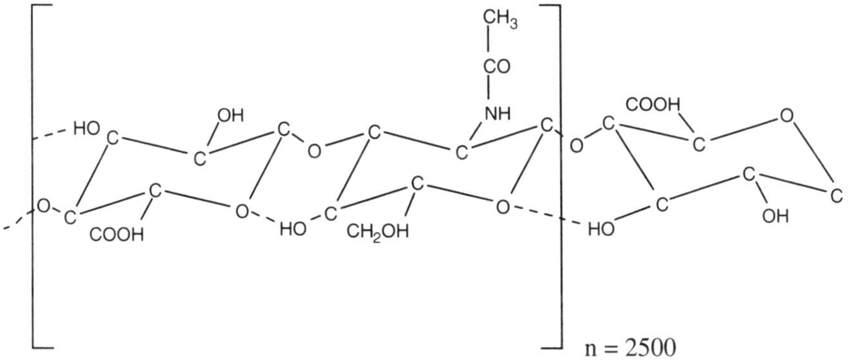

FIGURE 2.32. Repeat disaccharide of hyaluronan. Hyaluronan is composed of β-(1–3) linkage of glucuronic acid to D-N-acetyl glucosamine that is linked β-(1–4) to D-glucuronic acid. Both β-D-glucuronic acid and β-D-N-acetyl glucosamine are in the C₁ chair conformation. The 0°,0° conformation shown occurs when the atoms attached to the carbons connected to the oxygen linking the sugar units are along the axis of the unit (axially) and eclipse each other when viewed along the chain.

and if it involves the first and fourth carbon we refer to the linkage as either α or β 1–4.

The repeat disaccharide shown in Figure 2.32 is composed of D-glucuronic acid β (1–3) linked to D-N-acetyl glucosamine linked β (1–4) to D-glucuronic acid. Both sugar units are shown in the C1 chair conformations. The 0°,0° conformation that is illustrated occurs when the atoms attached to the 1 and 3 or 1 and 4 positions (i.e., the axial-oriented side chains) eclipse each other when viewed from a plane perpendicular to the bonds. The stereochemical plot that is produced when the dihedral angles are rotated through 360 degrees is shown in Figure 2.31. The stereochemical plot for HA shows a limited number of conformations, that is, about 4% of the theoretical total compared to the stereochemical plots for proteins. It is interesting to note that as we discuss further below, hyaluronan behaves to a first approximation at high shear as a flexible molecule as opposed to collagen, which is more rigid and attains its flexibility in another manner. Hyaluronan is flexible because even though the conformational plot has what appears to be a limited number of allowable conformations that center around 0,0, flexibility is associated with the continuous range of allowable conformations around the 0,0 position. What this tells us is that polymers of glucose and glucose derivatives such as cellulose have inherent chain flexibility; however, introduction of a hydrogen bond can limit this flexibility as has been postulated for HA. It turns out these polymers are also thixotropic; that is, at low shear they are rigid and at high shear they are flexible. This property of thixotropy of high molecular weight polysaccharides is a reflection of their stereochemistry and the ability to form hydrogen bonds along the chain backbone that are broken at high strain rates.

2.3.2.1 Supramolecular Structure

A number of helical structures have been proposed for HA in the solid state based on results of X-ray diffraction studies. Although these reports are more than isolated studies, the majority of solution data suggests that HA exists in solution and probably in the interfibrillar matrix as flexible coils that form entangled domains at high molecular weight. Therefore HA and other polysaccharides do not appear to form higher-ordered structures but exist as a series of flexible domains that are characterized by viscosities that are low enough to promote fluid flow during shearing of collagen and elastic fibers. In cartilage the high charge density of negatively charged GAG side chains of proteoglycans (see below) prevents tissue compression and limits flow outflow during locomotion. A diagram of the extended structure of HA is found in Figure 2.32.

2.3.3 Structure of Glycosaminoglycans

Most of the polysaccharides of interest in this text are termed glycosaminoglycans, polymers that contain an amino sugar in the repeat unit. Glycosaminoglycans that are abundant in mammalian tissues include hyaluronan, chondroitin sulfate, dermatan sulfate, keratan sulfate, and heparin–heparan sulfate (see Table 2.4). Most of these glycosaminoglycans,

TABLE 2.4. Glycosaminoglycans found in mammalian tissues

Glycosaminoglycan	Repeat disaccharide
Hyaluronan	β-D-Glucuronic acid + β-D-N-acetyl glucosamine
Chondroitin sulfate	β-D-Glucuronic acid + β-D-N-acetyl galactosamine
Dermatan sulfate	β-D-Glucuronic acid or β-L-Iduronic acid + α-D-N-acetyl galactosamine
Heparin, heparan sulfate	β-D-Glucuronic acid or β-L-Iduronic acid + α-D-N-acetyl glucosamine
Keratan sulfate	β-D-Galactose + β-D-glucosamine

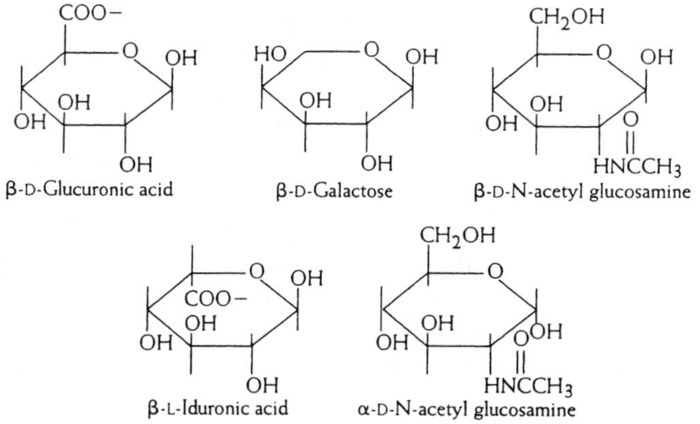

β-D-Glucuronic acid

β-D-Galactose

β-D-N-acetyl glucosamine

β-L-Iduronic acid

α-D-N-acetyl glucosamine

with the exception of hyaluronan, are composed of short polysaccharide chains and have limited secondary structure. However, extensive studies on hyaluronan have indicated that it likely has some limited secondary structure. The rate of oxidation (reaction with oxygen) of hyaluronan and chondroitin sulfate indicate that the C(2)–C(3) glycol group in the glucuronic acid moiety is very slowly oxidized. This could be explained by steric hindrance of the glycol group (OH) in hyaluronic acid and chondroitin sulfate (which could hydrogen bond to N-acetyl glucosamine) but not in dermatan sulfate (which contains N-acetyl galactosamine and cannot hydrogen bond to the same glycol groups). The proposed hydrogen bond scheme is shown in Figure 2.32 and is consistent with a highly extended ribbonlike helix with an axial rise per disaccharide of 9.8 Å. The stiffening of hydrogen bond arrays over short segments of the molecule explains the large expanses (space-filling role) that are occupied by HA molecules. A conformational plot for the repeat units found in HA is shown in Figure 2.33 showing the inherent flexibility of the β(1–3) and β(1–4) linkages.

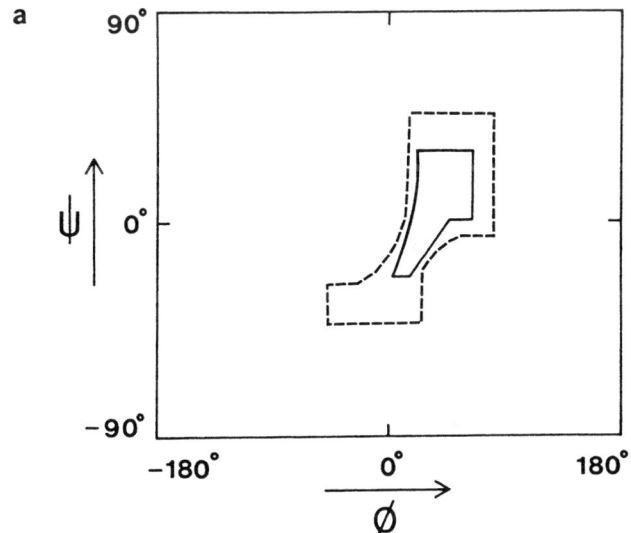

FIGURE 2.33. Conformational plot of β (1→3) and β (1→4) linkages in hyaluronic acid: plots of fully allowed (—) and partially allowed (---) conformations of φ and ψ for (a) D-glucuronic acid which is linked β (1→4) to N-acetyl glucosamine and (b) N-acetyl glucosamine which is linked β (1→3) to D-glucuronic acid. Allowed conformations center around (0°, 0°) and indicate that stereochemically the backbone of hyaluronan has some flexibility.

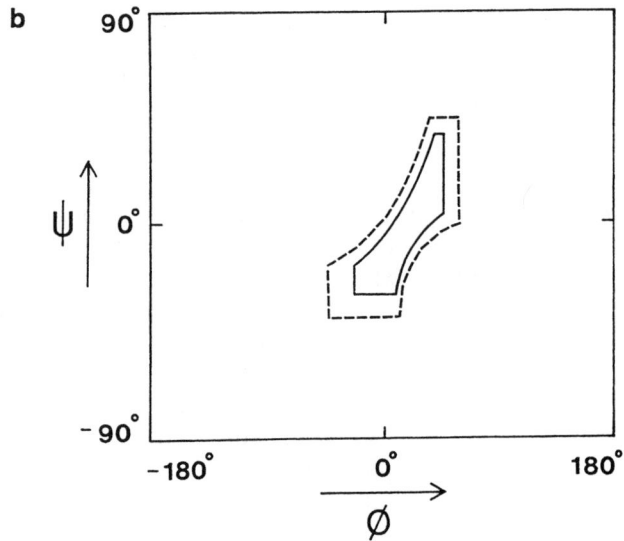

FIGURE 2.33. *Continued*

2.4 Glycoprotein and Proteoglycan Structure

There are other macromolecules found in tissues that are combinations of polypeptides and polysaccharides. These include glycoproteins that are composed of polypeptides to which sugar chains are attached and proteoglycans that are composed of a protein core onto which polysaccharides are grafted. Fibronectin and collagen are actually glycoproteins because both contain sugar rings and sugar polymers that are attached to these molecules.

Proteoglycans are a diverse family of glycosylated proteins, which contain sulfated polysaccharides as a principal constituent. Diverse structures are found for these molecules that depend on the type of glycosaminoglycan side chains, length and net charge. Aggrecan found in large amounts in cartilage (50 mg/g of tissue) is highly glycosylated with 200 chains containing chondroitin sulfate and keratan sulfate. The central core protein is about 220,000 in molecular weight with the overall aggrecan weight reaching 2 to 3 million (Figure 2.34). It interacts with hyaluronan via link protein. In contrast decorin and biglycan have relatively small core proteins (about 40,000) that have a leucine rich repeat and have one (decorin) and two (biglycan) chondroitin/dermatan sulfate side chains (Figure 2.34). Decorin binds to collagen fibrils whereas biglycan does not (Figure 2.35). A unique heparan sulfate proteoglycan, perlecan, is the major PG found in basement membranes. It has a large protein core containing about 3500 amino acids, consisting of multiple domains. Perlecan is able to self-associate or interact with several other basement membrane macromolecules including laminin and type IV collagen. In the kidney, heparan sulfate PG contributes a negative charge to the basement membrane and is thought to exclude serum

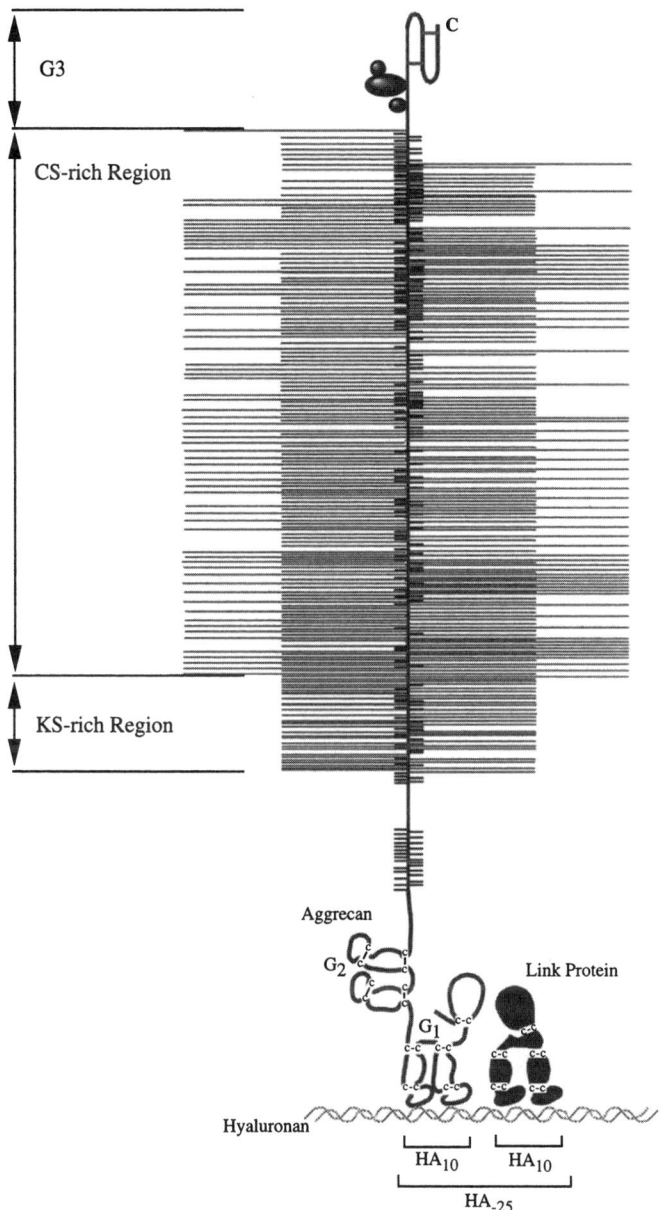

FIGURE 2.34. Interaction between aggrecan and hyaluronan. The diagram illustrates the interaction between large aggregating proteoglycan (aggrecan), link protein, and hyaluronan.

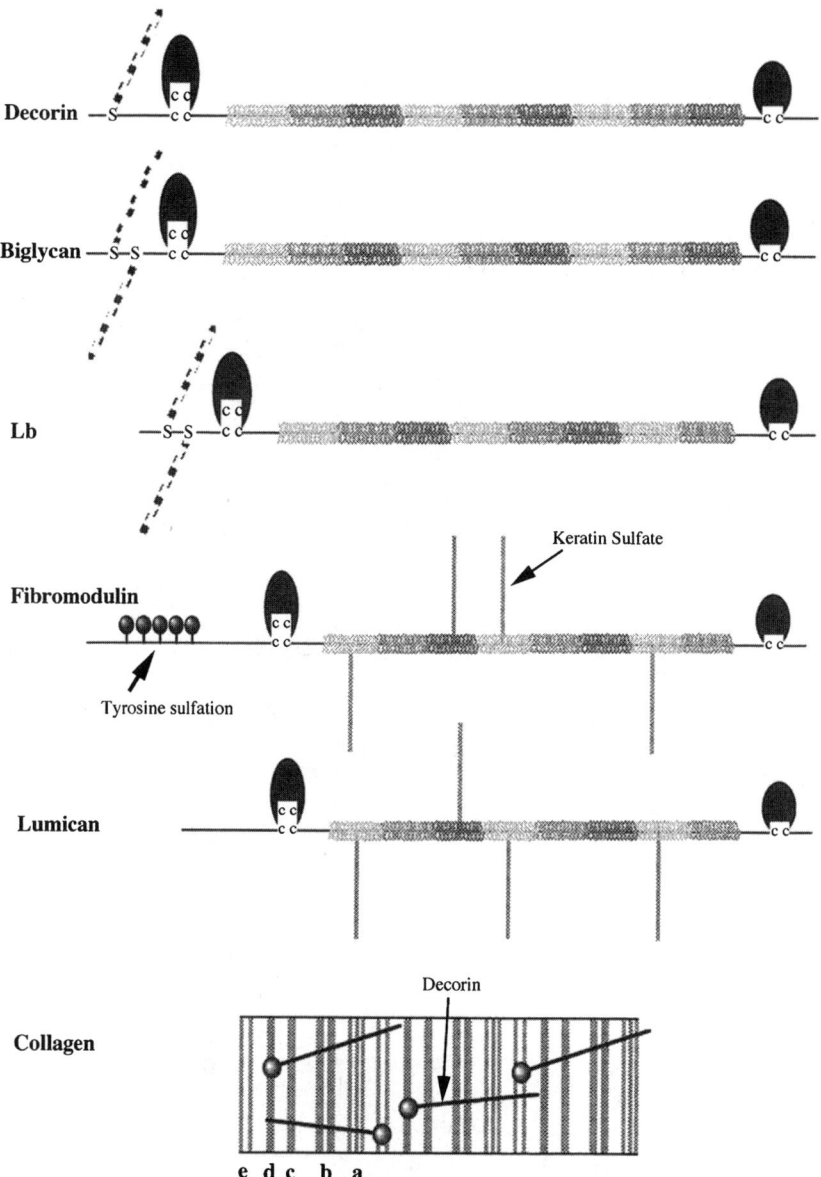

FIGURE 2.35. Comparison between proteoglycan structures. The diagram illustrates structures of decorin (one glycosaminoglycan side chain), biglycan (two glycosaminoglycan side chains), proteoglycan-Lb, fibromodulin, and lumican. Also shown is the specific binding of decorin to the d and e bands on collagen fibrils.

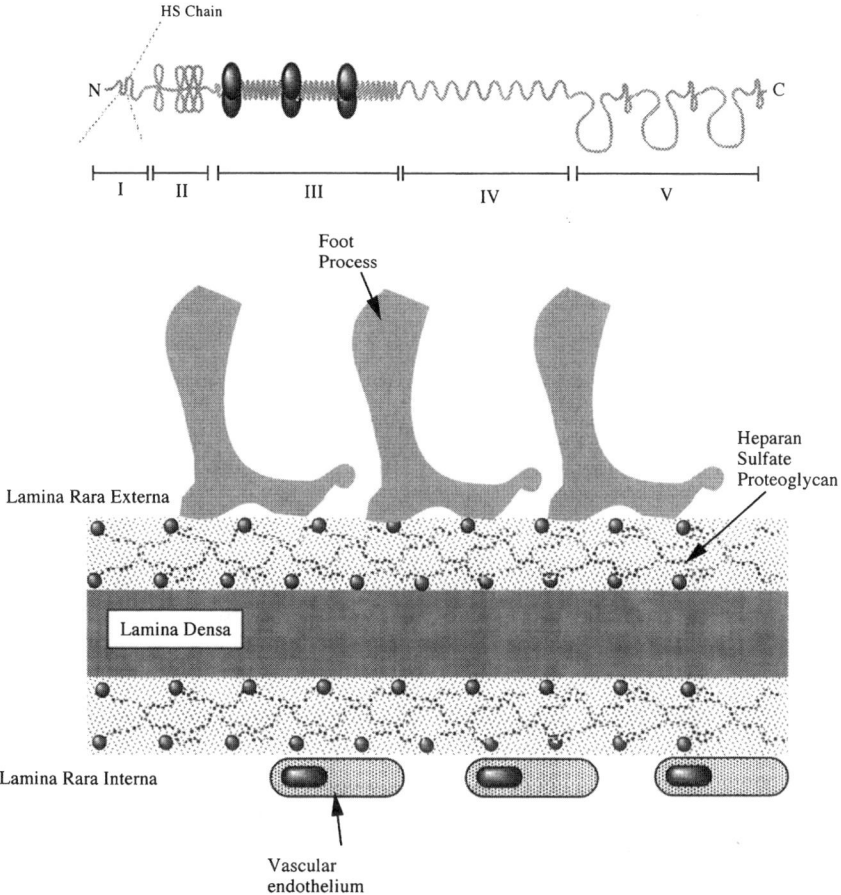

FIGURE 2.36. Model of perlecan in basement membranes. The illustration shows the structure of perlecan, a basement membrane proteoglycan, and localization of proteoglycans in the basement membrane.

proteins from being filtered out of the blood. Although the structure of perlecan is still not completely understood a model has been developed (Figure 2.36).

Virtually every mammalian cell has heparan sulfate proteoglycans as a plasma membrane component. They are inserted into the cell membrane either through a transmembrane domain in their core protein (syndecans), or via a modified glycosaminoglycan region that is linked to the cell membrane (glypican). Heparan sulfate PGs interact with numerous molecules such a growth factors, cytokines, extracellular matrix proteins, enzymes, and protease inhibitors. They are believed to act to transduce signals that emanate from the interplay between components in the extracellular matrix. Syndecan is associated with epithelial cell differentiation after migration and with the intracellular actin cytoskeleton; expression of syndecan by cells appears to inhibit epithelial cell invasion into collagen.

Syndecan consists of a core protein that is inserted through the cell membrane and contains both heparan sulfate and chondroitin sulfate chains. Glypican is covalently linked to the head group of membrane phospholipids in the plasma membrane and contains only heparan sulfate side chains.

2.5 Stereochemistry of Lipids

Lipids are the major components of cell membrane and are also found in blood. The stereochemistry of lipids is very similar to that of poly(ethylene) chains and proteins with small side chains. A freely rotating polymer chain (lipids without double bonds in the backbone) has a stereochemical plot that is very similar to Figure 2.11. Therefore most hydrocarbon chains are quite flexible and adopt a number of conformations. In order to pack hydrocarbon chains efficiently into the cell membrane, the hydrocarbon component is in a planar zigzag. The reason that the hydrocarbon component is only 14 or so carbon atoms long in cell membranes is that longer chains would crystallize as a result of the van der Waals bonds that form between hydrogen atoms between the chains. That is exactly what happens with poly(ethylene) when it is polymerized into chains with molecular weights exceeding 100. This is easily illustrated by examining the physical form of hexane which is a low molecular weight poly(ethylene). At room temperature hexane is a liquid. As the chain length of the hydrocarbon is increased the material is first a wax (i.e., it will flow under pressure and heat such as paraffin) and then at higher molecular weights the material becomes a solid, although if we heat it enough the solid will melt. Therefore, when the chains get long enough the secondary forces between them prevent the molecules from being flexible and allow the chains to crystallize. The flexibility of the chains can be assessed by noting the viscosity of gasoline is very low compared to that of a 2% solution of carboxy methylcellulose (CMC), a food additive used to make food products thick.

The small nature of the hydrogen atoms attached to the backbone and the similarity in size (the chain is composed of a carbon backbone with hydrogen side chains) allow these polymer chains to pack efficiently. They pack efficiently because of the stereochemistry of the side chains. If we look at a drawing of a hydrocarbon chain in a zigzag conformation, the hydrogens as viewed from the side of the chain are staggered and alternately fall between the other hydrogens (see Figure 1.7) or they can be positioned on top of the previous ones (eclipse each other). As we discussed above when we referred to Equation (2.2), the Gibbs free energy should be minimized in the most favorable conformation. The free energy is minimized when the hydrogens are staggered with respect to the backbone of the hydrocarbon chain.

2.6 Stereochemistry of Nucleic Acids

DNA and RNA are the blueprints for making proteins and other polymers involved in cell and tissue metabolism. Examples of the repeat units in DNA and RNA are given in Figure 2.4. The stereochemistry is similar to that of polysaccharides in that a sugar unit is linked through oxygen to another sugar unit. The difference is that the oxygen linkage has several other atoms attached to it between the sugars giving additional flexibility to the chain backbone. Therefore the stereochemical plot for nucleic acid polymers will be similar to that of polysaccharides except there will be additional allowed conformations. This flexibility is necessary for DNA so that it can fold into a double helix as we discuss below.

2.6.1 Primary and Secondary Structure of DNA and RNA

The structure of DNA was originally thought to contain equal amounts of the four purines and pyridines. By the late 1940s it was found that the ratios of adenine and thymine were always very close to unity; the same was true for guanine and cytosine. This implied that for some reason, every molecule of DNA contained equal amounts of adenine and thymine and also equal amounts of guanine and cytosine. Using this chemical information together with X-ray diffraction patterns of DNA, a model was proposed for the structure of DNA in the early 1950s. It was proposed that a molecule of DNA consists of two helical polynucleotides wound around a common axis to form a right-handed "double helix". In direct contrast to the arrangements in helical polypeptides (where the amino acid side chains are directed to the outside of the helix), the purine and pyrimidine bases of each polynucleotide chain are directed towards the center of the double helix in a manner such that they faced each other. Based on stereochemical considerations, it was further suggested that the only possible way that the nitrogen bases could be arranged within the center of the double helix that was consistent with the predicted dimensions was that in which the purine always faced the pyrimidine. Based on consideration of the possible hydrogen bond patterns between purines and pyrimidines it was concluded that adenine must be matched with thymine and guanine with cytosine.

The parameters of the double helix that is formed by DNA include a diameter of 20 Å, a rise per nucleic acid residue of 3.4 Å with ten residues per complete turn. The two chains that make up the molecule are antiparallel; the chains grow by adding repeat units to the 3′ group on ribose or the 5′ group on ribose and therefore one chain is joined 3′ to 5′ by phosphodiester bonds and in the other chain the riboses are joined by 5′ to 3′. The two polynucleotides are twisted around each other in such a way as to produce two helical grooves in the surface of the molecule. The structure of RNAs are different than DNA and do not form a double helix but fold to form different structures such as the large and small subunit of the ribosome.

2.7 Relationship Between Higher-Order Structures and Mechanical Properties

We have learned about the types of macromolecular structures found in tissues. These include helices, extended structures, and random coils. Ultimately, the properties of biological polymers are dictated not only by the macromolecular structure, but also by the levels of structural hierarchy that are found in tissues. For instance, mechanical loading very easily deforms elastin. In contrast, structures containing keratin are less easily deformed not only because they are made up of α helices, but also the α helices are packed into higher-ordered structures. It is the higher-ordered structures that in the case of keratin and even collagen dictate whether a tissue is soft and pliable or hard and rigid. Therefore although helical molecules are in general more difficult to deform (require more force per unit area) than random coils, the presence of higher-order structure gives biological systems the ability to tailor structures and therefore physical properties. The relationship among macromolecular structure, energy storage, and mechanotransduction is explored further in later chapters. The purpose of this chapter is to underscore that force transfer is best accomplished using a polymer that is in an extended conformation. This allows stress to be transferred with a minimum of deformation.

2.8 Summary

Biological tissues contain a variety of different types of macromolecules that exist in α helices, single and double helices, extended chain structures, β pleated sheets, random coils, and collagen coiled-coils that form supramolecular structures that maintain cell and tissue shape, resist mechanical forces, act to transmit loads, generate contractile forces, provide a mechanical link between extracellular matrix and the cell cytoskeleton, and act in recognition of foreign cells and macromolecules. The key element needed to understand how these molecules form is: (1) the chemistry of the repeat unit, (2) the nature of the backbone flexibility, (3) the types of hydrogen bonds that form, (4) the nature of the secondary forces other than H-bonds that form between the chains, (5) the manner in which individual chains fold, (6) the way in which folded chains assemble with other chains, and (6) the manner that assembly is limited. Unfortunately, we are only beginning to understand the intricacies of how chain folding and assembly lead to biological form and function. However, the little we know makes this field so very exciting because nature has created some very clear structure–function relationships. At the very least we now know that structural materials that are aimed at reinforcing tissues have many levels of organization that pack molecules into a regular array. For instance, materials that provide tissue integrity (i.e., keratin and collagen) contain linear regions of highly ordered structure and crosslinks within the molecule to prevent extensive molecular slippage. Macromolecules that transmit force to cells such as actin in the cell cytoskeleton, and myosin in skeletal muscle contain more globular regions that are connected by turns and bends.

Suggested Reading

Laurent TC, Fraser RE. Hyaluronan, *FASEB J.* 1992;6:2397.

Pauling L, Corey RB. The polypeptide conformation in hemoglobin and other globular proteins, *Proc Natl Acad Sci.* 1951;37:282.

Pauling L, Corey RB. Configurations of polypeptide chains with favored orientations around single bonds: Two new pleated sheets, *Proc Natl Acad Sci.* 1951;37:729.

Ramachandran GN, Kartha G. Structure of collagen, *Nature.* 1955;176:593.

Ruoslahti E. Integrins, *J Clin Invest.* 1991;87:1.

Schultz GE, Shrimer RH. Patterns of folding and association of polypeptide chains. In: *Principles of Protein Structure*, New York: Springer-Verlag; 1979: Chapter 5, 66–107.

3
Microscopic and Macroscopic Structure of Tissues

3.1 Introduction

One of the most obvious changes that occur to vertebrate tissues exposed to external mechanical loading involves changes to gross tissue structure. These include thickening of the superficial layer of skin, the epidermis, as a result of mechanical loading of the skin surface or as a result of placing an expandable balloon below the surface of the skin. Increasing the tension in the epidermis by applying frictional forces to the surface or with a balloon expander below the epidermis in the dermis results in an increased epidermal thickness while at the same time the expansion of a balloon beneath the epidermis in the lower layer of skin, the dermis, actually causes the dermis to decrease in thickness. Clearly there is a direct relationship between the type of load and the location that influences tissue metabolism. In this chapter we analyze normal tissue structure and ways that mechanical forces alter it. How do loads applied to the surface of tissues end up affecting the synthesis of proteins in the cells of the tissue? This must involve several factors including mechanosensing (the sensing of mechanical forces) and mechanotransduction (transducing mechanical forces into the generation of new tissue or tissue loss). In order to understand these two concepts we must relate how external mechanical loading affects tissue structure. Therefore we must define what the normal structure of each tissue is in the presence of "everyday" external loads.

Our understanding of normal tissue structure comes from the accumulation of knowledge that includes gross and microscopic observations. Gross observations give information about the size, shape, and surface properties of a tissue. Microscopic structure usually includes images made with light and electron microscopes, which span low magnification (light microscope 4× to 1000×) and high magnification (electron microscope up to 300,000×). At low magnification we can determine the structural units that make up a tissue or organ (Figure 3.1). The structural units usually contain connective tissue and ECM in some 3-D arrangement that is characteristic to that class of tissue. The connective tissue contains specific cell types; the distribution

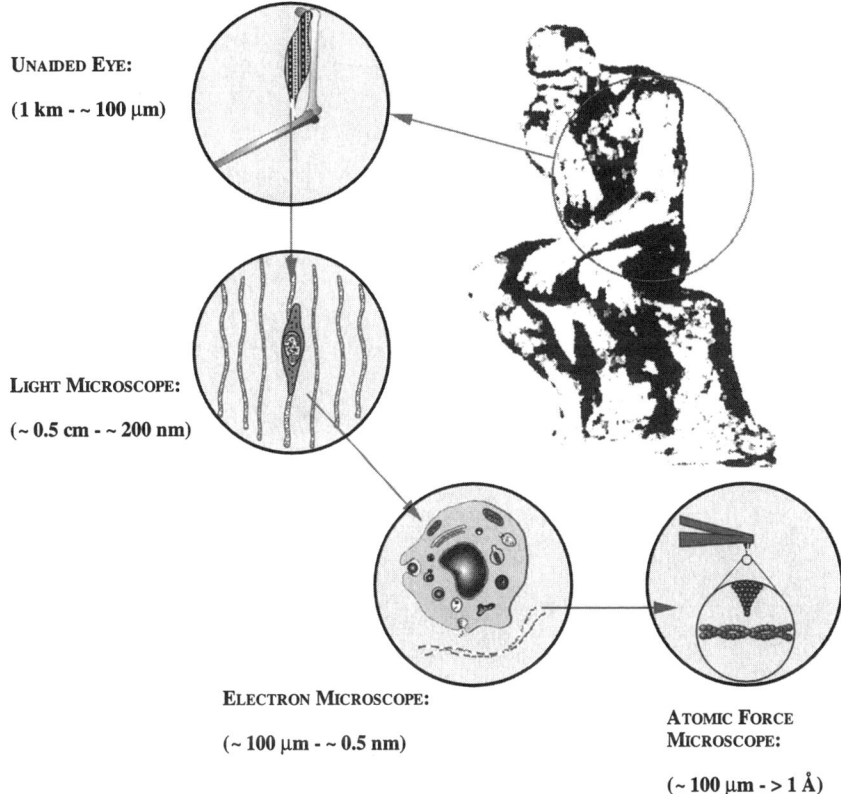

UNAIDED EYE:

(1 km - ~ 100 µm)

LIGHT MICROSCOPE:

(~ 0.5 cm - ~ 200 nm)

ELECTRON MICROSCOPE:

(~ 100 µm - ~ 0.5 nm)

ATOMIC FORCE
MICROSCOPE:

(~ 100 µm - > 1 Å)

FIGURE 3.1. Levels of structural analysis of tissues. The levels of structural hierarchy of tissues include levels viewed by eye, light microscope, electron microscope, and atomic force microscope.

and type of connective tissue components and cell types are a "fingerprint" for each specific tissue. At higher magnification in the electron microscope cell organelles and assemblies of different macromolecules can be identified (Figure 3.1). It is important to understand the arrangement of specific macromolecules inside cells and in the extracellular environment as well as how structural units fit together to make up gross tissue structure because it is the behavior of the macromolecules and their packing arrangement that ultimately dictate how a tissue will behave as is discussed in Chapter 2.

There are a number of techniques that are used to elucidate the structure of tissues at the light and electron microscopic levels. At the light level different techniques are used to process tissue for making standard "paraffin" sections that are the gold standard of light microscopic sample preparation. The steps used to prepare paraffin sections include: fixation, decalcification, dehydration, embedding, section preparation, and staining (Figure 3.2a). Fixation of tissue involves chemical crosslinking to stabilize the tissue to prevent post-mortem changes, preserve various cell constituents, and harden tissue for easier processing. Fixatives include formaldehyde also known as Formalin, glutaraldehyde, paraformaldehyde, and buffered solutions containing these crosslinking materials. In order to preserve tissue structure to reflect normal mechanical loading in vivo, the tissue should be fixed on a mandrel to prevent length, width, and shape

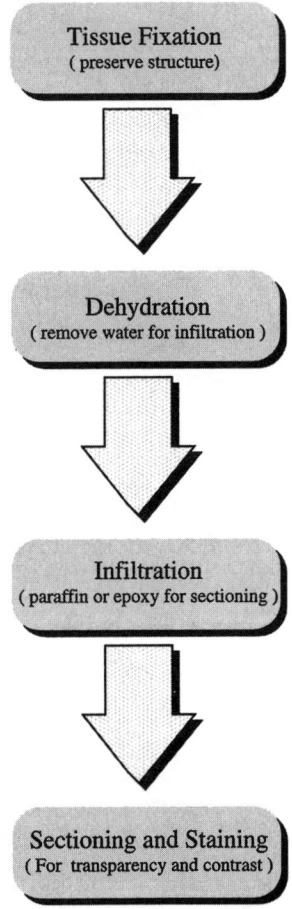

FIGURE 3.2(a). Processing of tissues for microscopy. When processing tissues for light and electron microscopy, tissues are isolated and fixed with solutions containing aldehydes and then dehydrated through alcohol solutions to remove all water. Dry tissue samples are infiltrated with either paraffin wax or epoxy to stiffen the sample. Samples are then cut with a knife into sections several millimeters thick, through which light or electron beams can pass.

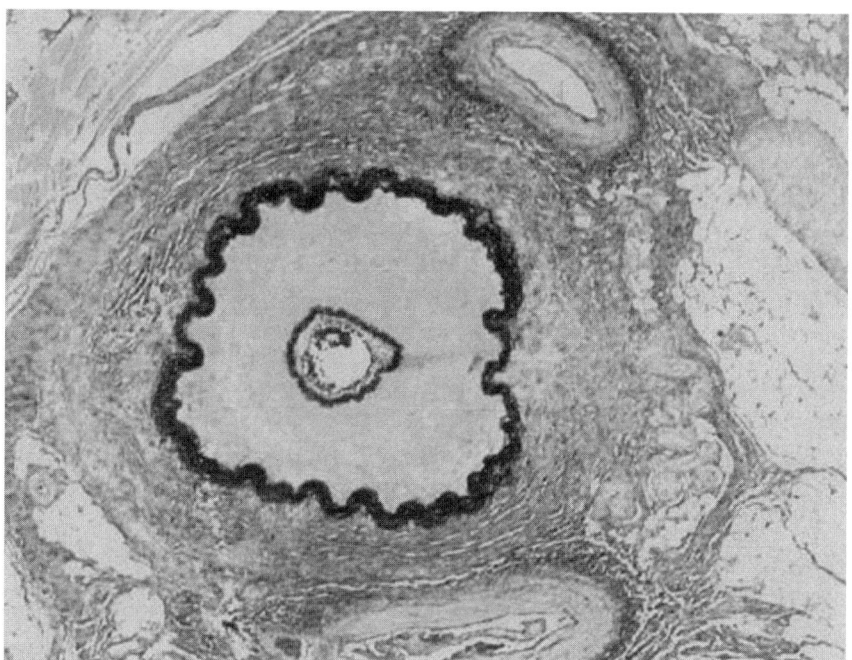

FIGURE 3.2(b). Micrograph of femoral artery in arteriotomy site in a pig stained with Van Giesen's and counterstained with eosin. Note artery in center of picture with dark staining elastic lamellae has fibrous tissue being deposited around it especially at about 4 o'clock where the fibrous tissue appears to be integrating with fat. The individual fat cells (to right of artery) appear to be fusing at this point and appear to becoming integrated into the fibrous tissue. (See color insert)

changes during dehydration. Typically, internal tension that is present in most ECMs is lost when the tissue is released from its contact with the surrounding ECM. Therefore it is important to record the initial tissue dimensions prior to removal from the host. In addition, fixation will cause further tissue shrinkage and alter the tissue structure.

Decalcification of hard tissues can be accomplished using nitric acid, formic acid–sodium citrate, or immersion in a solution containing a chelating agent (an agent that binds calcium) such as ethylene diamine tetraacetic acid (EDTA). Dehydration of the tissue is then accomplished by passing the tissue through a graded series of solutions with increasing concentrations of alcohol. It should be noted that during this process the tissue dimensions shrink; this must be accounted for if you wish to do quantitative light microscopy. The alcohol is replaced with a clearing agent such as xylene which is miscible in both the dehydrant (alcohol) and the embedding material (paraffin or rigid polymer). The clearing agent transforms the translucent tissue into a transparent material. The section is next infiltrated with liquid paraffin to provide support for the tissue block that is then cut into thin sections during sectioning. Sections about 3 to 6μm are cut using a knife mounted on an instrument termed a microtome. The cut sections are floated without wrinkles onto slides and attached to the section using an adhesive such as gelatin. Sections are deparaffinized and stained with dyes such as hematoxylin and eosin, van Gieson's, Mason's Trichrome, and alcian blue.

Hematoxylin is a natural dye substance that combines with a metal salt such as an aluminum salt. After staining with hematoxylin and eosin (H&E), the nuclear structures are stained dark purple or blue and most of the cytoplasmic and intracellular structures are stained various shades of pink. Collagen fibers in the extracellular matrix are stained light pink with eosin (see Figure 3.2b; note this image has been stained with Van Giesen's for elastic tissue, which stains black, and counterstained with eosin, which stains collagen pink). Mason's trichrome is used to stain collagen in tissues; collagen fibers stain blue in contrast to cells that stain purple using this stain. Elastic tissue stains black with van Gieson's solution with the rest of the extracellular matrix staining off-white as a background. If an eosin counterstain is used and the background is pink then other components of the ECM can be visualized (Figure 3.2b). Glycosaminoglycans stain blue with alcian blue and the surrounding extracellular matrix stains pink with eosin as a counterstain.

Sometimes tissues are frozen and then sectioned so that they can be cut directly without the need for embedding. In other instances tissues are embedded in a polymer such as methacrylate instead of paraffin after fixation. There are several variations that are used to make tissue sections to view under the light microscope. In addition to staining with dyes, fluorescent molecules can be specifically attached to macromolecules using antibodies.

At the electron microscopic level, there are two approaches to revealing tissue structure. The first approach involves the analysis of surface structure by bouncing electrons off the surface of a tissue fragment and then forming an image. This process is termed scanning microscopy because scanning the surface collects secondary electrons from the surface of the object. The power of this approach is that the composition of the surface can be determined by analysis of the energy spectrum of the electrons. The energy spectrum of electrons is characteristic of the atoms on the surface. Images obtained by scanning electron microscopy give good resolution up to a magnification of about 40,000×. In transmission electron microscopy, the electrons of the imaging beam go through the specimen (therefore the specimen must be sectioned) and impinge on photographic paper to make an image. The resolution obtained using this technique is much greater than that obtained using scanning microscopy and can be as high as several hundred thousand. Using these two techniques the structure of individual macromolecules as well their assemblies can be elucidated. However, in each tech-

COLOR PLATE

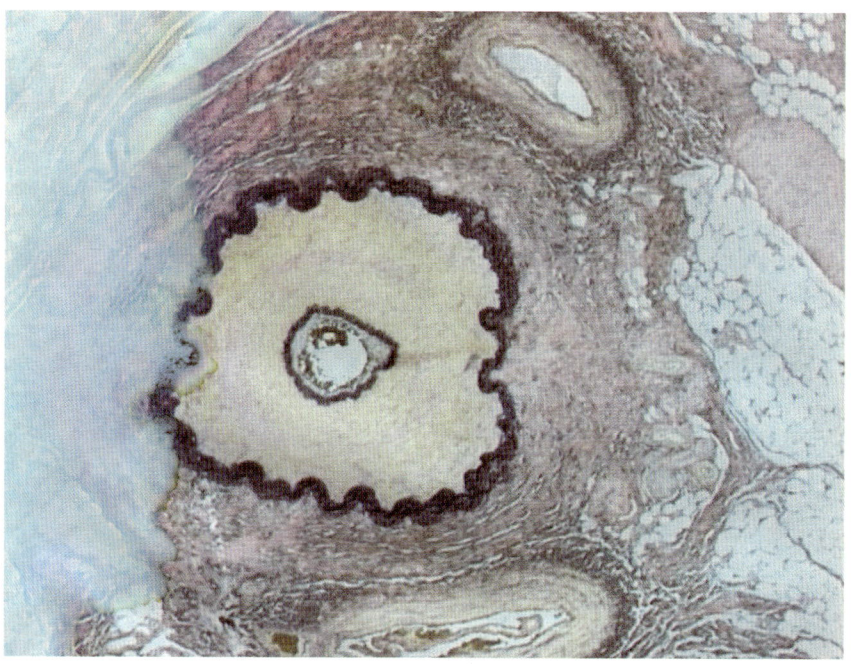

FIGURE 3.2(b). Micrograph of femoral artery in arteriotomy site in a pig stained with Van Giesen's and counterstained with eosin. Note artery in center of picture with dark staining elastic lamellae has fibrous tissue being deposited around it especially at about 4 o'clock where the fibrous tissue appears to be integrating with fat. The individual fat cells (to right of artery) appear to be fusing at this point and appear to becoming integrated into the fibrous tissue.

nique the macromolecule must be first fixed and then either coated with metallic ions by vapor deposition (scanning microscopy) or stained with salts of heavy metals (transmission electron microscopy) to scatter the electrons of the imaging beam. Below we discuss how light and electron microscopic techniques help us elucidate the structure of biological tissues.

3.1.1 Generalized Approach to Tissue Structural Analyses

Memorizing what each tissue looks like is the general method used to analyze tissue structure. The difficulty with this approach is that it is very difficult for the reader to quickly get a solid overview of the subject. To simplify the analysis of tissue structure, we arbitrarily categorize tissues in this chapter into four groups: surface lining structures, conduit and holding structures, musculoskeletal tissues, and specialized organs; three of these categories are tabulated in Table 3.1. The influence of external forces on all of these categories of biological tissues is important. For instance, frictional forces occur at the interface of the skin. These forces cause the epidermis

TABLE 3.1. Surface and internal lining structures

Structure	Composition of lining	Function
Alveoli	Squamous epithelium	Allows oxygen transport
Cornea	Stratified squamous epithelium	Protects eye from injury
Mouth	Stratified squamous epithelium	Protects oral tissues
Peritoneum	Mesothelial cells	Protects stomach organs
Pleura	Mesothelial cells	Protects chest organs
Skin	Stratified squamous epithelium	Protects body surface
Uterus	Columnar epithelium	Protects internal surfaces
Vagina	Stratified squamous epithelium	Protects internal surfaces

Conduit and holding structures

Structure	Composition	Function
Bladder	Mucosa, muscularis, serosa	Holds urine
Blood vessels	Intima, media, adventitia	Transports blood
Bronchiole	Mucosa, smooth muscle, adventitia	Distributes air
Bronchus	Mucosa, submucosa, adventitia	Distributes air
Esophagus	Mucosa, submucosa, muscularis, fibrosa	Collect air
Large intestine (colon)	Mucosa, submucosa, muscularis externa, serosa	Transports food
Rectum	Mucosa, submucosa, muscularis externa, serosa	Transports waste
Stomach	Mucosa, submucosa, muscularis serosa	Hydrolyzes food
Small intestine (duodenum)	Mucosa, submucosa, muscularis externa, serosa	Adsorbs food
Trachea	Mucosa, submucosa, fibrocartilage, fibrosa	Distributes air
Ureter	Mucosa, muscularis, fibrosa	Transports liquid waste

TABLE 3.1. *Continued*

Skeletal structures

Structure	Composition	Function
Articular cartilage	Superficial, intermediate, deep zones	Absorbs shock
Compact bone	Circumferential, concentric lamellar bone	Prevents bending of long bone
Cruciate ligaments	Collagen, proteoglycans	Stabilizes knee
Intervertebral disc	Nuclear pulposis, annular fibrosa	Supports spine
Muscular tissue	Smooth muscle	Constricts tubular walls
	Skeletal muscle	Allows for locomotion
Periodontal ligament	Collagen, proteoglycans	Connects tooth to bone
Spongy bone	Circumferential, concentric lamellar bone with cavities	Stores blood cell precursors

to thicken. In this book we focus on the influence of external forces on surface-lining structures, conduit structures, and skeletal tissues because the literature on mechanochemical transduction is extensive for these tissues. After reviewing the generalized structures of several tissues, we examine the detailed structure of several tissues that are further discussed in subsequent chapters.

3.2 Structure of External and Internal Lining Tissues

Surface and internal lining structures keep what is outside from entering the body and mixing with host internal organ structures. These tissues are normally loaded in tension under normal physiological conditions. The tension decreases with age and the tissues sag as a result. This group of tissues has specialized structures that give specific functions including: (1) selective adsorption of gases (alveoli); (2) prevention of external mechanical and chemical injury to the eye (cornea), oral tissues (muscosal lining), skin (cornified epithelium), and vagina (mucosa); and (3) prevention against internal mechanical injury to stomach and chest organs (peritoneum and pleura).

All of these structures have an epithelial lining that lies at the interface as well as extracellular matrix including basement membranes and loose connective tissue that supports the cellular layers (Table 3.2). These tissues are similar in their general structure: they all have an inner cellular layer, supportive connective tissue, and an outer cellular layer. It is important to be familiar with the structure of these tissues to be able to analyze how external and internal mechanical forces are transduced at both the macroscopic and microscopic level into and out of cells. The effect of mechanical loading on these tissues is complex, but as discussed above, with increased frictional forces on the epidermis, the surface layer of skin actually increases the thickness of the epidermis.

TABLE 3.2. Cellular and noncellular composition of lining structures

Structure	Cellular layers	Noncellular layers	Function
Alveolus	Squamous epithelium	Basement membrane	Lines inside of alveolus
	Capillaries and fibroblasts	Connective tissue	O_2 absorption
Cornea	Stratified squamous		Protects eye
	Columnar epithelium	Bowman's membrane	Supports epithelium
	Stromal fibroblasts	Collagen lamellae	Provides refraction
	Posterior epithelium	Descemet's membrane	Controls fluid transport
Mouth Mucosa	Stratified squamous		
	Columnar cells	Basement membrane	Protects mouth
	Capillaries	Connective tissue	Supports epithelium
Submucosa	Blood vessels	Connective tissue	Provides nutrition
	Nerves and fat		
Peritoneum Pleura	Squamous epithelium (Mesothelium)	Basement membrane	Protects organs
	Fibroblasts	Loose connective tissue	Supports epithelium
Skin Epidermis	Stratified squamous epithelium		Provides resistance to shear injury
	Stratum corneum		
	Stratum spinosum	Basement membrane	
	Stratum basale		
Dermis	Fibroblasts, histiocytes immune cells	Loose connective tissue	Supports epidermis
	Epithelium of accessory structures	Dense connective tissue	Contains accessory structures such as sweat glands and hair follicles
	Endothelium	Blood vessel walls	
	Adiposites		
	Smooth muscle cells		
Uterus Endometrium	Columnar epithelium	Basement membrane	Secretes substances
	Fibroblasts and glandular cells		
	Basal epithelium	Loose connective tissue	
	Endothelium	Blood vessel walls	
Myometrium	Smooth muscle cells	Loose connective tissue	Supports endometrium
	Fibroblasts	Smooth muscle fibers	
	Endothelium	Blood vessel walls	

3.2.1 Histology of Alveoli and Bronchus

External forces acting on the lung include the negative pressure present in the chest cavity as well as the air pressure that pushes against the inside of the alveolus. The alveolus is the gas exchange unit of the respiratory system. The negative pressure that is maintained in the chest cavity expands this unit, passively resulting in the alveolus being stretched even after air is expired. Therefore the alveolus is under passive tension even in the absence of external tension that occurs as the diaphram is loaded. Alveoli consist of mazes of air spaces surrounded by a thin layer of tissue that is lined with squamous epithelium (see Table 3.2 and Figure 3.3). Adjacent alveoli have a common wall (interalveolar septum) containing capillaries supported by small amounts of connective tissue containing fibroblasts. The capillaries are close to the squamous epithelium so that oxygen and carbon dioxide are exchanged by diffusion through the vessel and alveolar walls. External tension that results in stretching of the alveolus results in tension at the cellular level.

The bronchus is a tubular structure that brings air into the alveoli. It is lined with pseudostratified columnar epithelium surrounded by a thin lamina propria containing fine collagen and elastic fibers. A thin layer of smooth muscle surrounds the lamina propria. Glands are found in the submucosa and hyaline cartilage and the pulmonary arteries are found in the outer layer (adventitia).

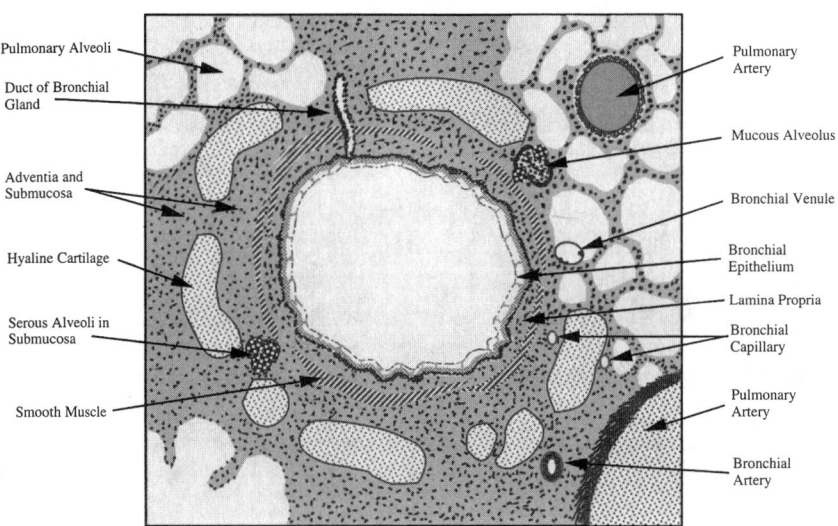

FIGURE 3.3. Structure of alveoli and bronchi. The alveoli absorb oxygen from the air and transport it into the capillaries. The bronchi are tubular structures through which air flows to the alveoli.

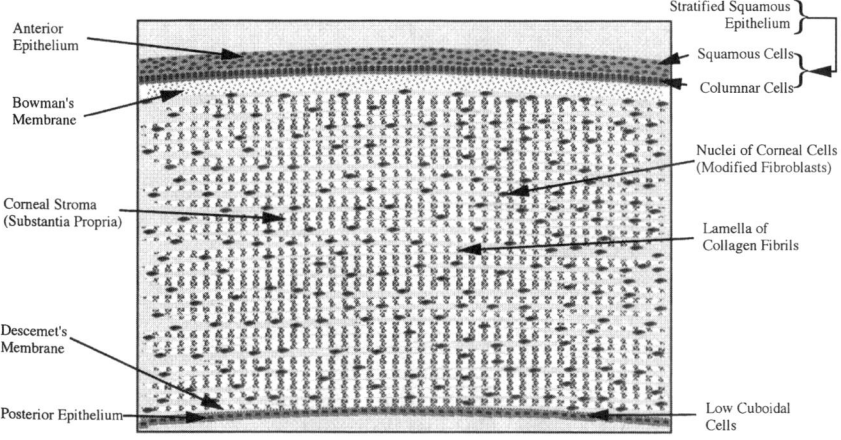

Anterior Epithelium

Bowman's Membrane

Corneal Stroma (Substantia Propria)

Descemet's Membrane

Posterior Epithelium

Stratified Squamous Epithelium

Squamous Cells

Columnar Cells

Nuclei of Corneal Cells (Modified Fibroblasts)

Lamella of Collagen Fibrils

Low Cuboidal Cells

FIGURE 3.4. Structure of cornea. The structural components of cornea include epithelium, corneal stroma, collagen lamellae and Bowman's and Descemet's membranes.

3.2.2 Histology of Cornea

The outer surface of the cornea is covered with a smooth layer of stratified corneal epithelium (Figure 3.4). The lower layer of cells is columnar in shape and rests on a basement membrane that sits on top of a thick limiting structure termed Bowman's membrane derived from the corneal stroma below. The corneal stroma is composed of parallel bundles of collagen fibrils termed lamellae and rows or layers of branching corneal fibroblasts termed keratocytes. The posterior of the cornea is covered with a low cuboidal epithelium with a wide basement membrane (Descemet's membrane) and rests on the posterior portion of the corneal stroma. The corneal epithelium is normally under tension due to the pressure that is present in the anterior chamber just behind the cornea. The intraocular pressure is between 10 and 20 mm of mercury and is enough to cause the cornea to contract about 5% when it is excised from the eye. Therefore this pressure must be transferred between epithelium via cell–cell junctions.

3.2.3 Oral Histology

The mouth and esophagus are composed of two layers, the mucosa and submucosa (Figure 3.5). The mucosa is lined on its outer surface by a stratified squamous epithelium with layers of polyhedral cells of the intermediate layers and low columnar cells of the basal layer. Below the cellular layer is the lamina propria containing loose connective tissue with blood vessels and small aggregates of lymphocytes. Smooth muscle within the mucosa (muscularis mucosal layer) is seen as small bundles. The submucosa

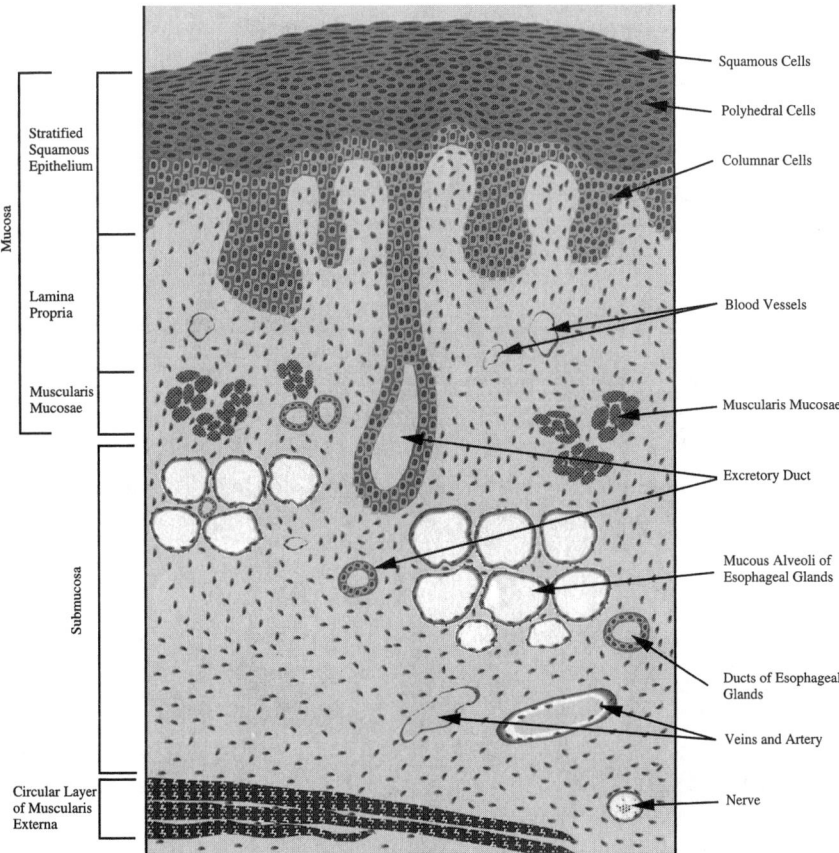

FIGURE 3.5. Structure of oral tissues. The mouth and esophagus are composed of two layers: the mucosa and submucosa. The mucosa is composed of layers of squamous epithelium, lamina propria, and muscularis. The submucosa contains blood vessels, nerves, adipose tissue, and skeletal muscle.

contains blood vessels, nerves, adipose tissue, and skeletal muscle similar to the subcutaneous tissue seen in skin. The mucosa and submucosa are loaded in tension under normal physiological conditions because they contract when they are removed from the underlying connective tissue.

3.2.4 Histology of Peritoneum and Pleura

The pleura and peritoneum are composed of a thin mesothelial cell lining that rests on a basement membrane and a thin layer of connective tissue that contains endothelium and smooth muscle cells (peritoneum) and fibroblasts (Figure 3.6). These cell linings are continuous with internal tissues that lie beneath and also contract when the tension is released but not as much as cornea or skin.

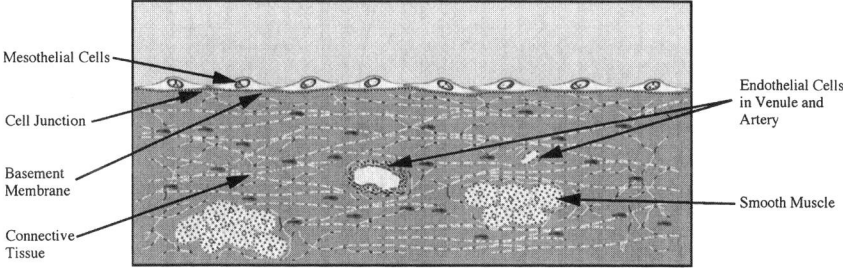

FIGURE 3.6. Structure of peritoneum and pleura. The peritoneum and pleura contain a thin mesothelial cell lining, a basement membrane, connective tissue, and smooth muscle cells.

3.2.5 Histology of Skin

Skin is composed of two layers, the epidermis and dermis (Figure 3.7). The epidermis is a stratified squamous epithelium composed of a number of cell layers including the keratinized outer cell layer (stratum corneum),

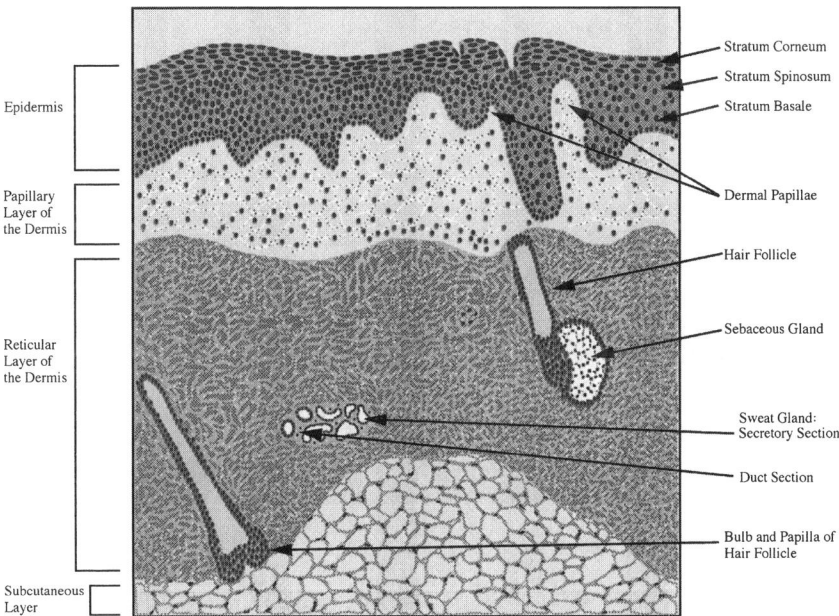

FIGURE 3.7. Structure of skin. Skin is composed of two layers: the dermis and epidermis. Epidermis is a stratified squamous epithelium containing stratum corneum, stratum spinosum, and stratum basale (basal cell layer). The dermis contains papillary and reticular layers and a subcutaneous layer containing fat, blood vessels, and skeletal muscle.

polygonal cells of the stratum spinosum, and the inner layer of basal cells (stratum basale). Basal cells sit on a basement membrane below which is found a fine-fibered dense connective tissue of the papillary dermis. The next layer contains thick collagen fibers and is termed the reticular layer. Reticular dermis is above the subcutaneous layer containing fat (adipose tissue), blood vessels, and skeletal muscle. The accessory structures in the skin are found in the dermis and include hair follicles, sweat glands, and glands that secrete a waxy substance (sebaceous glands). Skin contracts significantly when it is excised from the subcutaneous tissue; the amount of contraction depends on the skin location because the tension is higher on the limbs than on the stomach and back.

We examine the detailed structure and composition of skin below inasmuch as the ECM in skin is involved in sensing of external forces applied to the epidermis and transducing these forces into changes in skin thickness and composition. It is well known that increased friction on the surface of skin leads to thickening of the epidermis.

3.2.5.1 Detailed Structure and Composition of Skin

The epidermis consists of several cell layers beginning with a layer of viable basal keratinocytes that differentiate into a cornified nonviable layer of squamous epithelium that covers the surface (see Figure 3.8). The various

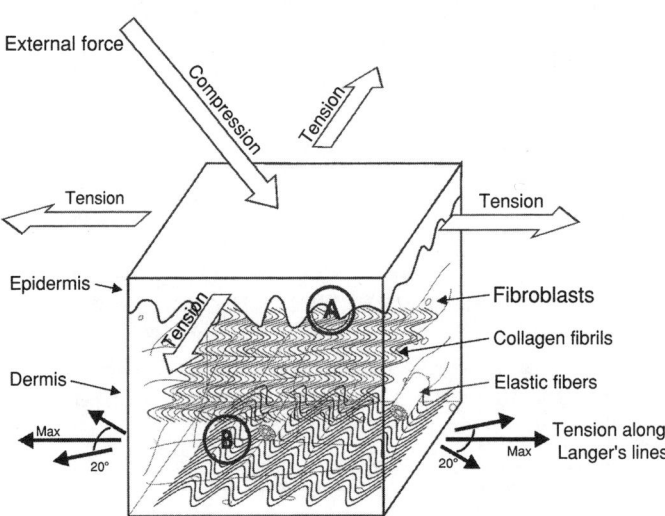

FIGURE 3.8. Diagram illustrating the structure of skin. Note orientation of collagen and elastic fibers with respect to Langer's lines. Both the epidermis and dermis are under tension.

cells of the epidermal layer differ in size, shape, position, and physical properties, which reflect the state of differentiation of the cells in their respective layers. In addition, the exact nature of the mechanical loading on individual cell layers differs depending on the distance from the basement membrane and the level of external loading acting at the air–epidermal interface. In the absence of external loading the stress is highest at the interface between the epidermis and dermis; in areas of skin under high levels of external loading, such as on the hands and feet, the epidermis thickens suggesting that epidermal thickness is controlled by the balance between external and internal forces acting on/in this layer.

Basal keratinocytes are joined together by desmosomal junctions and attach to the underlying basement membrane by hemidesmosomes and integrin receptors. Both types of junctions involve low molecular weight and keratin intermediate filaments (K14 and K15) that extend throughout the cytoplasm and insert into the attachment plaques of the junctions at the cell periphery. Forces are transmitted between epithelial cells via specific cell adhesion molecules including cadherins; mechanical loading leads to activation of secondary messengers that affect genetic expression and cell growth.

Together with actin microfilaments and microtubules, keratin filaments make up the cytoskeleton of vertebrate epithelial cells. Keratins belong to a family of intermediate filament proteins that form α-helical coiled-coil dimers that associate laterally and end to end to form 10 nm diameter filaments. Keratin and actin filaments and microtubules form an integrated cytoskeleton that preserves the shape and structural integrity of the keratinocyte as well as serves to transmit mechanical loads. Keratins account for about 30% of the total protein in basal cells.

Above the basement membrane and basal cell layer in the epidermis is found the spinous layer consisting of three to four cell layers of keratinocytes that are polyhedral in shape. Keratin accounts for up to 85% of the total protein in the cells of the spinous layer and is presumed to stiffen the cell cytoskeleton. Above this layer is the granular cell layer consisting of two to three cell layers where the keratin filaments are associated with a protein, profilaggrin, to form keratohyaline granules. Profilaggrin is a high molecular weight, histidine-rich, phosphorylated polymer composed of monomers joined by link proteins and binds calcium. The protein constituents of the cornified cell envelope become crosslinked by calcium-dependent, epidermal transglutaminases within granular cells. A cytosolic form of epidermal transglutaminase is present in granular cells, and a particulate membrane-bound form is found in spinous and granular layer cells. These transglutaminases are believed to form lysine-derived crosslinks between envelope protein precursors that make up the cell envelope that forms when granular cells flatten, become dehydrated, and cornify.

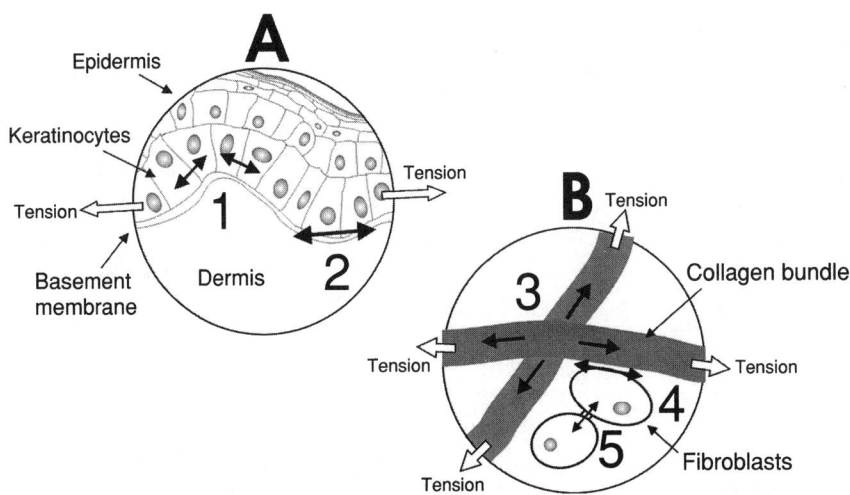

FIGURE 3.9. Diagram illustrating (A) the cellular structure of the epidermis and the location of the epidermis with respect to the basement that is found at the interface with the dermis: (B) the relationship between fibroblasts and collagen fibrils in the dermis.

The boundary between the epidermis and dermis is a basement membrane (see Figure 3.9); it can be described by four planes proceeding from the basal epidermal side to the dermal side: (a) the border of the basal keratinocyte; (b) the lamina lucida, an electron lucent layer that lies beneath the epidermis; (c) the lamina densa, an electron dense layer also known as the basal lamina; and (d) the reticular lamina or subepidermal zone consisting of connective tissue immediately below the epidermis. The mechanical continuity at the epidermal–dermal junction, as well as between keratinocytes, is key to normal transfer of internal and external mechanical forces between the epidermis and dermis.

The epidermis and dermis are connected through the hemidesmosome attachment plaques. The bullous pemphigoid antigen (BPA) is associated with the attachment plaque of the hemidesmosomes where it appears to colocalize with the $\alpha6\beta4$ integrin receptor. BPA may also be found in the lamina lucida adjacent to the hemidesmosome. Keratin filaments of the basal keratinocyte insert into attachment plaques connecting the cell cytoskeleton with the cell surface and to the matrix of the upper dermal layer, the papillary dermis. Anchoring filaments, composed of type VII collagen that span the lamina lucida, are usually increased in density beneath the hemidesmosomes. In most regions of the skin both ends of the anchoring fibrils are embedded into the lamina densa. In patients with a genetic connective tissue disease affecting type VII collagen, termed epidermolysis bullosa, the epidermis spontaneously separates from the dermis upon application of external forces. This observation underscores the importance of maintaining the mechanical continuity between the dermis and epidermis in order to maintain the normal mechanical force transfer.

Thus, external forces applied to the epidermis and/or internal forces present in dermis are transmitted through the cornified layer and lead to stretching of keratinocyte–keratinocyte cell junctions and the epidermal–dermal junction. These forces appear to be transmitted between epithelial cells via cadherins and may result in activation of secondary messengers which alter gene expression and protein synthesis.

Fibers composed of type I and III collagens are found in both the papillary and reticular dermis; the type III to type I ratio is somewhat higher in the papillary layer as compared to the reticular layer with type I collagen comprising about 80 to 90% of the total collagen content. The mean fractional volume of collagen fibers for both papillary and reticular dermis is about 70%. In the young adult, the collagen in the papillary dermis appears as a feltwork of randomly oriented thin fibers; and in the reticular dermis it consists of loosely interwoven, large, wavy, randomly oriented collagen bundles. Reportedly, the spaces between individual collagen bundles decrease with age, which is reflected as an increase in collagen fiber density. The fibers comprising the bundles appear to unravel. In general, collagen concentration in the skin of rats is known to increase up to six months of age, after which it decreases. The type III collagen content of rat skin falls from 33% at two weeks of age to 18.6% at one year.

Elastic tissue in skin forms a three-dimensional network of branched fibers of variable diameter and elastin content, which spans from the papillary layer to the deep dermis (Figure 3.8). Mechanically, the elastic fiber network of skin is in parallel with that of collagen and in skin from older individuals elastic fibers appear to fray and contain holes. Diameters of elastic fibers increase from about $1\,\mu$m to $2\,\mu$m in proceeding from the papillary to the reticular dermis. They form a continuous network that can be isolated by treatment with strong alkali and autoclaving after removal of other components. Individual elastic fibers are composed of what was originally believed to be an amorphous core of elastin, which constitutes up to 90% of the fiber and a microfibrillar component consisting of 10 to 12nm diameter fibrils. Later studies suggest that elastin may contain regions with different levels of order. The microfibrils are composed of fibrillins and microfibrillar-associated glycoproteins. Oxytalan fibers, found in the papillary layer, are composed of bundles of 12nm microfibrils with little associated elastin. They extend into the papillary dermis from the dermal–epidermal junction, where they merge with elaunin fibers, with a higher elastin content, and then join elastic fibers in the deeper dermis. Elastic fibers in the deep dermis contain about 90% elastin. The relative volume of elastic fibers increases from about 0.7% to about 2.5% in proceeding from the papillary to the recticular dermis.

Skin contains a number of glycosaminoglycans (GAGs) including hyaluronan, which is not connected to a protein core, and a number of proteoglycans consisting of a protein core to which GAG side chains are attached. Proteoglycans found in skin include heparin/heparan sulfate proteoglycan, chondroitin-6-sulfate proteoglycan found primarily associated with basement membranes, chondroitin sulfate/dermatan sulfate proteoglycan in dermal matrix, and low levels of keratan sulfate (Figure 3.10). The

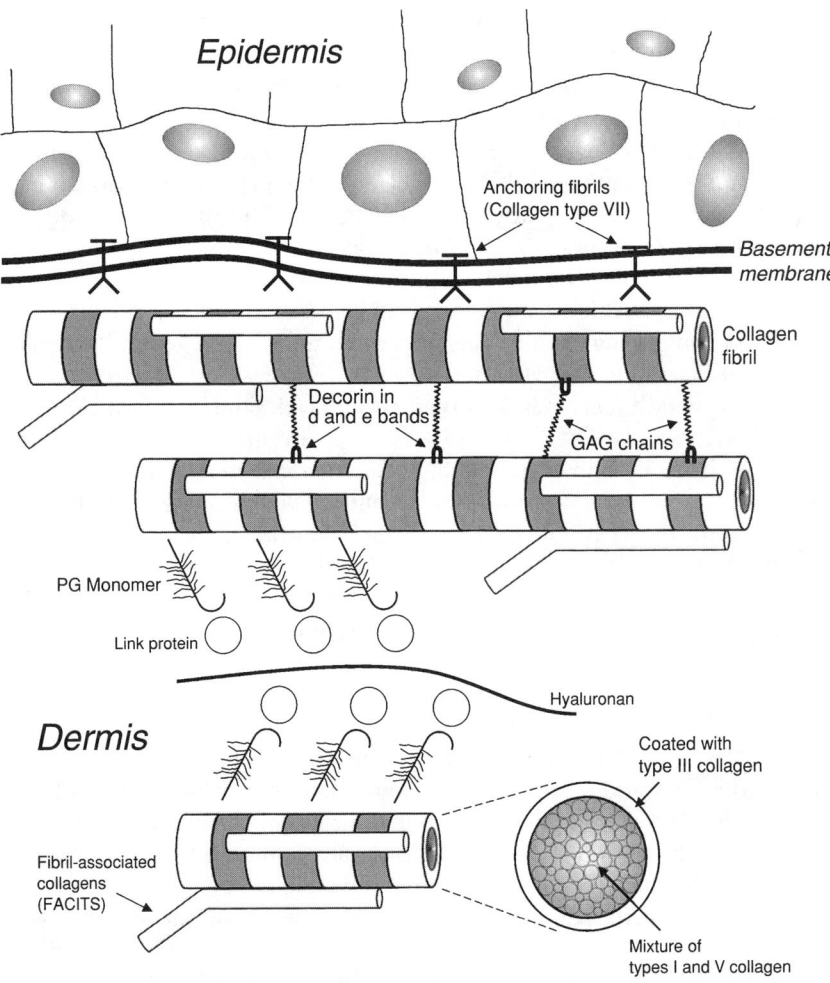

FIGURE 3.10. Diagram illustrating the components of the dermis. Collagen type VII fibrils anchor the epidermis to the dermis. Collagen fibrils in dermis are composed of types I, III, and V and are coated with proteoglycans (PGs).

high molecular weight chondroitin sulfate PG in skin is versican, whereas the low molecular weight keratan sulfate containing members of the leucine-rich proteoglycan family found in skin includes lumican, decorin, and biglycan. Decorin and biglycan have small molecular sizes and consist of one (decorin) or two (biglycan) GAG chains attached to a core protein with a molecular weight of about 40 kDa. Both decorin and biglycan share homologous core proteins containing 7 to 24 repeats of characteristic leucine-rich amino acid motifs. Decorin has been shown to bind to type I collagen via leucine-rich regions designated 4 to 5 although biglycan does not bind. Lumican is found associated with fibrillar collagens in skin and in its absence, animals develop abnormally thick collagen fibrils.

The hyaluronan content of skin has been estimated to be between 0.03% and 0.09%, and dermatan sulfate represents 30 to 40% of dermal proteoglycans. The GAG content is reported to decrease with respect to the amount of protein with increased age, and a recent report suggests that there is a decrease in the proportion of versican and an increase in decorin with increased age, which is associated with the appearance of a small PG that may be a catabolic fragment of decorin.

Thus, dermis contains a variety of macromolecular components that are involved in the transmission and distribution of forces. Collagen fibers appear to transfer internal tension and external loads to the surrounding ECM containing nonfibril-forming collagens, FACITs, other minor collagen types, proteoglycans, elastic fibers, and other macromolecular components as well as to fibroblasts. External forces applied to the skin not only cause stretching of keratinocyte–keratinocyte cell junctions and the epidermal–dermal junction but also appear to lead to stretching of elastic and collagen fibril networks as well as collagen–fibroblast and fibroblast–fibroblast interfaces.

3.2.6 Histology of Uterus

Uterus is composed of two layers termed the endometrium and myometrium (Figure 3.11). On the outside is a lining layer termed perimetrium that is a mesothelial layer similar to peritoneum. The endometrium is composed of two layers termed functionalis and basalis. The functionalis is composed of an outer layer of columnar secretory epithelium overlying a connective tissue layer termed laminar propria. Long tubular glands, the uterine glands are present in the laminar propria. The glands are usually straight in the superficial portion and are coiled in deeper portions.

The myometrium contains compactly arranged smooth muscle fiber bundles separated by thin partitions of connective tissue as well as the blood vessels that provide nutrition to the organ. Contraction of the myometrium results in forces being applied to the uterus and is another example of force application to a surface lining tissue.

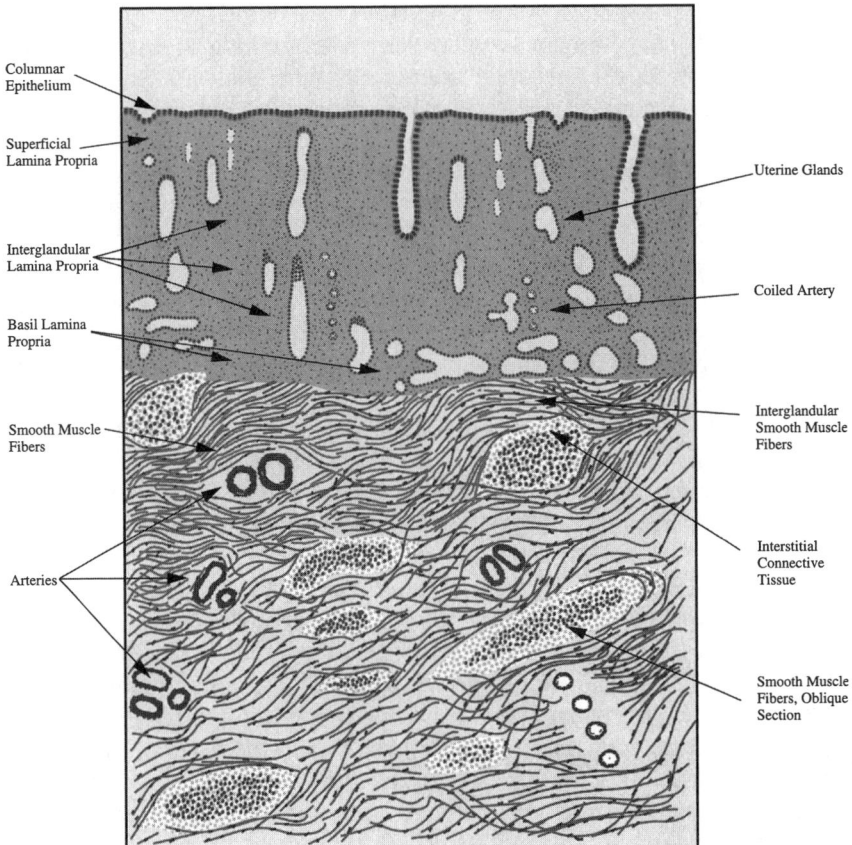

FIGURE 3.11. Structure of uterus. The uterus contains two layers: the endometrium and the myometrium. Endometrium contains columnar epithelium, lamina propria, uterine glands, and arteries. Myometrium contains smooth muscle fibers, interstitial collagen, and arteries.

3.3 Conduit and Holding Structures

Conduit and holding structures are a second category that we have used to introduce structures that are important in this text. Fluid is transported throughout the body through these structures and fluid flow is accomplished by muscular contraction of the wall. This increases fluid pressure as well as exposes cells to tensile and shear forces as the wall stretches. These structures include conduits for distributing blood (heart and blood vessels), food (stomach, small intestine, and large intestine), and waste products (bladder and ureter). Most of these structures contain three layers similar to blood vessels. They are composed of an inner cellular lining (intima), a muscular layer (media), and an outer layer (adventitia) containing connective tissue that anchors the tissue to surrounding structures. Below we examine each of these subclassifications of structures to determine the name of the arrangements of the cells and extracellular matrix (Table 3.3).

The influence of mechanical loading on conduit and holding structures is only partially understood. Increased intravascular pressure associated with hypertension leads to increased vessel wall diameters and thickness. This may be due to increased mechanotransduction by medial smooth muscle cells.

3.3.1 Structure of Blood Vessels and Lymphatics

The heart and blood vessels pump and distribute blood to the peripheral circulation. This is accomplished by contraction of the heart wall increasing the pressure exerted on the fluid inside, which increases the tension on cells and ECM that make the tissue. In situ, blood vessels are stretched along their axis and radially by about 40% to 50% during the low pressure (diastolic) part of the pressurization cycle. This "prestress" and strain provides a mechanical signal to maintain cellular synthetic properties that allow the cells in blood vessels and heart wall to quickly respond to increased resistance to blood flow and higher demand for oxygen. This cyclic stretching leads to a progressive increase of vessel wall diameters with increased age and ultimately can lead to aneurysm or ballooning out of vessel walls.

Blood vessels include capillaries, arterioles, arteries, and veins. Each of these structures is composed of three layers: the intima, media, and adventitia (Figure 3.12). For example, in elastic arteries the wall consists of an intimal layer containing endothelial cells and connective tissue; a media containing smooth muscle cells, collagen, and elastic fibers; and an adventitia containing collagen fibers, nerves, and blood vessels. In cross-section,

TABLE 3.3. Cellular and noncellular composition of conduit and holding structures

Structure	Cellular layers	Noncellular layers	Function
Blood vessels			
Intima	Endothelium	Basement membrane	Maintain blood flow
Media	Smooth muscle	Connective tissue	Prevents blowout
	Fibroblasts	Elastic fibers	Auxillary pump
Adventitia	Fibroblasts	Connective tissue	Anchors vessel
	Nerve	Connective tissue	
	Blood vessels	Connective tissue	
Stomach and intestines			
Mucosa	Columnar epithelium	Basement membrane	Food transport
	Glandular cells or pyloric cells	Connective tissue	Release of material
	Fibroblasts	Connective tissue	
Submucosa	Fibroblasts	Connective tissue	
	Glands (intestine)		
Muscularis	Smooth muscle	Connective tissue	Contractile motion
Serosa	Fibroblasts	Connective tissue	Part of peritoneum

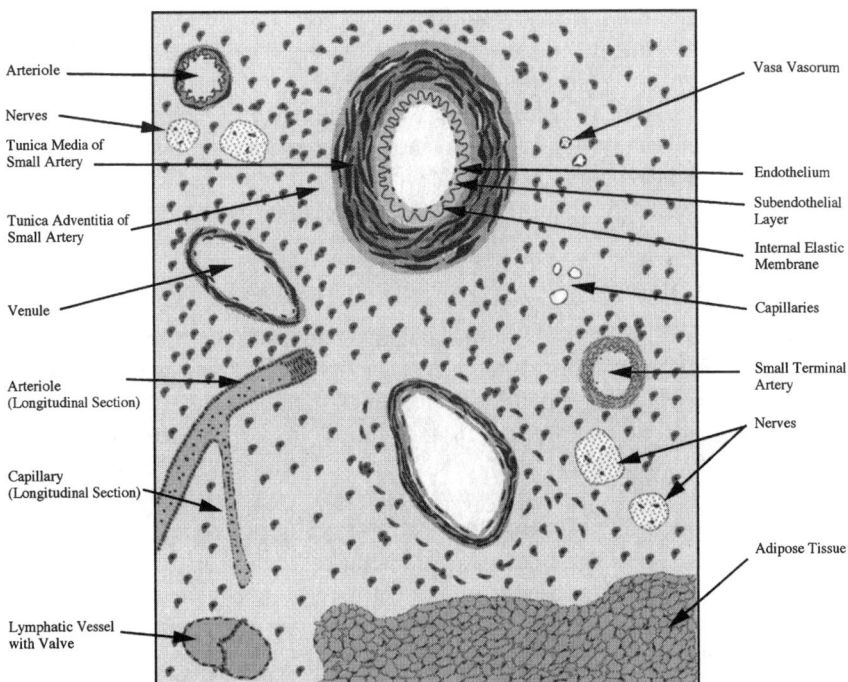

Arteriole

Nerves

Tunica Media of
Small Artery

Tunica Adventitia of
Small Artery

Venule

Arteriole
(Longitudinal Section)

Capillary
(Longitudinal Section)

Lymphatic Vessel
with Valve

Vasa Vasorum

Endothelium

Subendothelial
Layer

Internal Elastic
Membrane

Capillaries

Small Terminal
Artery

Nerves

Adipose Tissue

FIGURE 3.12. Structure of vessels. All vessels contain three layers: the intima, media, and adventitia. In large elastic arteries, the intima is found beneath the internal elastic membrane and interfaces with the lumen. The media is found between the internal and external elastic membranes, and the adventitia is found outside the external elastic membrane. The media is less prominent in the other types of vessels.

arteries and arterioles are circular; the only structural difference is that arteries are larger in diameter and have a more prominent media. Capillaries are also circular and usually contain red blood cells but are much narrower in diameter. In contrast, veins and venules are oval and contain a less prominent media. Lymphatic vessels (vessels that recirculate interstitial fluid to the vasculature) are very thin walled and typically are recognized by the valves within the lumen. Endothelium, fibroblasts, and smooth muscle cells must monitor the blood and tissue stresses and respond to increased tension by synthesizing new components of the ECM and by undergoing cell division because both the diameter and wall thickness of the elastic arteries are observed to increase with increasing age and blood pressure.

3.3.1.1 Detailed Structure and Composition of Blood Vessels

The arterial wall of animals contains three layers: the intima, media, and adventitia; however, the wall structure of different vessels varies somewhat. The intima extends from the blood vessel lumen to the internal elastic lamina, the first layer of concentric elastic tissue that surrounds the lumen (Figure 3.13). The internal elastic lamina is reported to be 100 mm in diameter. The dry weight of the vessel wall is about 30% of the total weight; collagen and elastin make up about 50% of the dry weight. About 20% of the dry weight of the ascending aorta is collagen and 41% is elastin whereas about 46% of the dry weight of the descending aorta is collagen and 30% is elastin. The weight fractions of pepsin-solubilized collagens type I and III in thoracic and abdominal aortas of humans decrease from the arch to the abdominal aorta.

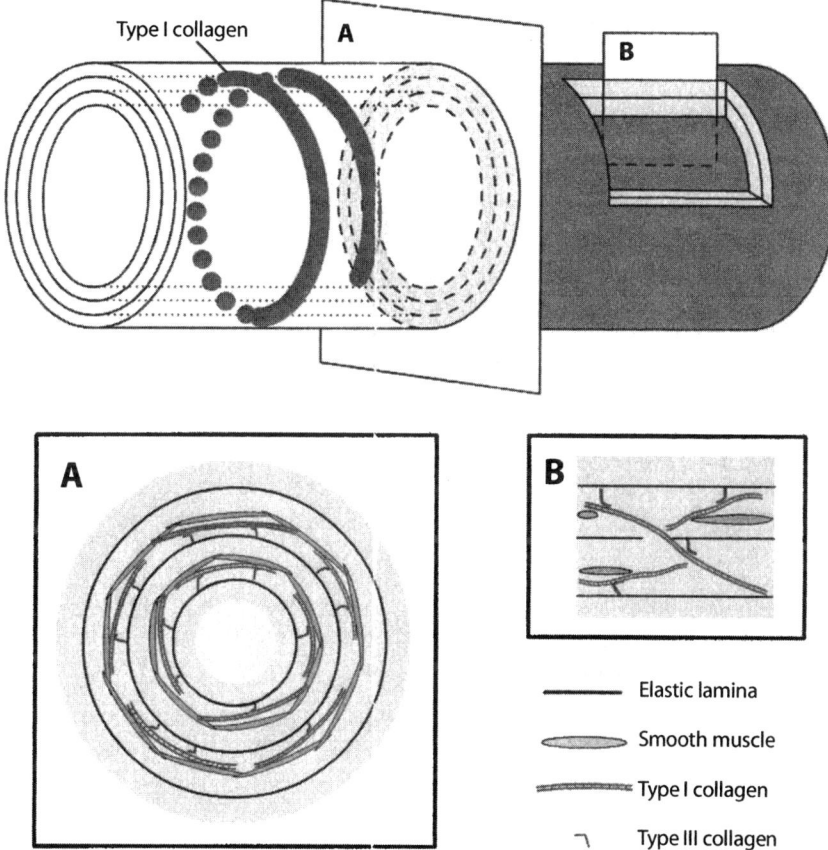

FIGURE 3.13. Diagrammatic representation of connection between elastic lamellae and collagen fibrils. The diagram depicts the circumferential arrangement of elastic lamellae in the aortic wall and the association between type I collagen fibrils and smooth muscle cells and type III collagen fibrils and the elastic lamellae. Type III collagen is shown on the surface of the type I collagen fibrils where it appears to provide attachments to the elastic lamellae. The low elastic spring constant for collagen fibrils in aortic tissue may be a result of the direct connections between the smooth muscle and the collagen fibrils.

The aortic diameter is seen to increase from about 28 mm at age 30 to about 35 mm at age 100. The increased diameter is associated with a decreased content of type III collagen with respect to type I, as well as an increased aortic circumferential modulus. Therefore, it appears that the lower type III collagen content is associated with larger values of the aortic stiffness.

The intima is lined with a continuous layer of polygonal endothelial cells that align with the flow direction and which form the interface with the blood. Endothelial cells are supported by a basement membrane composed of type IV collagen, proteoglycans, and laminin; the latter forms a continuous bond between the cells and the extracellular matrix. The basement membrane provides a cushion that allows bending and changes in diameter associated with changes in blood pressure. The intima in healthy young adults is very thin and is believed to play an insignificant role in the mechanical properties of the aortic wall. The intima thickens with age at the same time the circumference increases, which alters the mechanical behavior. A variety of collagen types are found in the intima including types I, III, IV, V, VI, and VIII, as well as fibronectin, proteoglycans, and hyaluronan. Besides water, which composes about 65% of the weight of the intima, the major components of this layer in the abdominal aorta are type I collagen (16%), type III collagen (8%), type IV collagen (2%), proteoglycans (2%), cellular components (5%), and other collagens (5%). The collagen fibrils in the intima are from 30 to 40 nm in diameter and 50 to 100 nm in diameter in the outer adventitia.

The mechanical behavior of the aorta primarily reflects the thickness and structure of the media. The media of the aorta in humans contains 50 to 65 concentric layers of elastic lamellar units. Each layer is about 2.5 nm thick and contains smooth muscle cells, elastic fibers, and collagen fibrils. The media is delineated on the inside by the internal elastic lamella and on the outside by the external elastic lamella. Structural models suggest that layers of smooth muscle cells are arranged circumferentially around the vessel lumen with the cell axis directed along the vessel axis. Sheaths of basement membrane and orthogonal layers of fine collagen fibrils surround the cell layers. The collagen fibrils are oriented at angles of approximately 45° to the vessel axis and are surrounded by elastic lamellae that are closely associated with the smooth muscle cells.

Compositionally, the media contains elastic fibers composed of elastin and fibrillin; collagen types I, III, IV, and V; proteoglycans; fibronectin; and hyaluronan. The media in the thoracic region contains 65% water, 8% type I collagen, 16% type III collagen, 2% other collagen types, 2% elastic fibers, 2% proteoglycans, and 5% cellular components. Type III collagen has been reported near elastic lamellae whereas type I collagen is located closer to the surface of smooth muscle cells. Under intraluminal pressures higher than 40 mm Hg, the collagen fibers of the media are oriented circumferentially with a 50% degree of orientation that increased to above 75% at pressures above 150 mm Hg.

Outside the external elastic membrane is found extracellular matrix consisting of fibroblasts, large-diameter collagen fibrils, elastic fibers, adipose tissue, branching vessels, and proteoglycans. Fibroblasts in this layer are mostly circumferentially oriented and the collagen fibers are longitudinally oriented. Compositionally, the adventitia contains 65% water, 14% type I collagen, 7% type III collagen, 2% other collagens, 2% proteoglycans, 5% elastin, and 4% cellular components. The adventitia is not believed to contribute extensively to the mechanical behavior except in tethering the aorta to surrounding tissues damping pulsatile flow. Collagen fibrils in the adventitia are about 52.1 nm in diameter.

3.3.2 Structure of Stomach and Intestines

Food is propelled through the stomach by wall contraction in a similar manner to the way blood is pushed through blood vessels. The stomach and intestines contain several layers including the mucosa, submucosa, muscularis, and serosa. In the stomach the inner lining is made up of a mucous layer with columnar epithelium that extend into the gastric pits about 1/4 of the thickness of the wall (Figure 3.14). Below the epithelium is the lamina

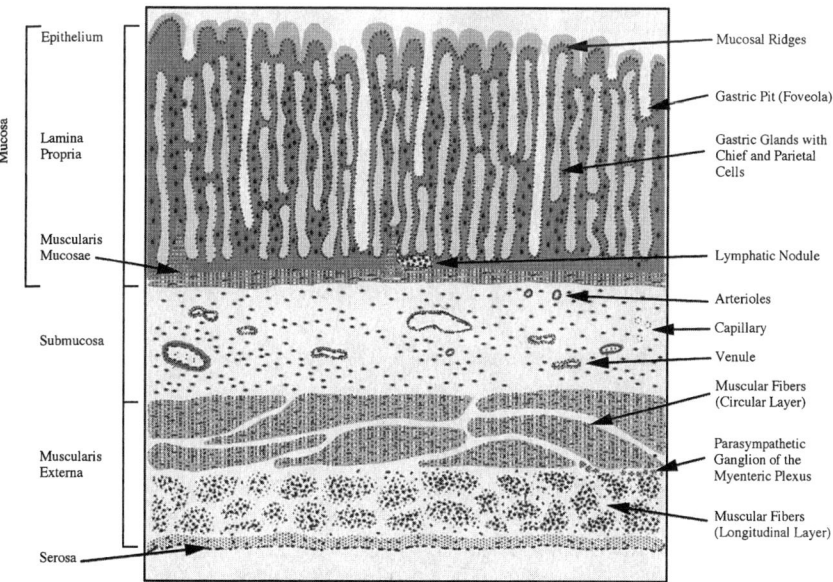

FIGURE 3.14. Structure of the stomach. Stomach contains four layers, including the mucosa, submucosa, muscularis, and serosa. The inner infolded mucous layer is made up of columnar epithelium, and gastric pits are found within the infoldings. Submucosa contains connective tissue and vessels, and the muscularis contains muscle fibers. The serosa is on the outside of the stomach and is continuous with the peritoneum.

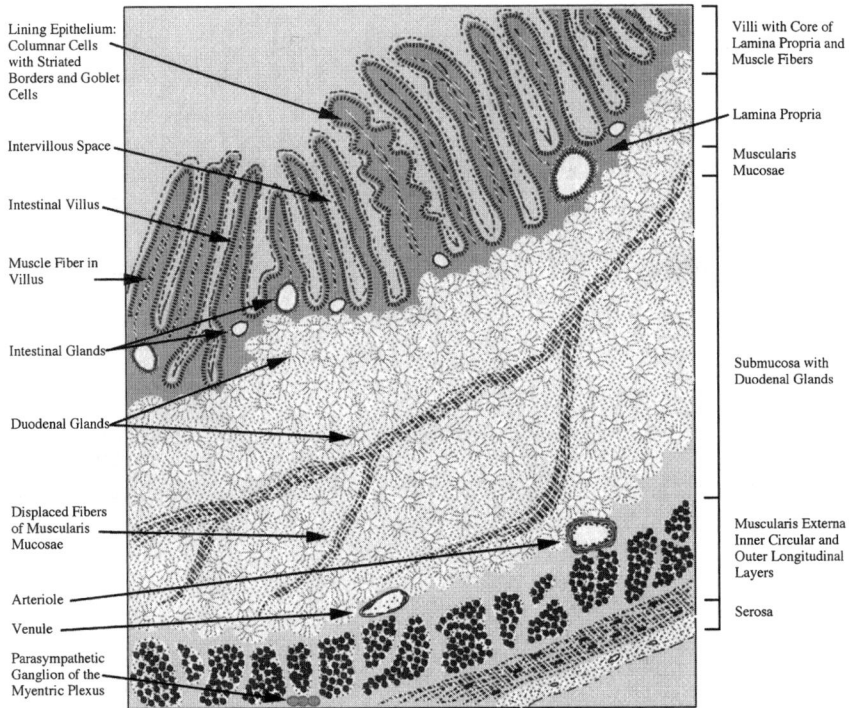

Lining Epithelium:
Columnar Cells
with Striated
Borders and Goblet
Cells

Intervillous Space

Intestinal Villus

Muscle Fiber in
Villus

Intestinal Glands

Duodenal Glands

Displaced Fibers
of Muscularis
Mucosae

Arteriole

Venule

Parasympathetic
Ganglion of the
Myentric Plexus

Villi with Core of
Lamina Propria and
Muscle Fibers

Lamina Propria

Muscularis
Mucosae

Submucosa with
Duodenal Glands

Muscularis Externa
Inner Circular and
Outer Longitudinal
Layers

Serosa

FIGURE 3.15. Structure of small intestine. In small intestine, the mucosa contains villi or surface projections, lamina propria, and muscle fibers. Submucosa contains duodenal glands, and the muscularis contains muscle fibers.

propria that includes the bases of the gastric pits where the gastric glands release their products. At the base of the mucosa is a band made up of smooth muscle termed muscularis mucosa below which is found the submucosa. The submucosa is made up of connective tissue containing small blood vessels.

The next layer is the muscularis externa, which contains circular muscle fibers on the top layers and longitudinal muscle fibers on the outer layers. Finally, the serosa is found which is continuous with the peritoneum. Food movement through the stomach into the intestines occurs by contraction of the muscle fibers in the musclularis.

In contrast, in the small intestine (Figure 3.15), the mucosa is composed of surface projections or villi with a core of lamina propria and muscle fibers. In the lamina propria itself are found intestinal glands, fine connective tissues, reticular cells, and lymphatic tissue. The submucosa is filled with glands in the duodenum whereas in the muscularis externa an inner circular layer and an outer longitudinal layer of muscle are present. Parasympathetic ganglion cells are seen in the thin layer of connective tissue between the two muscle layers. The serosa forms the outermost layer and is continuous with the peritoneum.

3.3.3 Structure of Bladder and Ureter

The structure of the bladder, ureter, and urethra are similar in that they contain three layers, the mucosa, muscularis, and serosa. In the bladder (Figure 3.16) the inner layer (mucosa) when empty is infolded and it is made up of transitional epithelium. The lamina propria that is found below contains collagen and elastic fibers in the deeper layer. The muscularis is prominent and contains muscle fibers that are arranged in branching bundles separated by connective tissue. Muscular contraction causes expulsion of fluid from the bladder into the ureter. The connective tissue between the muscle fiber bundles merges with the connective tissue of the serosa. The serosa is continuous with the peritoneal lining.

In comparison, the ureter is similar except the muscularis is less prominent (Figure 3.17). In the ureter the muscularis consists of an inner longitudinal layer and an outer circular layer of smooth muscle. The outer layer is termed the adventitia and contains fibroelastic and adipose tissue in which are found blood vessels and nerves.

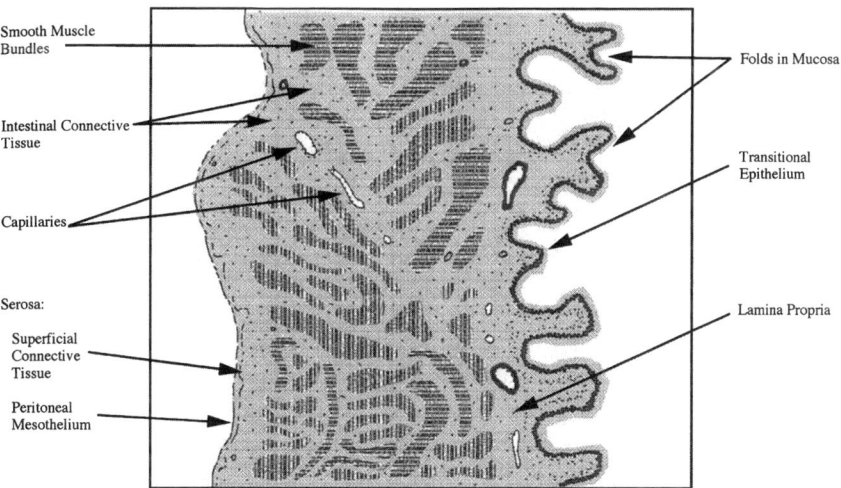

FIGURE 3.16. Structure of bladder. This structure is composed of mucosa, muscularis, and serosa. In the bladder, the mucosa is infolded when empty and is supported by connective tissue in the lamina propria. Muscularis contains smooth muscle bundles, and the serosa is continuous with the peritoneum.

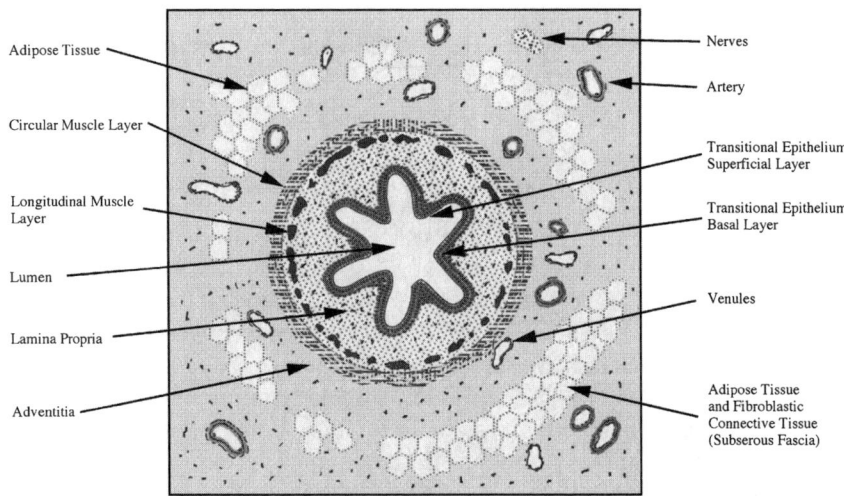

FIGURE 3.17. Structure of ureter. The structure of the ureter is similar to that of the bladder, except the muscularis is less prominent, and adventitia, the outer layer, contains fibroelastic and adipose tissue in which blood vessels and nerves are embedded.

3.4 Parenchymal or Organ-Supporting Structures

Internal organs such as the liver and pancreas have complex structures. They consist of cells that are functional units supported by connective tissue (the parenchyma). The relationships between the cells that make up the functional units vary from organ to organ and so does the geometry of parenchyma. However, in each organ there is usually an afferent (going into the organ) blood supply and an efferent (going out) one. The location of the blood supplies and the flow of blood through each unit are important to understand. Parenchymal structures do appear to be under tension under normal physiological conditions; however, these structures have not been studied as extensively as have other tissues.

3.5 Skeletal Structures

The skeletal tissues of importance include cancellous bone, cortical bone, cartilage, intervertebral disc, skeletal muscle, tendon, and ligament. Both bone and tendon appear to be directly affected by mechanical loading. Both of these tissues appear to stiffen and become denser upon loading. The effect of loading on cartilage structure is more complex, because mechanical loading appears to contribute to cartilage wear. Histologically these are made up of the cellular and noncellular components described in Table 3.4. Cancellous bone consists of slender trabeculae made up of oriented layers of mineralized collagen fibers that connect and form irregular cavities that contain bone marrow (Figure 3.18). The outer layer is composed of

TABLE 3.4. Cellular and Noncellular components of skeletal tissues

Structure	Cellular layers	Noncellular layers	Function
Cancellous bone	Periosteum		Covers surface
	Osteocytes	Bony trabeculae	Mechanical support
	Osteocytes	Haversian systems	Mechanical support
	Blood cells	Marrow cavities	Source of cells
Compact bone	Periosteum		
Circumferential	Osteocytes	Circumferential lamellae of collagen and hydroxyapatite	Mechanical support
Osteon	Osteocytes	Concentric lamellae	Mechanical support
	Endothelium	Connective tissue (haversian canal)	Blood supply
Cartilage	Perichondrium		Covers surface
Hyaline	Chondrocytes	Connective tissue	Mechanical support
	Chondrons	Interterritorial matrix	Connects cells
Intervertebral disc	Annulus fibrosis Nucleus pulposis	Connective tissue	Covers surface
	Chondrocytes	Connective tissue	Shock absorber
Muscle	Fibroblasts	Endomysium (connective tissue)	Holds fibers together
	Muscle cells	Actin and myosin	Contractile elements
Tendon and ligament	Epitendineum		Covers surface
		Fascicle	Separates units
	Fibroblasts	Dense, aligned connective tissue	Supports loads

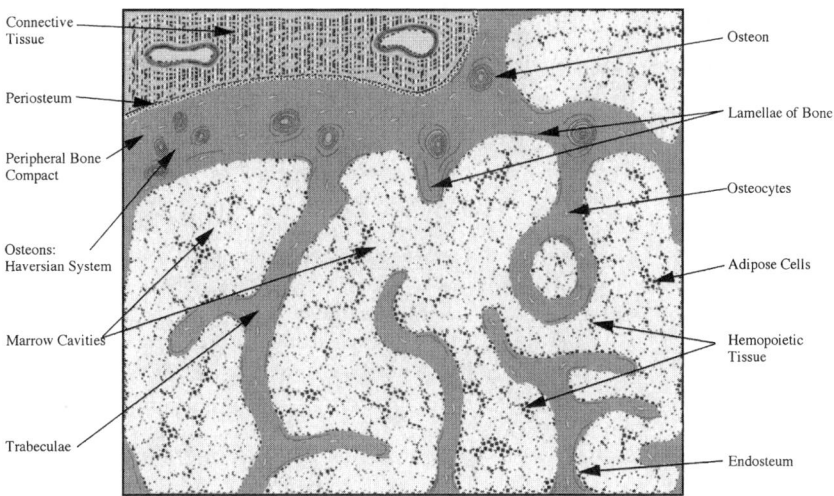

FIGURE 3.18. Structure of cancellous bone. Cancellous bone is composed of periosteum, an outer layer apposed to the peripheral compact bone. The layer beneath peripheral compact bone contains bone trabeculae, marrow cavities, adipose tissue, hemopoietic cells, and the endosteum.

periosteum containing undifferentiated connective tissue producing cells that merges with the surrounding tissue. Underneath is peripheral bone composed of parallel sheets of mineralized collagen fibers in which circumferentially wrapped collagen fibers containing mineral in the form of osteons are inserted. Bone-making cells (osteocytes) are observed in lacunae. The marrow cavity contains adipose tissue and hematopoietic tissue. The endosteum covers the interface between the marrow and the trabeculae.

Compact bone that forms the outer layer of long bones is composed of lamellae that are thin plates of bony tissue containing osteocytes in lacunae (Figure 3.19). The outer layer of bone beneath the periosteum is formed from circumferential layers of mineralized collagen (lamellae) that are parallel to each other and the long axis of the bone. Similarly the inner layer of bone in the shaft is also circumferential lamellae. Between these two layers are found osteons also known as Haversian systems. Each osteon consists of a number of concentric lamellae (plied layers of mineralized collagen containing osteocytes) surrounding a central canal that contains reticular connective tissue and blood vessels. The boundary between each osteon is a thin layer of matrix, the cementing line. Osteons constantly are being formed and resorbed at the same time the balance is dictated by the mechanical loading to the tissue. Repeated loading of long bones leads to bone formation suggesting that osteoblast synthetic metabolism is up-regulated by mechanical loading.

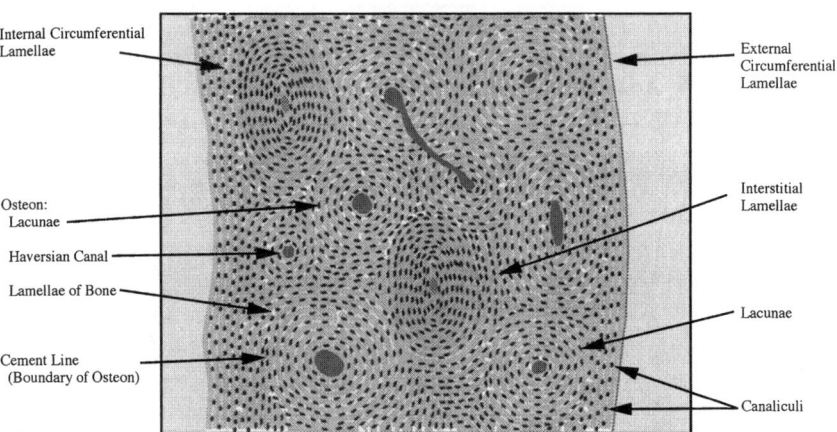

FIGURE 3.19. Structure of compact bone. Compact bone consists of an outer layer of mineralized lamellae that are wrapped around the shaft of the bone. Beneath the outer layer are concentric rings of mineralized collagenous lamellae. Each concentric unit is termed an osteon and has a vessel running through its center (haversian canal).

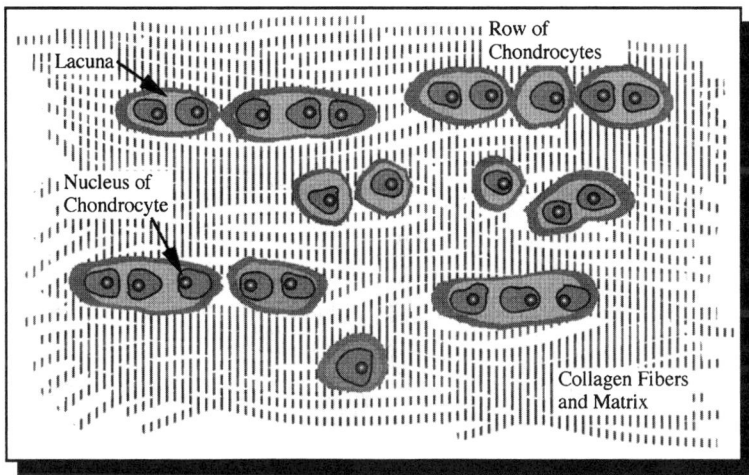

FIGURE 3.20. Structure of hyaline cartilage. Hyaline cartilage contains chondrocytes within lacunae separated by a matrix containing collagen fibers and proteoglycans.

Hyaline cartilage is found on the ends of long bones surrounded by a connective tissue sheath termed perichondrium containing undifferentiated cells (Figure 3.20). It is loaded in tension in the joint and appears to contain cells termed chondrocytes that are up-regulated by increased loading. Beneath the perichondrium is found connective tissue rich in type II collagen, proteoglycans, and cells termed chondrocytes that make matrix. The chondrocytes are formed into vertical groups termed chondrons that are surrounded by extracellular matrix termed territorial (surrounding individual chondrocytes in their lacunae) and territorial matrix (separating different chondrocytes and chondrons).

In fibrocartilage such as the intervertebral disc, the outer layer consists of circumferential lamellae of plied collagen and fibroblasts that supports an inner layer of chondrocytes that make type II collagen fibers. Small chondrocytes in lacunae usually lie in rows within the collagen matrix. Elastic cartilage is composed of chondrocytes embedded in a matrix containing collagen and elastic fibers (Figure 3.21).

Skeletal muscle is composed of muscle cells, thin and thick filaments, and endomysium, a connective tissue containing fibroblasts that holds the fibers together. Mechanical loading regulates normal muscle metabolism: in the absence of normal tensile loading muscle atrophy results. Interactions between thick (myosin) and thin (actin) filaments result in lines or bands containing one or more of the muscle fiber components. The H band represents overlap of only the thick filaments whereas the I band represents the area of overlap of thin and thick filaments. The z lines are the points at which the sarcomere repeats itself.

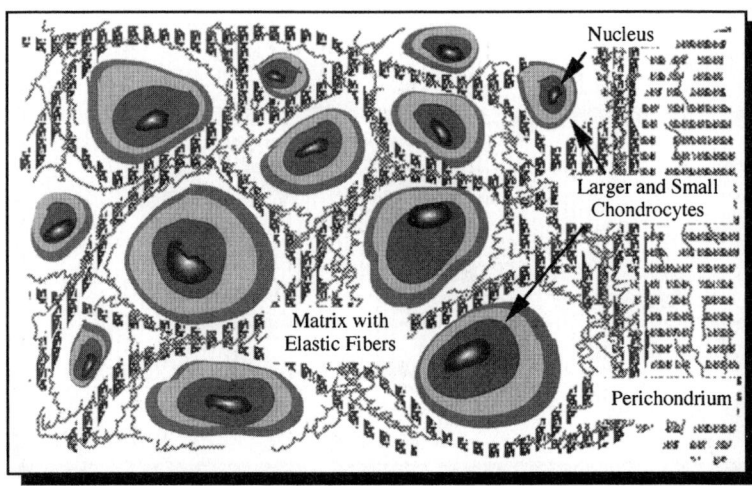

FIGURE 3.21. Structure of elastic cartilage. Elastic cartilage is composed of chondrocytes in a matrix containing collagen and elastic fibers.

Tendons and ligaments are made up of units termed fascicles that are bound into functional units by a sheath termed epitendineum (Figure 3.22). Individual fascicles are composed of rows of fibroblasts that alternate with bundles of collagen fibrils parallel to the tendon axis. These structures are normally loaded in tension to maintain joint stability.

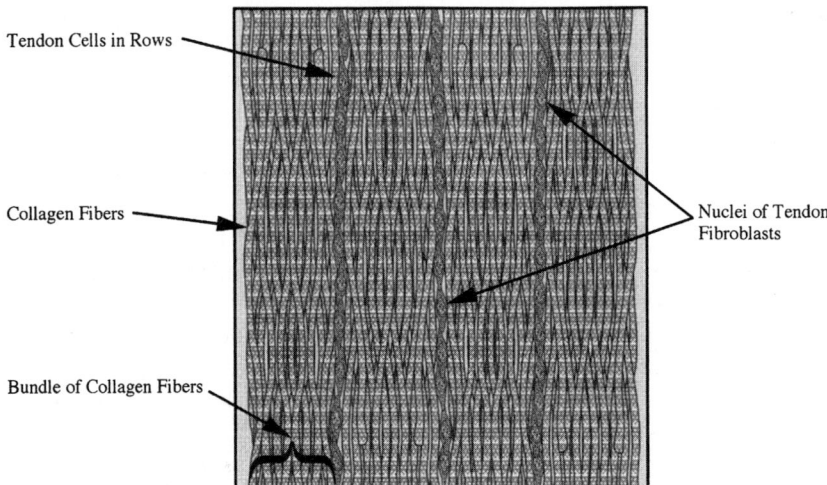

FIGURE 3.22. Structure of tendon. Tendon is composed of rows of tendon cells that are in columns aligned parallel to the axis and parallel bundles of collagen fibers.

3.5.1 Detailed Structure and Composition of Hyaline Cartilage

Articular cartilage, which lines the surfaces of diathrodial joints, is a specialized connective tissue that plays an important role in dissipating loads in joints due to translational and rotational motion. All cartilaginous tissues contain cells (chondrocytes and chondroblasts) that secrete an extracellular matrix composed of proteoglycans and collagen fibrils. It is the amount of extracellular matrix and its organization that gives rise to normal mechanical functioning of the joint surface.

Articular cartilage is a specialized hyaline cartilage that has been characterized structurally, biochemically, and mechanically. It is divided into four zones termed the superficial (or tangential), intermediate (or transition), deep (or radial), and calcified going from the joint surface to the underlying bone (Figure 3.23). The collagen and proteoglycan composition and structure of each of these zones is different, giving rise to differences in their mechanical properties.

The composition of articular cartilage differs with age, site in the joint, and depth from the surface. Normal articular cartilage contains 60 to 70% collagen and 5 to 15% proteoglycans on a dry weight basis. The collagen is highest at the superficial layer, which comprises about 10% of the cartilage thickness, and then is constant throughout the thickness of the noncalcified zones. The middle zone comprises about 60% of the cartilage thickness and the deep zone comprises about 30% of the thickness. The hexosamine

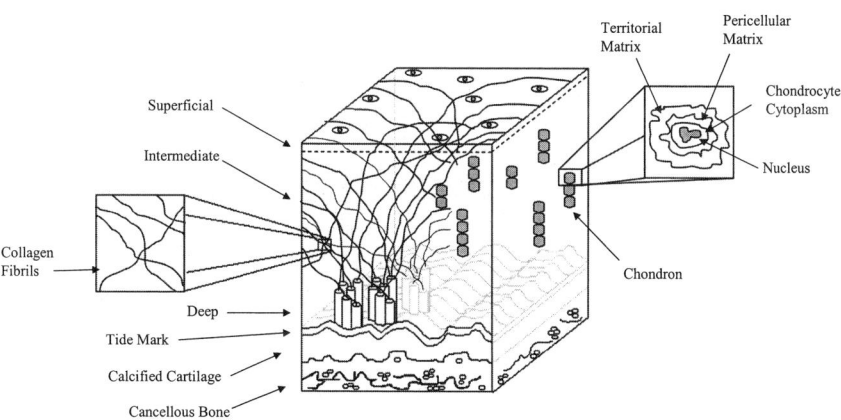

FIGURE 3.23. Diagram illustrating the zonal structure of articular cartilage. The superficial zone contains aligned collagen fibrils; the intermediate zone contains unoriented collagen fibrils; the deep zone contains collagen fibrils perperdicular to the subchondral bone.

content, which reflects the composition of proteoglycans, reaches a maximum percentage at about 1/3 of the total depth and then decreases with increasing thickness. The fluid content is about 85% at the surface and remains constant for about 25% of the depth; it then falls to below 75% for the remaining portion of the cartilage.

Cartilage is composed of units termed chondrons that consist of chondrocytes surrounded by a pericellular matrix enriched in hyaluronic acid and proteoglycans, and a pericellular capsule composed of a fine network of collagenlike fibrillar material associated with a dense accumulation of PGs (Figure 3.24). The pericellular matrix is a fine network of thin, branching, filamentous material interspersed with granules typical of condensed PGs, and the pericellular capsule is a basketweave network of fine fibrillar material 15 to 20 nm in diameter with a 54 nm periodicity. At the light microscopic level the chondrocytes appear to reside within spaces or compartments called lacunae. Groups of chondrocytes line up to form a chondron. Chondrons in cartilage are embedded into a collagenous matrix containing

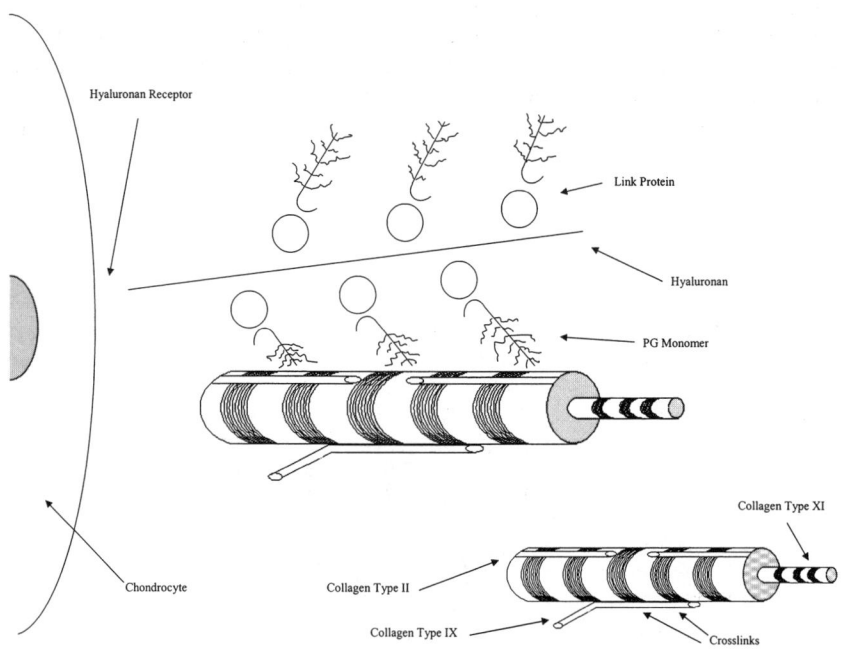

FIGURE 3.24. Diagram of the collagen fibrils in cartilage, collagen fibrils in cartilage are composed of a core of types XI and II collagen with zone type IX on the surface. Collagen fibrils are connected to each other and to the cell surface by proteoglycans.

collagen fibrils that run approximately vertically throughout most of the layers of cartilage, except at the joint surface where they are parallel to the surface.

On the basis of electron microscopy, a lighter staining material 5 to $10\,\mu m$ wide, immediately outside the pericellular capsule is defined as the territorial matrix. This zone consists of bundles of collagen fibers ranging in width from 0.2 to $0.8\,\mu m$ interspersed with electron lucent spaces 0.13 to $1.0\,\mu m$ wide that stain for PGs. The interterritorial matrix is characterized by closer collagen fiber packing. In general, the collagen fibrils are thicker than those in the territorial matrix compartment, and the granules are more abundant and are observed along and between collagen fibers.

3.5.1.1 Macromolecular Components of Articular Cartilage

Articular cartilage contains a number of macromolecular components, including collagens, proteoglycans, and glycoproteins that form the extracellular matrix. The predominant collagen found in articular cartilage is type II. It is composed of three alpha chains, designated $\alpha 1(II)$ which have chromatographic characteristics similar to those of $\alpha 1(I)$. The latter chain is one of the two alpha chains that are found in type I collagen, the form of collagen found in tendon and skin. Type II collagen is also characterized by the high content of bound carbohydrates and can contain up to fivefold more of these residues than the homologous chain from skin. Collagen types that form fibrils are composed of alternating flexible and rigid domains; the flexible domains have been proposed as the sites of elastic energy storage (Figure 3.25). Types II and III collagen have been shown to be more flexible than type I collagen, suggesting that types II and III are involved in elastic energy storage and dissipation.

Cartilage also contains type XI collagen aggregates that form the fibrillar core of fibrils containing type II collagen. When hybrid collagen fibrils containing type XI and type II collagen reach a limiting diameter, growth is hypothesized to be terminated by interaction with type IX collagen molecules that are arranged in an antiparallel direction. Type X collagen is found in the calcified zone and appears to be associated with type II/type XI fibers and with fine fibers surrounding hypertrophic chondrocytes. Type XIII collagen mRNA and type XIIA have been detected in the cartilage growth plate.

Several different types of proteoglycans are found in cartilage. Aggrecan, the most abundant large PG in cartilage, is thought to provide resistance to compressive deformation in weight-bearing joints. In tension, PGs retard the rate of stretch and alignment of collagen fibrils. PGs consist of a protein core with attached glycosaminoglycan side chains that bind to hyaluronan at its amino terminal end in the presence of link protein. Aggrecan is rich in chondroitin sulfate and heparan sulfate GAG chains. As many as 100

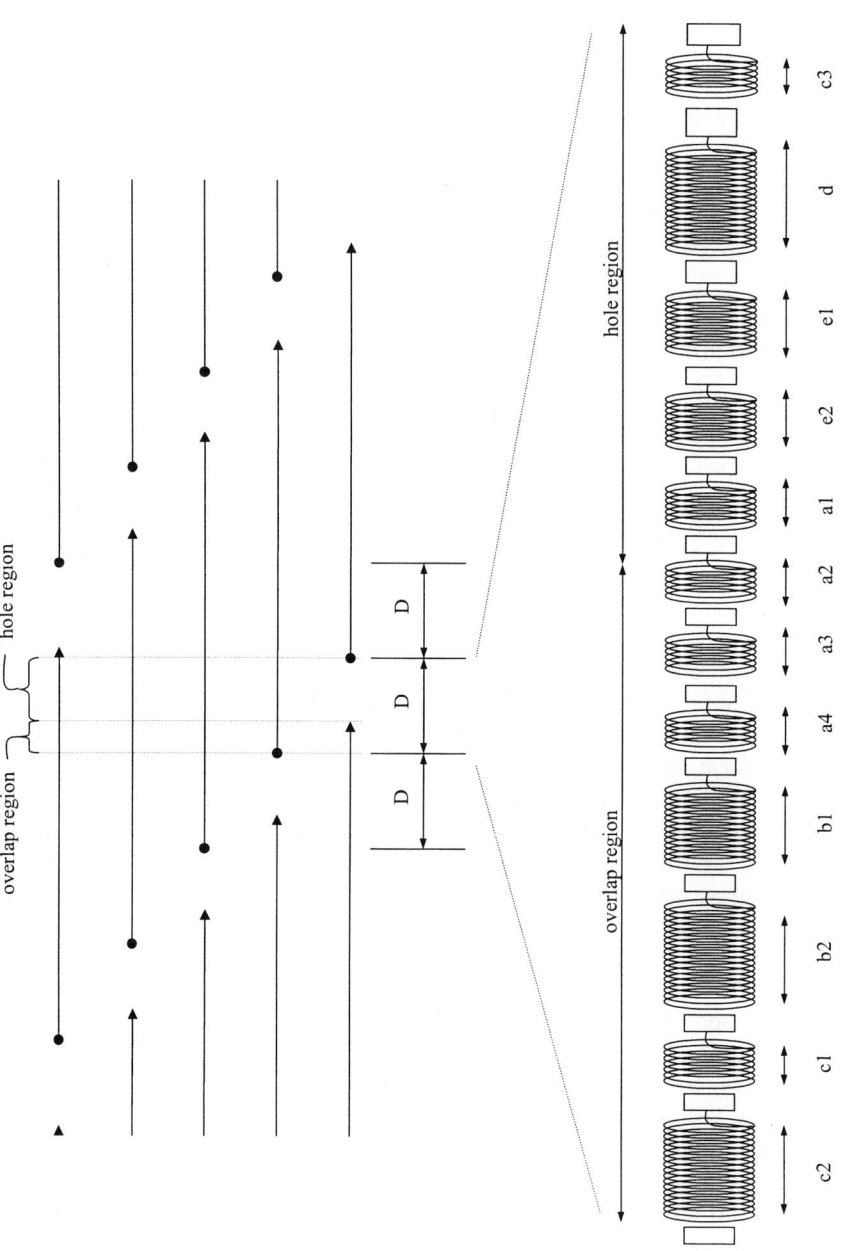

FIGURE 3.25. Diagram illustrating the relationship between quarter-staggered collagen molecules in the microfibril (top) and the flexible and rigid regions of the amine acid sequence (bottom). The "D" period is one repeat containing 4 molecules (hole) followed by 5 molecules in cross-section (overlap). Mineral is deposited in the hole region in bone.

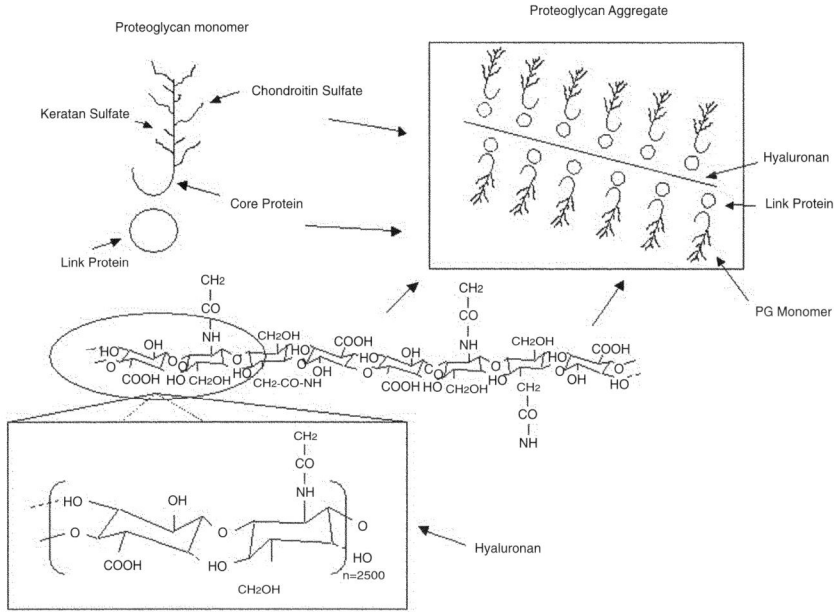

FIGURE 3.26. Diagram illustrating the relationship between hyaluronic acid (hyaluronan), link protein, and proteoglycans in proteoglycan aggregates.

aggrecan and link protein molecules can bind a strand of hyaluronan, forming a complex with a molecular weight as high as 2×10^8 (Figure 3.26). Each aggrecan molecule has a molecular weight greater than 10^6. The high net negative charge associated with the GAG chains provides a swelling pressure that resists compressive forces associated with weight bearing.

Cartilage contains two forms of dermatan sulfate proteoglycans (DSPGs), called biglycan and decorin. These PGs are rich in dermatan sulfate and bind to collagen, transforming growth factor-β, and other macromolecules of the extracellular matrix. DSPGs are believed to inhibit tissue repair processes and may explain the difficulty associated with healing superficial cartilage lesions. Biglycan and decorin are small proteoglycans with molecular weights of 10^6 and 0.7×10^6, respectively. The core proteins of both of these DSPGs have a molecular weight of about 37,000. Biglycan has two attached glycosaminoglycan chains, whereas decorin has a single one. Decorin core protein binds to specific regions in collagen fibrils. Decorin is mainly found in the superficial and intermediate zones of adult cartilage.

The core protein of biglycan does not specifically bind to collagen. Compressive loading of cartilage in tissue culture results in increased synthesis of decorin whereas biglycan synthesis is unchanged. Biglycan is localized in the pericellular region and its function is unclear. Fibromodulin is another collagen-binding PG found in cartilage. It inhibits collagen fibril formation in a manner similar to that of decorin. Small cartilage PGs may function as regulators of collagen fibril diameter. PGs play a role in increasing energy storage and dissipation through collagen fibril extension and slippage during mechanical deformation.

3.5.1.2 Collagen Fibril Orientation in Articular Cartilage

When a pin, the tip of which is covered with dye, is inserted into the surface of the joint, the orientation of the resulting split lines or cleavage pattern can be used to study the macroscopic structure of cartilage (Figure 3.27). When histological sections are made of the surface, the principal orientation of the collagen fibrils is along the split lines. Scanning electron microscopy of cartilage shows that collagen fibrils are parallel to the surface in the superficial layer whereas in deeper layers they run in a direction that is perpendicular (radial) to the subchondral bone. Cartilage cells, or chondrocytes, exist in cavities, termed lacunae, within the fibrillar network and appear not to be anchored to the surrounding matrix.

Although the structure of cartilage has been interpreted as arising from the behavior of several distinct zones or layers, one model of articular cartilage depicts collagen fibrils in the form of bundles of fibrils with different orientations. In the intermediate and superficial zones, the fibril bundles are composed of a fine meshwork of collagen fibrils that run at angles or parallel with respect to the surface. In the calcified and deep zones, the fibrils are arranged in vertical bundles with adjacent bundles closely connected with bridging fibrils.

In the superficial layers of cartilage, the collagen fibrils are smaller in diameter and the network is finer than in the deeper zones. Closely adherent to the horizontally oriented collagen fibrils is the surface layer or "lamina splendens", a morphologically distinct layer that appears to lack the fibrillar structure of the deeper layers of the superficial zone. The collagen fibrils in cartilage are about 50 nm in diameter and distinct collagen fibers cannot be visualized in cartilage at the light microscopic level.

3.5.1.3 Zonal Structure of Articular Cartilage

Although cartilage contains a continuous network of collagen fibrils in the form of interconnected fibrils, much of our understanding of the structure comes from analysis of the different "zones" contained within this network. The superficial zone consists of fine collagen fibrils running parallel to the articular surface, as discussed above. These fibrils are reported to have a banding pattern that is characteristic of collagen. The lower part of the superficial zone contains flattened polarized chondrocytes aligned parallel to the surface. Immunostaining reveals PGs and link protein, although cationic dyes do not stain this zone as intensely as the deeper zones. Scanning electron microscopy reveals that the collagen of the superficial zone

Anterior

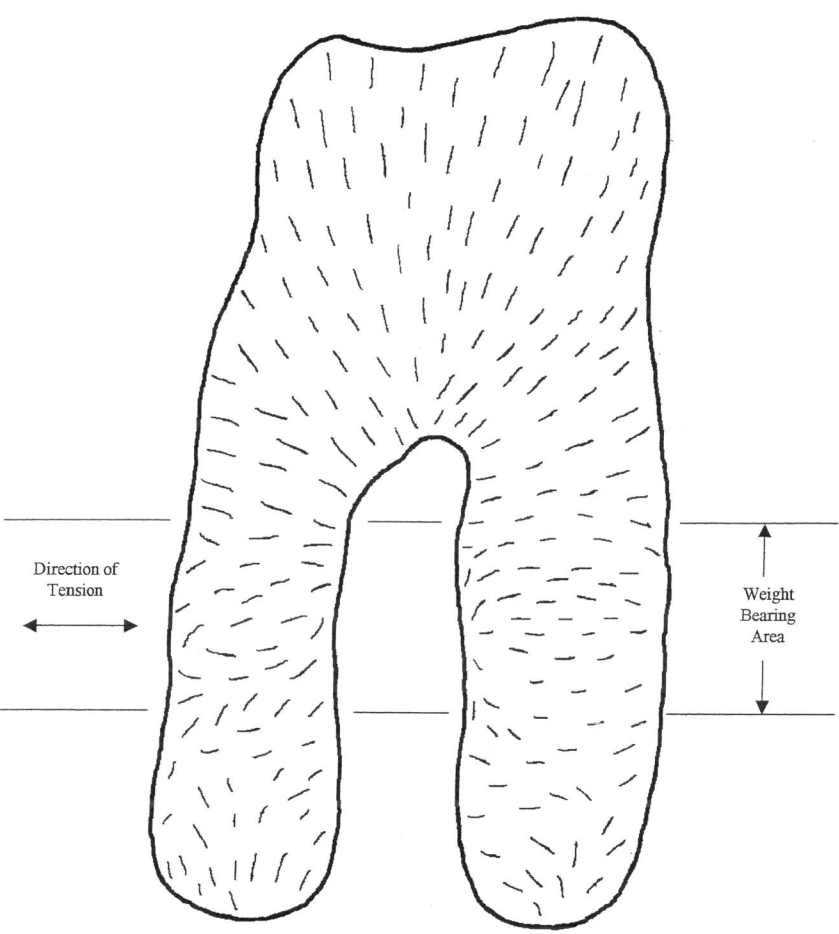

Posterior

FIGURE 3.27. Diagram illustrating split line pattern in articular cartilage. Collagen fibrils are oriented approximately parallel to the split lines in cartilage.

has two distinct morphologies. The upper layer, the lamina splendens, of the superficial zone is observed to form three or four horizontal layers. The meshwork of fibrils forming these layers is finer than that observed in the intermediate zone. The collagen fibrils that originate in the intermediate zone become almost parallel to the surface in the superficial zone. The collagen fibers in the superficial zone are continuous with fibers of the deep zone.

The chondrocytes of the intermediate zone are more spherical than, and differ from, the cells of the superficial layer, which show surface polaritity. Chondrons are found in the intermediate layer. They consist of chondrocytes, pericellular matrix, and pericellular capsule. In this zone, the chondrocytes do not form vertical columns as extensively as is observed in the deep zone. At high magnification the collagen fibrils have a beaded appearance, which may reflect the high level of PGs seen in this region.

In the deep zone, the cells and collagen fibrils tend to align themselves perpendicular to the joint surface. Collagen fibrils in the calcified zone are also arranged vertically with adjacent bundles closely linked by bridging fibrils. The deep and calcified zones are separated by the tidemark, which is a narrow band of aggregates of mineral associated with matrix vesicles. Matrix vesicles appear to bud off from the ends of chondrocyte cytoplasmic processes and may play a role in cartilage mineralization. The calicifed zone forms the interface between cartilage and underlying subchondral bone.

3.5.2 Detailed Structure and Composition of Tendon, Ligament, and Joint Capsule

Tendons, ligaments, and joint capsule are all dense regular fibrous connective tissue and are made up of cells, aligned collagen and elastic fibers, proteoglycans, and water (Figure 3.28). Although elastic fibers are present in ligament and capsule, they are only minor components of these tissues. Collagen is the most abundant protein in dense fibrous connective tissue and forms essential mechanical building blocks in the musculoskeletal system. It can be found in both fibril- and nonfibril-forming forms.

Tendon, ligament, and joint capsule are composed predominantly of type I collagen, although they contain small amounts of types III and V. The type III collagen content has been reported to be 10% for ligament as opposed to 5% for tendon. Collagen types II, VI, XII, and XIV have also been reported to be in ligament and capsule; however, these collagens appear to associate with fibrocartilage found at the junction with bone and not in the midsubstance. The high content of type I collagen in these joint tissues not only leads to mechanical stability but it also supports elastic energy storage.

In mature tendon, collagen fibril bundles (fibers) have diameters between 1 and 300 mm and fibrils have diameters from 20 to over 280 nm (Figure 3.28). The presence of a crimp pattern in the collagen fibers has been established for rat tail tendon as well as for patellar tendon and anterior cruciate ligament; the specific geometry of the pattern, however, differs from tissue to tissue. It is not clear that the crimp morphology of tendon is actually present in tendons that are under normal resting muscular forces.

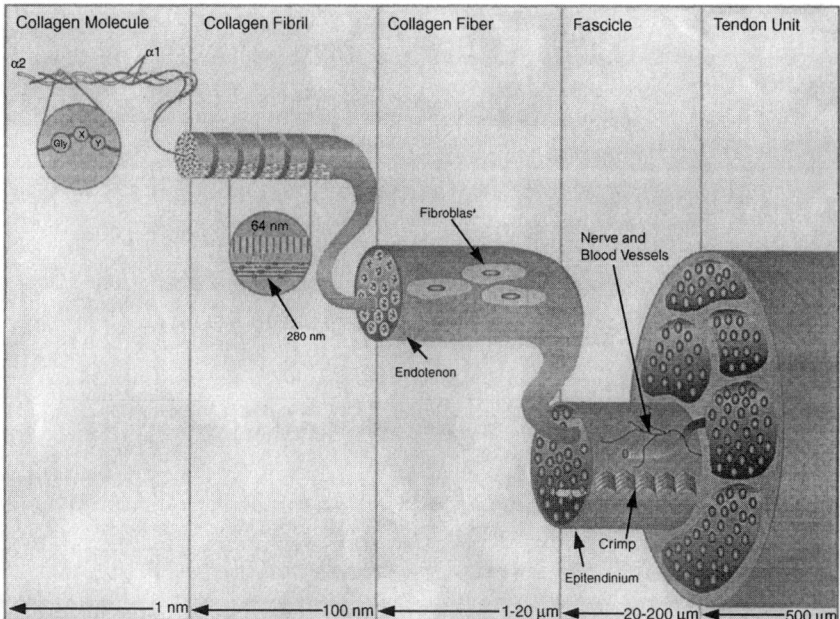

FIGURE 3.28. Diagram illustrating the structural hierarchy in tendon. Collagen molecules (left) form microfibrils (not shown) that make up collagen fibrils, fibers, fascicles, and tendon units.

3.5.2.1 Role of Proteoglycans (PGs) in Tendon

Tendon contains a variety of PGs including decorin, a small leucine-rich PG that binds specifically to the d band of positively stained type I collagen fibrils as well as hyaluronan, a high molecular weight polysaccharide. Other small leucine-rich PGs include biglycan, fibromodulin, lumican, epiphycan, and keratocan. In mature tendon the PG(s) are predominantly proteoder-mochondran sulfates. PGs are seen as filaments regularly attached to collagen fibrils in electron micrographs of tendon stained with Cupromeronic blue (Figure 3.29). In relaxed mature tendon, most PG filaments are arranged orthogonally across the collagen fibrils at the gap zone, usually at the d and e positively staining bands. In immature tendon, PGs are observed either orthogonal or parallel to the D period and the amount of PGs associated with collagen fibrils in tendon decreases with increased fibril diameter and age.

Animal models employing genetic mutations lacking decorin demonstrate collagen fibrils with irregular diameters and decreased skin strength, and a model lacking lumican shows abnormally thick collagen fibrils and skin fragility. Down-regulation of decorin has been shown to lead to development of collagen fibrils with larger diameters and higher ultimate tensile strengths in ligament scar. Models lacking thrombospondin 2, a member of a family of glycoproteins found in extracellular matrix, exhibit abnormally

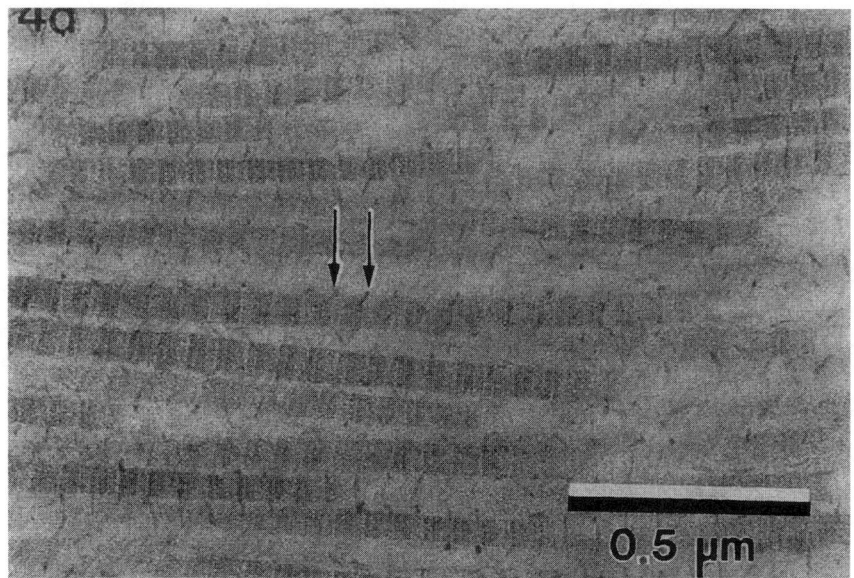

FIGURE 3.29. Relationship between proteoglycans and collagen fibrils in tendon. Transmission electron micrograph showing positive staining pattern of type I collagen fibrils from rabbit Achilles tendon stained with quinolinic blue. Proteoglycan filaments (arrows) are shown attached to collagen fibrils.

large fibril diameters and skin fragility. These observations suggest that PGs such as decorin and other glycoproteins found in the extracellular matrix are required for normal collagen fibrillogenesis. Decorin also appears to assist in alignment of collagen molecules in tendon as well as to facilitate sliding during mechanical deformation.

3.5.3 Detailed Structure of Mineralizing Tendon

Mineralization of the tendons of some species of vertebrates is a normal occurrence as well as pathological calcification of some human tendons. The tendons in the legs and wings of certain birds are examples of this phenomenon although not all such tendons mineralize in a particular bird and not all avians have mineralizing tendons. Mineralization of turkey tendon in some avians is believed to be a means to increase the ultimate strength of the tendon while preserving its ability to store elastic energy. Precise identification of the factor or factors responsible for tendon mineralization has not been made, but it is currently thought that tendon mineral formation is mediated by a combination of biological, physical-chemical, and biomechanical effects on the tissue.

In particular, the gastrocnemius, or Achilles tendon structure of the domestic turkey and manner of mineralization tendon have been described in detail. From an anatomic perspective, the gastrocnemius is a large, thick, cylindrically shaped tendon located at the rear of the turkey leg and extending from the claws of the animal to insertion points in two major hip muscles of the bird limb. The tendon is a single segment at the gastroc muscle that bifurcates into segments each having smaller diameter than its distal portion. The shape of these segments becomes more flattened and fanlike on their insertion into the medial and lateral hip muscles. Mineralization of this region of the gastrocnemius occurs in the vicinity of its bifurcation and proceeds in a distal-to-proximal direction along the two smaller segments (Figure 3.30).

The tendons in the turkey are comprised of specific cells that synthesize and secrete an extracellular organic matrix. The cells of the tendon, variously called tenocytes or tendonoblasts, are found in columnar arrangements that lie generally parallel to each other and to the tendon long axes. Microscopic examination shows that the cells undergo phenotypic changes in size and shape as a function of their location along the tendon. In the gastrocnemius, those cells closest to the tendon–muscle junctions are the most immature and resemble flattened fibroblasts. At sites more distal to tendon–muscle insertions, the cells become progressively larger and more oblate and elaborate a complex system of interconnecting processes. Adja-

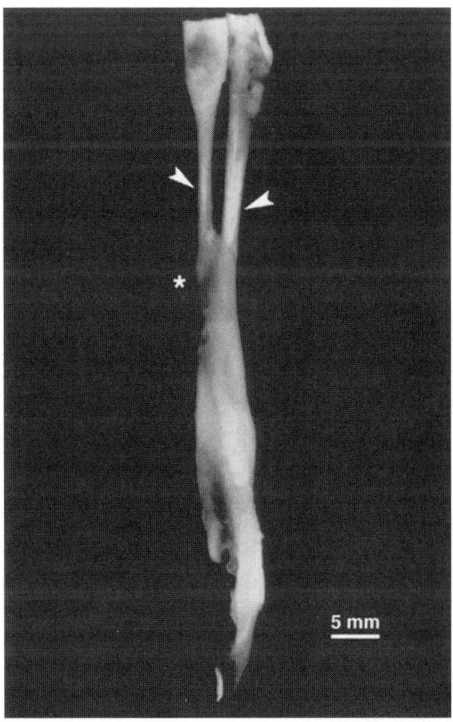

FIGURE   3.30. X-ray   showing mineral deposition in turkey leg tendon. Light areas reflect mineral that is electron dense forming on the collagen fibrils in tendon (Landis and Silver, 2002).

cent to the zone of tendon mineralization, hypertrophy of tenocytes reaches a maximum and their processes form extensive cell–cell and possible cell–matrix interconnections. Within and distal to the front of mineralization, mature tenocytes remain viable within lacunae but are obscured by the increasing mineral mass of the gastrocnemius. This columnar organization and orderly phenotypic transition to mature tenocytes resemble the morphological characteristics of chondrocytes in vertebrate growth plate cartilage, and the presence of viable tenocytes in lacunae surrounded by mineral has a similarity to osteocytes embedded in the mineralized matrix of bone. Thus, the mineralizing avian tendon has certain correlates in its morphology with that of growth plate cartilage and bone.

Tenocytes are responsible for secreting the extracellular organic matrix of the tendon. The principal matrix constituent of the turkey leg tendon, as all vertebrate tendons, is type I collagen. It comprises approximately 90 to 95% of a tendon matrix, the remainder being types V and XII collagen and certain noncollagenous proteins including osteopontin, bone sialoprotein, osteocalcin, and proteoglycans. Decorin, biglycan, lumican, and fibromodulin contribute to the latter. This tendon matrix composition is likely to be incomplete, and putative changes in the known constituents with respect to their location in the tendon, maturation of the tenocytes, and aging of the tissue have not been well characterized.

On the other hand, as noted above, the mineralization of the turkey gastrocnemius has been studied extensively. Mineral deposition appears to involve two distinct components of the tendon: type I collagen and small vesicles that are found in the extracellular spaces between the collagen fibrils organized in long, parallel, and highly aligned arrays.

Microscopically, vesicles initially appear conspicuous in the region of the tendon containing larger tenocytes and undergoing the transition to mineralization, just proximal to the mineral front. The early appearance of mineral in the tendon is found in association with these vesicles. Later, that is, closer to the mineral front, collagen also mediates mineral deposition and ultimately the bulk of the mineral is associated with this matrix component (Figure 3.31).

The mineral, itself, regardless of its relation with either vesicles or collagen, is a poorly crystalline apatite and is exceedingly small, measured as ~40–170 × 30–45 × 4–6 nm in length, width, and thickness, respectively. The pattern of mineral deposition associated with vesicles is one of randomly disposed crystals in radial-shaped spherulitic clusters of the size of vesicles, approximately 10 to 200 nm, in diameter. This pattern is distinct from that observed with collagen, in which the crystals are deposited in a specific and highly organized manner in the hole and overlap zones of the fibrils leading to ~64 to 70 nm periodicity. Furthermore, the c-axial planes of the crystals are generally aligned parallel to each other and to the collagen long axis with which they associate. Apatite crystals are also found deposited on collagen fibril surfaces in tendon, in which the disposition of apatite is disperse and without apparent preferred orientation.

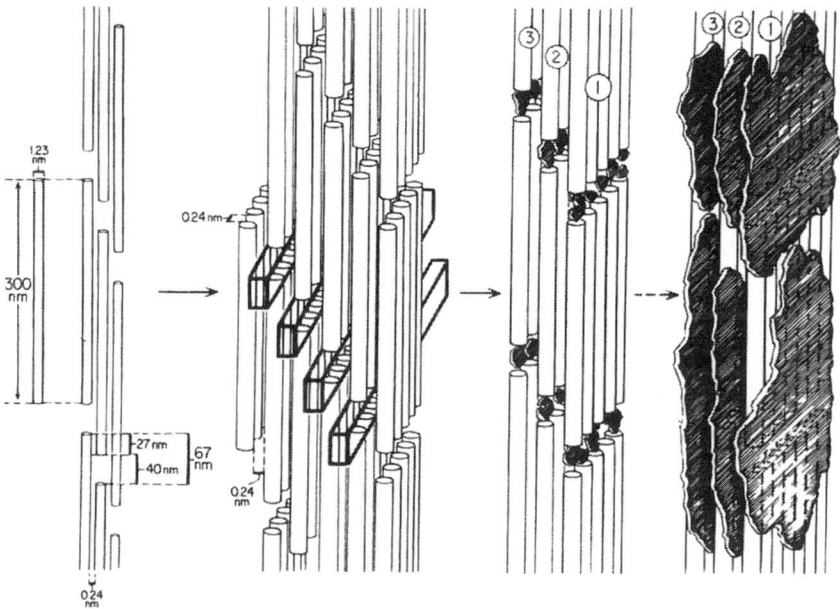

FIGURE 3.31. Diagram illustrating the relationship between collagen fibrils (cylinders) and mineral rectangles that grow into plates between the fibrils (Landis and Silver, 2002).

3.6 Summary

In this chapter we have attempted to sample the structure of a variety of tissues. Although tissue structure is quite complex, we do know that collagen and elastic fibers as well as smooth muscle are the predominant load-bearing materials in all tissues. We know from our personal experiences that mechanical loading affects the structure of different types of tissues, however, it is difficult to separate these effects from the effects of aging on tissue structure. It is clear that our skin thins and our blood vessel walls thicken as we get older, but are these effects due to increased duration of mechanical loading or due to aging alone? Although studies on athletes show that training increases muscle, tendon, and bone load-bearing ability, it is still unclear how mechanics influence tissue microstructure. The understanding of how training affects tissue microstructure is important in understanding how mechanical loading and physical therapy may positively influence and even modulate the effects of aging.

Suggested Reading

Bloom W, Fawcett DW. *A Textbook of Histology.* Philadelphia: W.B. Saunders; 1975.

di Fiore MSH. *Atlas of Human Histology.* Philadelpia: Lea & Febiger; 1981.

Landis WJ, Silver FH. The structure and function of normally mineralizing avian tendons. Comparative Biochemistry and physiology 2002;133:1135–1157.

Silver FH, Christiansen DL. *Biomaterials Science and Biocompatibility.* New York: Springer-Verlag; 1999; Chapter 4.

4
Determination of Physical Structure and Modeling

4.1 Introduction

In Chapter 3 we introduced the structure of a variety of tissue types and found that collagen is the major structural component of most mammalian tissues. Because of its triple-helical extended structure it can transmit mechanical loads and store energy with minimum deformation. To achieve these properties, a macromolecule must adopt an extended conformation, either the β sheet or collagen triple helix, because these are the only conformations that do not allow extensive conformational freedom inasmuch as they are almost totally extended in space. The collagen conformation has enough axial extensibility that it allows energy storage, which is critical for efficient locomotion and it can also transmit forces to cells through changes in conformation that are in turn transduced into changes in gene expression and protein synthesis. The relationship between structure and mechanical properties of collagen and other macromolecules has been elucidated in part from physical studies on the structure of individual macromolecules in solution. This holds true for other macromolecules of biological interest besides collagen.

In Chapter 2 we discussed the relationship between primary chemical structure (i.e., the sequence of monomeric units in a macromolecule) and the common structural elements that are found in tissues. Biological macromolecules are composed of combinations of structural units such that a protein may contain regions with random coil structure linking regions containing helical segments. The overall structure of a biological macromolecule is reflected in its physical properties including its flow behavior. In the absence of the analysis of the physical behavior of isolated macromolecules it would be difficult to understand the behavior of complex macromolecular networks such as those found in tissues. The purpose of this chapter is to describe the information about the physical properties of isolated macromolecules that have given us information about the structure and mechanical behavior of complex tissues.

Much of our understanding of the way cells and tissues behave reflects to a first approximation the behavior of isolated single macromolecules. For instance, the resistance of tendon to deformation reflects the high axial ratio of the collagen molecule and the packing of these molecules into parallel arrays of collagen fibrils and fibers. Although much information is gained from studying molecular structure by X-ray diffraction from which the average atomic coordinates of molecules in a crystal can be determined, other information is obtained by studying the behavior of isolated macro-molecules in solution. The flow behavior of isolated macromolecules is related to frictional forces that exist between a macromolecule and the sur-rounding environment, which, in turn is related to frictional forces that exist between structural units in tissues. We attempt to explain how the size and shape of macromolecules dictate the viscous behavior in solution as well as the viscosity obtained from macroscopic mechanical measurements.

In Chapter 2 we described the types of structures that are most com-monly found in proteins, polysaccharides, and lipids. These structures include the α helix, β sheet, collagen triple helix, extended chain structures in polysaccharides, lipids and nucleic acids, and random chain structures. Most macromolecules contain domains that are characterized by one of these structures connected to other types of structural domains. Although the collagen molecule is mostly composed of a triple helix it contains other flexible domains that alternate with the rigid triple helical domains. There-fore even though it is largely made up of a rigid triple helix, other domains in collagen make the structure depart from an ideal helical structure that gives the molecule unique physical and chemical properties. This departure from ideality makes understanding macromolecular behavior more diffi-cult; however, it also endows macromolecules with a wealth of both physi-cal and chemical properties.

Many of the methods used to extract information related to the structure of macromolecules come from studying the behavior of isolated macro-molecules in solution. These techniques are based primarily on the flow behavior in a velocity gradient, the rate of Brownian motion of a particle, or osmotic effects associated with the size of individual molecules. The tech-niques that have been employed to study size and shape of macromolecules most extensively include viscometry, light scattering, analytical ultracen-trifugation, and electron microscopy.

4.2 Viscosity

The property of viscosity is a very useful measure of the size and shape of a particle; long thin molecules give rise to increased solution viscosity as opposed to small spherical molecules. This is because the drag force exerted by one macromolecule on another or on a neighboring water molecule is proportional to the surface area. The surface area of rods is greater than the surface area of an equivalent sphere and therefore rodlike molecules have a higher surface-to-mass ratio than do spherical ones. This also implies that force transfer by shear between macromolecules or bundles of macro-molecules found in tissues would occur more efficiently with rodlike macro-molecules because they have a higher surface area for shear to occur. However, viscosity is a very inefficient way to transfer tension in aqueous systems inasmuch as energy transferred by friction is lost through viscous slippage. In nonaqueous systems, where the frictional forces are much higher, tension can be transferred through a matrix by shear forces. Tension is transferred in biological systems by formation of direct covalent linkages. We know this because defects in tissue crosslinking lead to premature tensile failure of components in tissues such as collagen.

4.2.1 What Is Viscosity?

Viscosity is a measurement of the resistance to flow of a fluid. A simple way by which the viscosity η is measured is by determining the time required for the fluid to flow from one point to another under the influence of gravity in a capillary tube with a fixed radius and length. A solution with a high viscosity requires a longer time to flow than one with a low viscosity. For instance, it is much easier to pour water out of a container than to pour maple syrup, therefore maple syrup has a higher viscosity than water. During simple fluid flow, layers of molecules slide past each other and experience molecular forces such as electrostatic interactions, hydrogen bonding, and dispersive forces leading to energy losses that arise from friction. In the simplest case the shear stress σ between adjacent layers of fluid is dependent on the viscosity η times the velocity gradient across the tube, dv/dr as shown by Equation (4.1). The fluid layer nearest to the wall moves more slowly than that in the center of the tube because the friction at the wall is higher than on other molecules.

$$\sigma = \eta \, (dv/dr) \tag{4.1}$$

According to Equation (4.1) the viscosity is the proportionality constant between stress σ and dv/dr. A fluid is defined as Newtonian if η is constant. For large macromolecules in solution the viscosity at a fixed concentration can be a function of the shear rate dv/dr. This occurs due to alignment of macromolecules with the direction of fluid flow at high shear rates. Therefore, the true viscosity of macromolecular solutions can only be measured at very low shear rates or by extrapolation of measurements at a number of shear rates to zero shear rate. In addition, Equation (4.1) also tells us that in a viscoelastic solid such as a tissue, the shear stress developed during mechanical loading is proportional to the viscosity and the strain rate. Therefore, tissues appear to require more stress to deform at high strain rates even though the viscosity may be constant. In actuality many tissues such as skin show a decrease in viscosity at high strain rates (thixotropy). This makes the stress–strain behavior of skin depend on how fast the load is applied and complicates investigation of the actual material properties of skin.

A further complication of determination of the viscosity of a macromolecular solution is that the viscosity depends on the concentration of macromolecules. Newton developed a formula for predicting the viscosity of a solution of macromolecules and solvent η', knowing the solvent viscosity η, shape factor for the macromolecule v, and the volume fraction of macromolecules φ. Newton's law of viscosity is given in Equation (4.2) and provides a relationship among viscosity, shape factor, and volume fraction of polymer.

$$\eta' = \eta \ (1 + v\varphi) \tag{4.2}$$

Using this formula it is possible to predict the viscosity of a macromolecular solution knowing the shape factor of a macromolecule and its volume fraction. However, to do this we must first determine the shape factor. This equation also tells us that macromolecules with large shape factors have high viscosities and show increased strain-rate dependence of the viscous component of the stress in tissues. Therefore, we need to be able to evaluate the shape factor for different macromolecules.

4.2.2 Determination of the Shape Factor

The shape factor of a macromolecule can be determined from theoretical treatments developed by Simha for prolate ellipsoids (cigar-shaped molecules) and oblate spheroids (disc-shaped molecules) (see Silver, 1987). Einstein determined that the shape factor was 2.5 for spheres. The shape factor for prolate and oblate ellipsoids is given in Figure 4.1 as a function of axial ratio a/b. The shape factor increases as the ratio of the surface area to volume increases suggesting that the viscosity increases rapidly as the surface area increases. Values of a and b in Figure 4.1 are the dimensions of the semimajor and semiminor axes for the ellipsoids. Knowing the shape factor we can calculate the axial ratio for the equivalent rod (prolate ellipsoid) or disc. It turns out the shape factor can be estimated by determining the intrinsic viscosity by viscometry. In addition, we can estimate the shear forces that exist between macromolecules in the solid state from knowledge of the shape factor as discussed in Chapters 8 and 9. As the shape factor increases (i.e., a/b increases), the viscous contribution due to shearing of macromolecules by each other increases. Therefore, macromolecules with large values of the axial ratio a/b are good candidates for supporting high stresses through intermolecular interactions.

4.2.3 Determination of Intrinsic Viscosity

The shape factor can be estimated by measuring the time t required for a solution of macromolecules to flow through a capillary. If this is measured

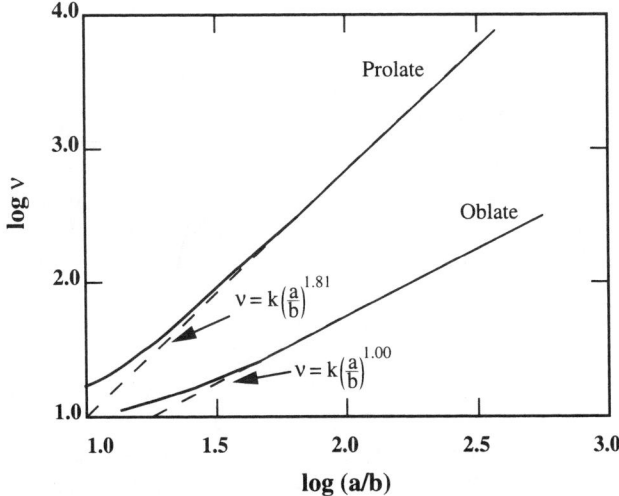

FIGURE 4.1. Theoretical values of the shape factor for ellipsoids. This is a plot of the shape factor, log ν versus log of the axial ratio (a/b) for prolate and oblate ellipsoids. Note that intrinsic viscosity can be approximated by the shape factor for ellipsoids.

for a series of solutions with increasing concentrations of macromolecules, then the intrinsic viscosity is obtained by calculating the reduced specific viscosity η_{sr} extrapolated to zero concentration as shown in Figure 4.2. The intrinsic viscosity [η] is defined by Equation (4.3) as the limit of the ratio of the flow time measured for the macromolecular solution minus the flow time of the solvent divided by the solvent flow time to, times the weight concentration c.

$$[\eta] = \lim_{C \to 0} \{(t - to)/to)1/c\} \tag{4.3}$$

Once [η] is determined from a plot similar to that shown in Figure 4.2, values for a/b can be obtained from Figure 4.1 for prolate and oblate ellipsoids. These values must be checked by determining a and b by other techniques discussed below.

By calculating the intrinsic viscosity of a macromolecule using Equation (4.2) we can determine how the viscosity is affected. For instance, the shape of the red blood cell influences the viscosity of blood. If the shape of the red blood cell changes from a disc to a prolate ellipsoid the viscosity of the blood will increase as will the energy required to pump blood through the capillaries. This occurs in sickle cell anemia when the disclike red blood cell changes in shape; as a result blood flow and oxygen absorption are impaired.

4.2.4 Intrinsic Viscosity of Biological Macromolecules

We can determine the intrinsic viscosity experimentally for biological macromolecules in a variety of different solvents and then model the size and shape of a macromolecule and whether the shape changes as a function of solution conditions. A good example of how shape affects intrinsic viscosity is that of collagen. In the triple-helical form collagen molecules have an intrinsic viscosity of about 1100 ml/g that decreases to 40 to 74 ml/g when the collagen molecule is uncoiled by heating. A shape change in the absence of molecular weight changes leads to a decrease by a factor of over

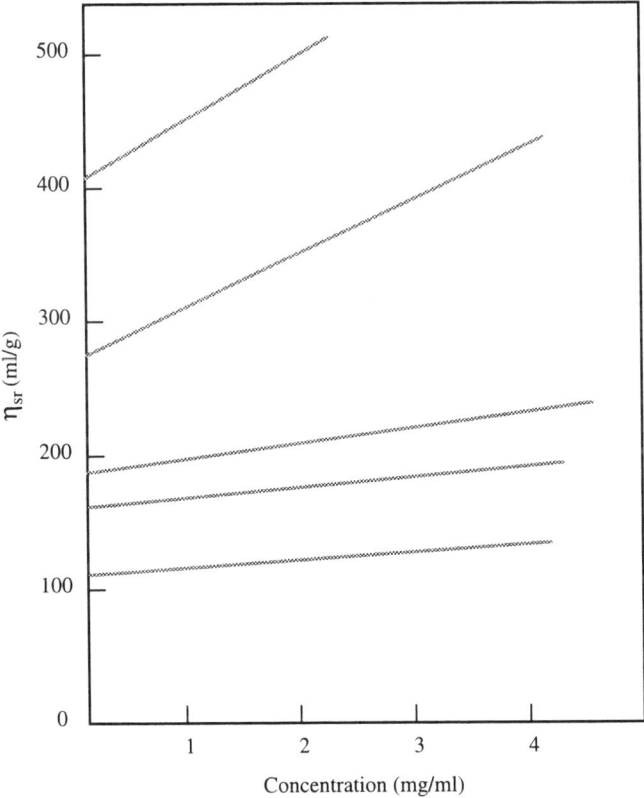

FIGURE 4.2. Determination of intrinsic viscosity. Intrinsic viscosity is determined by extrapolation of the reduced specific viscosity (η_{sr}) to zero concentration. The reduced specific viscosity is calculated by taking the difference between the flow time in a capillary for polymer in a solvent and that of the pure solvent and dividing the difference by the flow time of the pure solvent times the polymer weight concentration. Data are shown for hyaluronan for various fractions isolated from bovine vitreous.

TABLE 4.1. Intrinsic viscosity of connective tissue macromolecules reproduced from Silver, 1987

Macromolecule	$[\eta]$ (ml/g)
Collagen	
Type I	1200,1100,1150
Type II	1530
Type IV	75
α-chains, β- and γ-components	40–74
Hyaluronan	
Rooster Comb	2400,2670,7100
Bovine Vitreous Humor	556,945–970,280,718
Proteoglycans (Cartilage)	236,133–150,184
Glycoproteins	
Lubricin	92,120

25 in the intrinsic viscosity. Table 4.1 lists the intrinsic viscosities of some biological macromolecules.

4.3 Light Scattering

Another method by which the size and shape of macromolecules can be determined is based on the ability of large molecules to scatter light. As early as the 1940s Debye and coworkers recognized that light was scattered by a solution of macromolecules and that the scattered intensity was related to the molecular weight (see Silver, 1987). Because light is an electromagnetic wave characterized by electric and magnetic fields, the interaction between light and the electric and magnetic fields around a macromolecule alters these fields. As light moves through space, the magnitude of the electric and magnetic vectors changes as a function of time and position and from the magnitude of these changes the size of macromolecule can be determined.

It turns out that the intensity of scattered light Is is proportional to the molecular weight of the macromolecule M as well as the scattering angle ϕ, light wavelength λ, distance from the scattering source r, Avogadro's number No, change in refractive index with weight concentration dn/dc, the molecular polarizability α, and the refractive index n as shown in Equation (4.4). Note that the intensity of the incident beam is Io.

$$I_s = \frac{2\pi^2 n^2 (1 + \cos^2 \phi)(dn/dc)^2 M}{N_o \lambda^4 r^2}(cI_o) \qquad (4.4)$$

A simple way to determine the relationship between molecular weight and intensity of scattered light is to recast Equation (4.4) by defining an optical constant K using Equation (4.5) and Ro, the Rayleigh factor (Equation (4.6)).

$$K = \frac{2\pi^2 (dn/dc)^2 n^2}{No\lambda^4} \tag{4.5}$$

$$Ro = \frac{r^2 Is}{Io(1 + \cos^2 \phi)} = Kc/(1/M) \tag{4.6}$$

In the simplest case Equation (4.6) relates the Rayleigh factor to the molecular weight in the absence of solvent–macromolecule and macromolecule–macromolecule interactions. The molecular weight is determined from the slope of a plot of the Rayleigh factor versus weight concentration of macromolecules, c.

In the case of macromolecule–macromolecule interactions, the relationship between the Rayleigh factor and weight concentration of macromolecules is complicated by the fact that the coefficient B is needed to account for this effect (Equation (4.7)).

$$Kc/Ro = 1/M + 2Bc \tag{4.7}$$

Operationally the value of the Rayleigh factor for the solution is subtracted away from that of the solvent plus macromolecule and plotted versus weight concentration to get $1/M$ from the intercept and B from the slope. B is also known as the second virial coefficient that reflects the state of attraction between macromolecules (negative value of B) or repulsion (positive value of B). The weight of each macromolecule contributes to the scattered light intensity, therefore the molecular weight determined by light scattering is a weight average.

For macromolecules that have one or more dimensions larger than 1/20 of the light wavelength, less light actually is measured than is predicted by Equation (4.7). The observed scattered light intensity is less than that predicted from theory as a result of interference between light scattered by different portions of a macromolecule. This fact allows us to evaluate the shape factor from light scattering measurements. The light waves scattered by different portions of a macromolecule are in some instances one-half wavelength out of phase with each other and when summed cancel each other. The fraction of the theoretically predicted light that is actually scattered is termed the particle scattering factor $P(\phi)$. $P(\phi)$ is dependent on the scattering angle and the shape of a macromolecule. Correcting Equation (4.7) for the particle scattering factor, we now get Equation (4.8).

$$Kc/Ro = 1/(P(\phi))[1/M + 2Bc] \tag{4.8}$$

The molecular weight of a macromolecular solution is now obtained by plotting $Kc/(Ro)$ versus weight concentration at a series of decreasing angles to form a graph referred to as a Zimm plot. The molecular weight is obtained by double extrapolation, that is, to a weight concentration of zero and scattering angle of zero.

This double extrapolation procedure proves very time consuming and requires many hours of measurements in the laboratory. In the case of macromolecules that self-associate the determination of molecular weight by classical multiangle light scattering is a significant task. The development of low-angle laser light scattering devices allowed the measurement of Rayleigh factors at low scattering angles obviating the need for measurements at multiple angles (Figure 4.3). The particle scattering factor can be obtained by dividing the Raleigh factor measured at an angle between 90° and 180° by that measured at angles less than 4°. The ratio of the Raleigh factor at 90° or 180° to that at 4° gives us a measure of the shape factor.

Theoretical relationships between the particle scattering factor and molecular shape have been developed for rods, spheres, and random coils. Figure 4.4 illustrates the relationship between the particle scattering factor and ratio of length to width for rods as a function of scattering angle. The value of rod length L and molecular weight from light scattering can be compared with the value of a or a/b obtained from viscometry to get a better picture of the size and shape of a macromolecule. By comparing values of molecular weight and particle scattering factor one can determine the relative flexibility of different rodlike macromolecules. For instance rods, with similar molecular weights but different particle scattering factors have different flexibilities. Molecular weights and particle scattering factors for biological macromolecules of interest in this book are listed in Table 4.2.

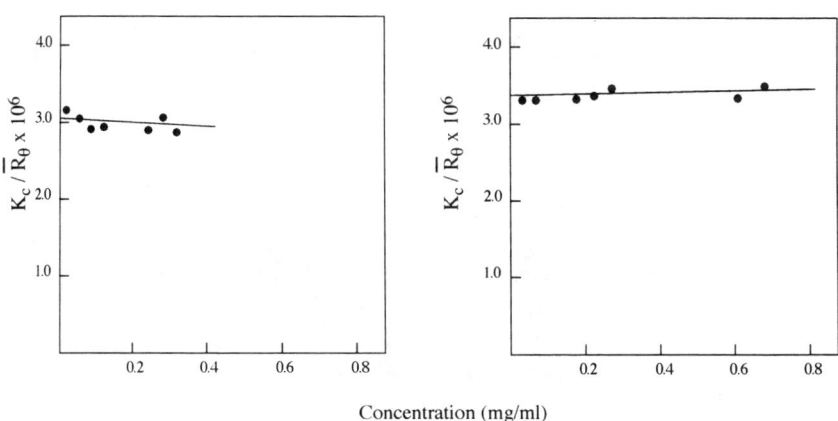

Concentration (mg/ml)

FIGURE 4.3. Determination of molecular weight from light scattering. The weight average molecular weight is determined from light scattering by determination of the Rayleigh ratio R_θ and plotting Kc/R_θ versus concentration. The weight average molecular weight is taken as the reciprocal of the intercept at zero concentration. Plots shown are for type I collagen from rabbit cornea (left) and sclera (right). Reproduced from Silver (1987).

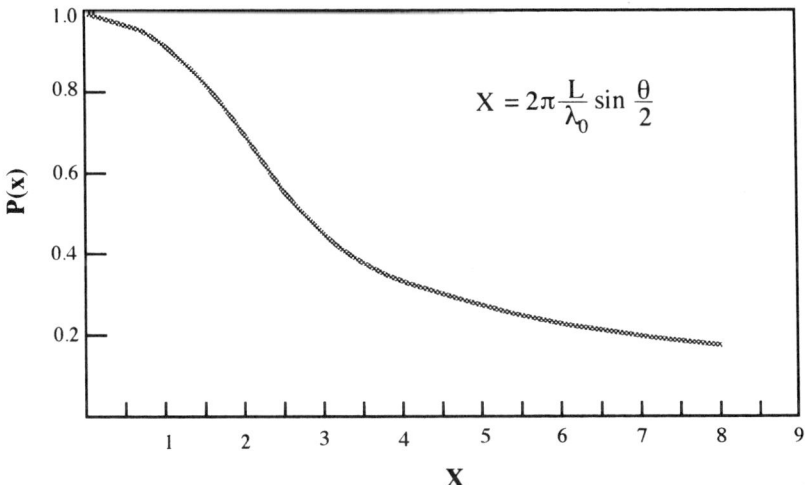

FIGURE 4.4. Determination of particle-scattering factor. Particle-scattering factor $P(X)$ as a function of angle (θ) and wavelength of light in solution (λ_o), versus X for rodlike molecules. Note that as the ratio of molecular length to light wavelength increases or the scattering angle increases, the particle-scattering factor decreases.

TABLE 4.2. Determination of weight average molecular weight and particle scattering factor by light scattering for connective tissue macromolecules reproduced from Silver, 1987

Macromolecule—Source	M^a	B^a	$P(175°)^a$
Collagen—carp swim bladder	345,000		
Collagen—rat skin (type I)	$320,000 - 4.2 \times 10^6$		
Collagen (type I)—rat tail tendon	282,000	-2.60×10^{-3}	
Collagen (type I)—rabbit cornea	328,000	0	0.394
Collagen (type I)—rabbit sclera	295,000	0	0.382
Collagen (type Ip)—bovine skin	280,000	NL^b	0.430
Collagen (type IIp)—bovine nasal septum	285,000	NL	0.450
Collagen (type IIIp)—bovine skin	289,000	NL	0.460
Collagen (type Vp)—bovine skin	307,000	-2.85×10^{-3}	0.440
Collagen (type Vp)—chick embryo	314,000	-2.95×10^{-3}	
Collagen (type IV)—EHS tumor	532,000	0	0.492
Procollagen (type I)—chick tendon	476,000	NL	0.40
Hyaluronan—bovine vitreous humor	370,000		
Hyaluronan—vitreous humor	1.27×10^6		
Hyaluronan—human umbilical cord	3.40×10^6		
Hyaluronan—bovine vitreous humor	340,000–500,000		
Hyaluronan—rooster comb	$4.4–4.6 \times 10^6$		
Hyaluronan—synovial fluid	12.8×10^6	3.3×10^{-3}	
Hyaluronan—bovine vitreous humor	320,000		
Hyaluronan—rooster comb	1.7×10^6		
Hyaluronan—bovine vitreous humor	150,000	-4.13×10^{-3}	
Hyaluronan—rooster comb	4.4×10^6	2.05×10^{-3}	
Fibronectin—bovine blood	444,000	0	0.760
Lubricin—bovine synovial fluid	206,000		
Lubricin—human knee synovial fluid	166,000	-2.21×10^{-3}	0.61
Proteoglycan—pig articular cartilage	$2.5–3.0 \times 10^6$		
Proteoglycan—synovial fluid	$0.704–0.938 \times 10^6$		0.791–0.822

[a] M, B and P (175.5) stand for molecular weight second virial coefficient and particle scattering factor at 175.5 degrees, respectively. All measurements were made at a wavelength of 633 nm. P signifies pepsin extract.
[b] NL—nonlinear and therefore B is a function of concentration.

4.4 Quasi-Elastic Light Scattering

In classical light scattering theory the average scattered light intensity is measured as a function of time and the molecular weight and particle scattering factor are determined. In addition to the average light intensity, the time dependence of the scattered light intensity gives information concerning the rate of movement (Brownian motion) of the macromolecules in solution. Although the average wavelength of the scattered light remains the same, the scattering process broadens the distribution of light wavelengths. In addition, the time dependence of fluctuations of the light intensity are related to the size and shape of the macromolecules in solution. This is due to the diffusion of macromolecules into and out of the control volume illuminated by the light. At any time the probability that there will be more than or less than the average density of molecules that are within the volume illuminated by the light is related to the diffusion coefficient and the rate at which molecules diffuse into and out of the light path. The diffusion coefficient can be calculated from the rate of decay of the product of two intensities that are tabulated into a mathematical function called the autocorrelation function.

Mathematically the autocorrelation function (Figure 4.5) of the scattered light $G(Ndt)$ in Equation (4.9) is related to the product of the two light intensities that are separated by a time interval dt. If dt is chosen correctly, that is, is within the time period required for diffusion to occur into and out of the control volume, then the diffusion coefficient can be calculated from the decay of the autocorrelation function over an observation interval of Ndt (Figure 4.5).

$$G(Ndt) = \sum_{N=1}^{\infty} [I(t)][I(t+dt)]/N \qquad (4.9)$$

The autocorrelation function decays to a value of $<I^2>$av that is the average squared scattered light intensity. Theoretically $G(Ndt)$ is related to a constant α, the experimental baseline B, and the decay constant Γ, as shown in Equation (4.10). The decay constant Γ is related to the translational

$$G(Ndt) = B(1 + \alpha e^{-2\Gamma t}) \qquad (4.10)$$

diffusion coefficient Dt, and the scattering vector Q, as shown in Equation (4.11). A typical normalized autocorrelation function for collagen is shown in Figure 4.6.

$$\Gamma = DtQ^2$$
$$Q = [4\Pi n/\lambda][\sin \phi/2] \qquad (4.11)$$

Taking the natural logarithm of Equation (4.10) and substituting Equation (4.11) we find that a plot of $\ln[(G(Ndt)/B) - 1]$ versus time has a slope equal

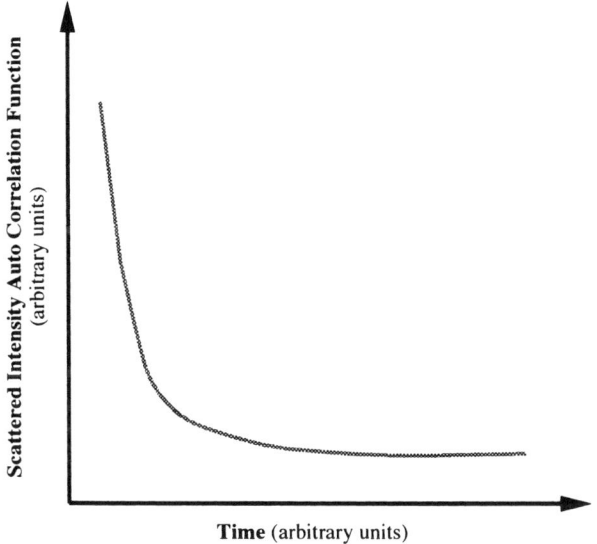

FIGURE 4.5. Autocorrelation function of scattered light. Schematic diagram showing the decay of the autocorrelation function to a baseline. The autocorrelation function is related to the product of two intensities separated by a time interval. As the time interval increases, the function decays to a baseline. The rate of decay is proportional to the translational diffusion coefficient.

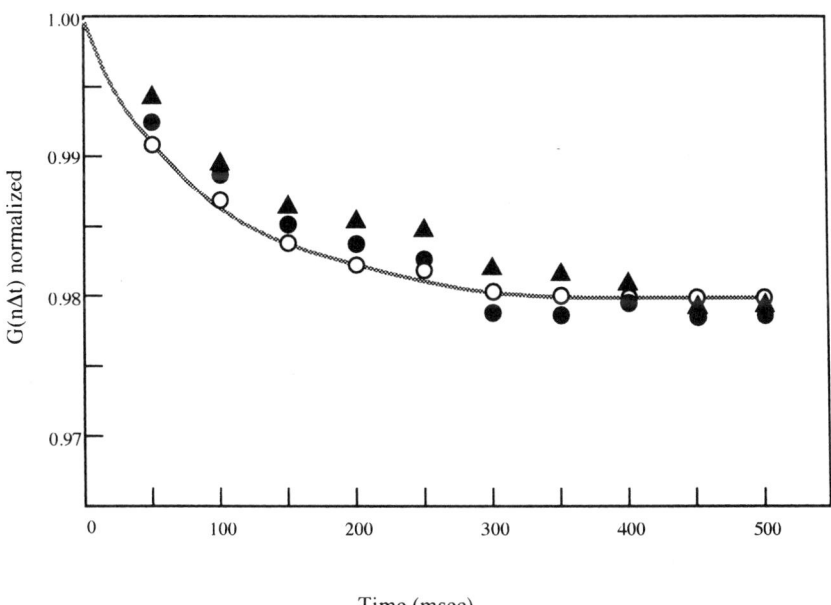

Time (msec)

FIGURE 4.6. Normalized autocorrelation function. Autocorrelation function for collagen single molecules. The autocorrelation function $G(n\Delta t)$ is normalized by dividing all points by the first experimental point $G(1)$. The autocorrelation function decays to a value of the average squared intensity of scattered light divided by $G(1)$. The average squared intensity is proportional to the weight average molecular weight, whereas the rate of decay is related to the translational diffusion coefficient. Reproduced from Silver, 1987.

to $-DtQ^2$ and Dt can be determined knowing the scattering angle, wave-length of light, and solution index of refraction. Dt is determined from such a plot for collagen α chains in Figure 4.7.

It is important to note that the translational diffusion coefficient can be determined this way for macromolecules that are less than $2\,\mu m$ in their largest dimension at scattering angles less than $4°$. Macromolecules larger than this or measurements made at higher scattering angles need to be corrected to get accurate values of Dt.

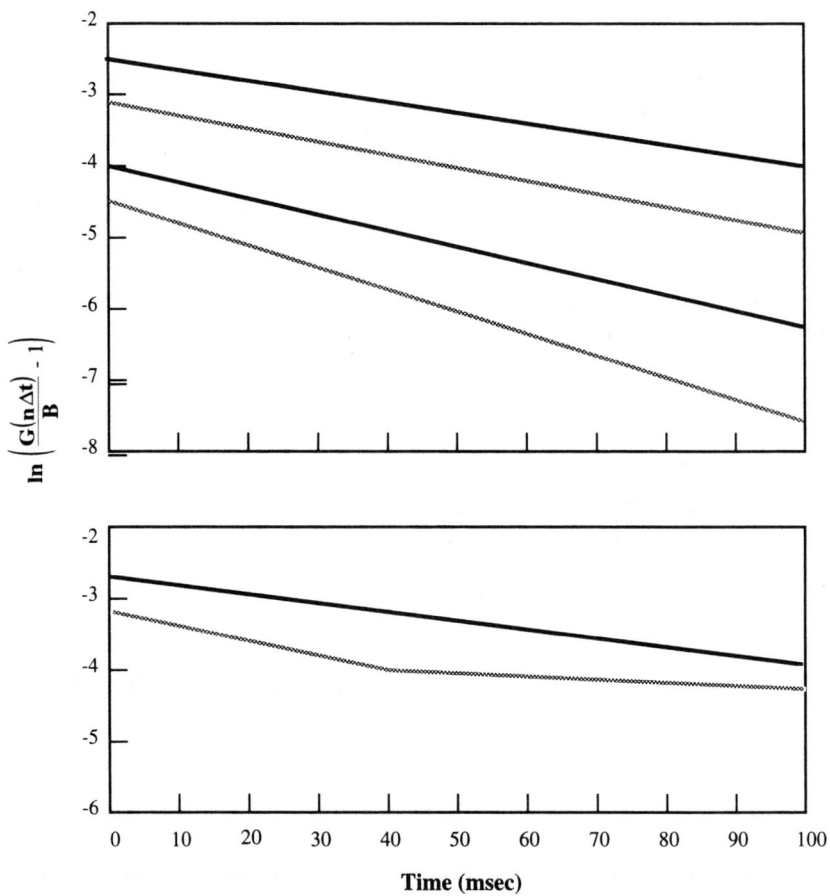

FIGURE 4.7. Determination of translational diffusion coefficient. Plots of $\ln[(G(n\Delta t))/B) - 1]$ versus time for collagen α chains (top) and mixtures of α chains and β components (bottom). The translational diffusion coefficient is obtained by dividing the slope of each line by $-2Q^2$, where Q is the scattering vector. B is the baseline of the autocorrelation function. Note for single molecular species, the slope is constant, and for mixtures with different molecular weights, the slope varies (molecular weight for α chains is 95,000 and for γ components is 285,000). (reproduced from Silver, 1987).

Once Dt for a macromolecule is determined, it can be compared to theoretical values for prolate ellipsoids and spheres using Equations (4.12) and (4.13) for standard conditions of 20°C and the viscosity of water and a value for the axial ratio can be estimated.

$$\text{Prolate ellipsoids } Z = a/b > 10$$
$$Dt_{20,w} = 2.15 \times 10^{-13} \text{ cm}^3/\text{sec} \left[\ln(2a/b)\right]/a \tag{4.12}$$

$$\text{Sphere of radius } R$$
$$Dt_{20,w} = 2.15 \times 10^{-13} \text{ cm}^3/R \tag{4.13}$$

Using Equations (4.12) and (4.13) and the experimental measurement of Dt, values of a and b or R can be determined for different macromolecules. These values can be compared to values generated from viscometry and classical light scattering. The large difference between the values of the translational diffusion coefficient for spheres and rods suggests that measurement of this parameter is useful in determining shape and size differences. For example, albumin is a protein present in blood that regulates osmotic pressure, that is, ions diffuse from the interstitial spaces into blood in order to lower the osmotic pressure of the blood. If albumin were rodlike rather than spherelike not only would the viscosity of blood be much higher, but the rate at which albumin diffuses in the blood would be much lower.

Values of Dt for various biological macromolecules are given in Table 4.3. From Dt we can get an idea of what the axial ratio of a macromolecule might be.

TABLE 4.3. Determination of translational and rotational diffusion coefficients for connective tissue macromolecules from quasi-elastic light scattering reproduced from Silver, 1987

Macromolecule—source	$D°_{20,w} \times 10^{+7}$ (cm^2/sec)
Collagen (type I)—rat tail tendon (pr)[a]	0.86[c]
Collagen (type I)—chick and rat skin	0.80,0.85
Collagen (type I)—rat tail tendon, skin	0.78
Collagen (type I)—rat tail tendon	0.78
Collagen (type I)—rabbit cornea	0.85
Collagen (type I)—rabbit sclera	0.86
Collagen (type I)—bovine skin (p)[b]	0.82
Collagen (type II)—bovine nasal septum (p)	0.85
Collagen (type III)—bovine skin (p)	0.86
Collagen (type IV)—EHS tumor	0.66
Collagen (type V)—bovine skin (p)	0.86
Collagen (type V)—chick embryo (p)	0.78
Hyaluronan—bovine vitreous humor	0.86
Hyaluronan—rooster comb	0.36
Lubricin—bovine synovial fluid	1.1
Proteoglycan—bovine nasal cartilage	3.32

[a] (pr) = Pronase treated.
[b] (p) = Pepsin treated.
[c] = Maximum value.

4.5 Ultracentrifugation

The size and shape of macromolecules in solution can be studied using two techniques termed equilibrium and velocity ultracentrifugation. These techniques use an ultracentrifuge to rotate solutions of macromolecules and place them under a centrifugal force to study their physical properties. The size and shape of the macromolecules can then be determined from the solution physical properties. The ultracentrifuge is equipped for direct measurement of the solution as it spins at high speed.

At speeds above several thousand rpm, macromolecules in the cell of the ultracentrifuge settle towards the periphery of the rotor, and thus the axial ratio can also be estimated from ultracentrifugation. At the same time the molecules tend to diffuse towards the center of the rotor because of the concentration gradient. At equilibrium, the centrifugal force due to spinning of the rotor exactly offsets the tendency for the molecules to diffuse towards the center of the rotor. Measurement of the solution optical density as a function of the concentration c, at different distances r, from the center of the cell, gives an idea of the apparent molecular weight $Mapp$, of the macromolecules. Using Equation (4.14), the apparent molecular weight is obtained knowing the values of V_2, R, ω, ρ, and T, the volume fraction of macromolecule, the gas constant, angular velocity, solution density, and temperature, respectively.

$$\ln c(r) = (1 - v_2\rho)/2RT\left[\omega^2 Mapp\, r^2\right] + \text{constant} \qquad (4.14)$$

At high speeds greater than 40,000 rpm in the ultracentrifuge, macromolecules settle towards the rotor periphery. Under these conditions the sedimentation coefficient s, is determined from the speed of sedimentation divided by the angular acceleration. The sedimentation coefficient is related to molecular weight using Equation (4.15).

$$M = RTs/\left[Dt(1 - V_2\rho)\right] \qquad (4.15)$$

Using M and s from ultracentrifugation experiments Dt can be calculated and compared to Dt determined from quasi-elastic light scattering.

Values of M and s determined by analytical ultracentrifugation are given in Table 4.4 for some biological macromolecules. From these values we can get an idea of the shape factor for a given biological macromolecule.

TABLE 4.4. Determination of molecular weights and sedimentation coefficients for connective tissue macromolecules using ultracentrifugation reproduced from Silver, 1987

Macromolecule—source	M	$s°_{20,w} \times 10^{13}$ (sec)
Collagen—rat skin	310,000–340,000	2.92
Collagen—rat and bovine skin	294,000–307,000	
Collagen—carp swim tunic	—	2.96
Collagen—calf skin	262,000	3.4
Collagen type II—chick xyphoid cartilage	262,000–307,000	
Procollagen I—chick embryo tendon cells	350,000	3.0
Hyaluronan—bovine vitreous	220,000–313,000	2.98–3.13
Hyaluronan—bovine vitreous	178,000	3.10
Hyaluronan—bovine vitreous	86,600	2.8
Lubricin—bovine synovial fluid	196,600	4.84
Proteoglycan—bovine nasal cartilage	$1.78–2.99 \times 10^6$	20.4–28.0
Proteoglycan—bovine articular cartilage	43,000–45,000	2.91–3.5
Proteoglycan—bovine articular cartilage	1.28(PGI), 1.07(PGII), 0.24 $\times 10^6$(PGIII)	
Proteoglycan core—bovine nasal cartilage	190,000–214,000	
Proteoglycan—human articular cartilage	1.6×10^6	10.5
Proteoglycan—human uterine cervix	73,000	2.1

4.6 Electron Microscopy (EM)

The identification of size, shape, and axial ratio can also be done by direct observation in the electron microscope (EM). This is accomplished by depositing single molecules (if they can be obtained) directly on polymer-coated copper grids and then shadowing them with heavy metals or making a replica of the molecular surface on mica. The sample can then be viewed in the transmission EM and photographs can then be taken after calibration of the magnification factor.

From these images complex shapes can analyzed using bead models (Figures 4.8 and 4.9) to calculate Dt and $P(\phi)$. The equations for calculating Dt for a series of N beads of diameter d and a radius between beads r_{ij} are given by Equations (4.16) and (4.17). In Equation (4.16) the summations are made for N from 1 to j and i from 1 to N as long as i does not equal j.

$$f = 6\pi\eta d/2\,N \Bigg/ \left[1 + d/N \sum_{i=1}^{N}\sum_{j=1}^{N} <r_{ij} - 1> \right] \qquad (4.16)$$

$$Dt = kT/f \qquad (4.17)$$

In Equation (4.17) k is Boltzmann's constant.

FIGURE 4.8. Electron microscope images of single molecules. Schematic images of
collagen molecules obtained by
shadowing the molecules with
heavy metals and viewing
either the molecules or replicas
of the molecules under the
electron microscope. Molecules shown include collagen
types I (A), II (B), III (C), and
V (D), and each molecule is
roughly 300 nm in length.

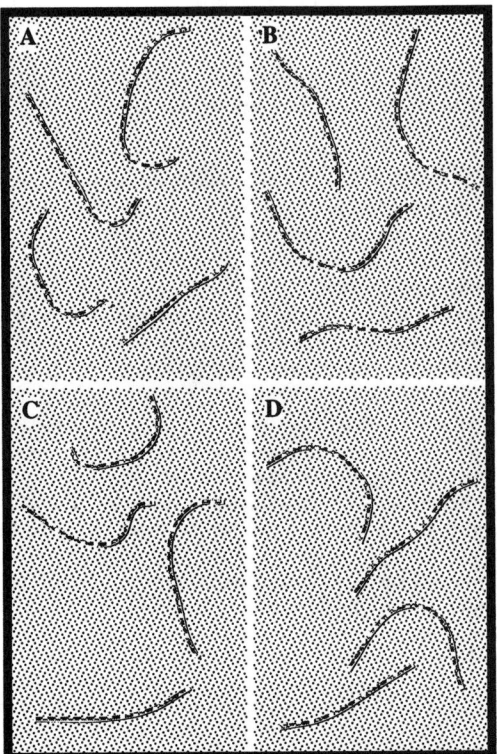

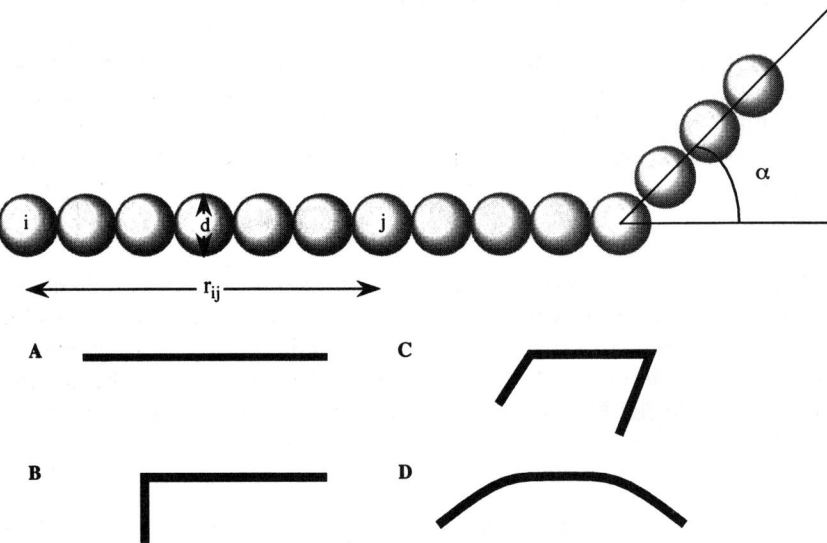

FIGURE 4.9. Bead models used to simulate macromolecular structures. The drawing
shows a model used to simulate a collagen molecule with a single bend at an angle
α. Each molecule is represented by a series of N beads of diameter d that are separated by a distance r_{ij}. (A) through (D) illustrate the typical shapes observed for
collagen type I by rotary electron microscopy. The molecule depicted has a total
length of 300 nm.

4.7 Determination of Physical Parameters for Biological Macromolecules

The values of intrinsic viscosity, molecular weight, translation diffusion coefficient, and sedimentation coefficient have been determined for a variety of different macromolecules and are listed in Table 4.5. As we can see from these data, molecules with high values of the intrinsic viscosity, that is, greater than 1000 ml/g and relatively low translational diffusion coefficients (1×10^{-7}) and sedimentation coefficients (3×10^{-13}) are rigid or semiflexible rods. These molecules are good candidates for supporting and transferring stress in biological systems. However, they must be assembled into continuous networks to make this possible. Molecules with low translational diffusion coefficients and intrinsic viscosity and high sedimentation coefficients are flexible rods. Finally, molecules with high translational diffusion coefficients and low intrinsic viscosity are roughly spherical.

A measurement of physical parameters in solution for isolated macromolecules provides a manner by which the shape of a macromolecule can be determined. The approximate dimensions and axial ratio or radius can be calculated by applying Equations (4.3) through (4.17). As shown in Figure 4.10, the particle scattering factor for collagen molecules depicted in Figure 4.9 is more sensitive to bends than is the translational diffusion coefficient.

TABLE 4.5. Physical constants for biological macromolecules

Macromolecule	M	$D_{20,w}$[a]	$S_{20,w}$[a]	$[\eta]$	Shape
Collagen I	285,000	0.85	2.96	1,100	Semiflexible rod
II	285,000	0.85		1,530	Semiflexible rod
IV	535,000	0.66		75	Flexible rod
Fibrinogen	330,000	2.02	7.9	25	Flexible rod
Hyaluronan (vitreous)	178,000	0.86	3.10	718	Wormlike coil
Myosin	570,000	1.0	6.4	217	Flexible rod
Serum albumin	66,000	5.94	4.31	3.7	Roughly spherical
Keratin	110,000	4.3	4.3	154.1	Flexible rod
Tubulin	110,000	4.7	6.0		Roughly spherical
G-Actin	42,000	7.9–8.13		3.0	Roughly spherical

[a] $D_{20,w}$, $S_{20,w}$ = translational diffusion constant and sedimentation coefficient back calculated to a temperature of 20°C and the viscosity of water.

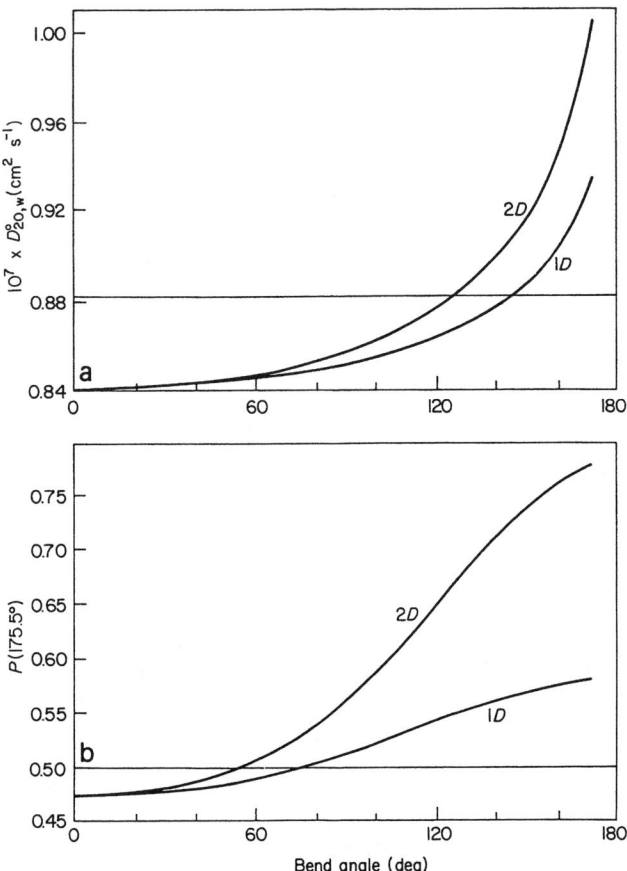

FIGURE 4.10. Theoretical dependence of translational diffusion coefficient of rods on location of bends. Theoretical relationships shown between translational diffusion coefficient (a), particle scattering factor at a scattering angle of 175.5° (b), and bend angle obtained using bead model with bends at $1D$ and $2D$ from the end. Points above horizontal lines shown for $D^{\circ}_{20,w}$ and $P(175.5°)$ are significantly different from those values for a straight rigid rod. The total rod length is $4.4D$. (reproduced from Silver, 1987).

4.8 Summary

The size and shape of a macromolecule can be determined by measuring the physical properties of isolated macromolecules in solution. Large rigid macromolecules that are derived from extended structures including the collagen triple helix result in rodlike rigid or semirigid structures. The size, shape, and physical parameters for macromolecules discussed in this book are shown in Table 4.5. This is important because molecules with high intrinsic viscosities and shape factors tend to have high moduli and are packed into strong stiff networks in tissues. As can be seen from Table 4.5, of the molecules listed collagen has the highest intrinsic viscosity and as we show later it has the highest axial ratio making it the best candidate for stress bearing and energy transfer. Flexible molecules tend to form from structures that have helices connected by flexible segments. These molecules are not only mobile in solution but form flexible low modulus structures in tissues. However, the behavior of individual macromolecules is only one measure of a candidate molecule's ability to support loads and transfer energy; the next requirement is that the macromolecule must be assembled in tissues into continuous crosslinked stress-bearing networks.

Suggested Reading

Birk DE, Silver FH. Corneal and scleral type I collagens: Analyses of physical properties and molecular flexibility. *Int J Biol Macromol* 1983;5:209.

Silver FH. *Biological Materials: Structure, Mechanical Properties, and Modeling of Soft Tissues*. New York: NYU Press; 1987: Chapter 4.

Silver FH, Trelstad RL. Type I collagen structure in solution and properties of fibril fragments. *J Biol Chem* 1980;255:9427.

5
Self-Assembly of Biological Macromolecules

5.1 Introduction

The ability of tissues to store energy, transmit and distribute loads, as well as dissipate energy requires that individual macromolecules be assembled into continuous networks. Although solutions of macromolecules can form gels like Jello that can bear small loads, it is important that the volume fraction of macromolecules be above about 0.5 and that there be chemical crosslinks so that large tensile loads can be borne by polymer networks. What this means is that although pulling on the ends can stretch macromolecules and as a result the molecules undergo conformational changes, the ability to transfer loads from one molecule to another requires intermolecular interactions. As we saw in Chapter 2 covalent interactions are much stronger than electrostatic bonds and they are both stronger than individual van der Waals attractions. Therefore, in an aqueous environment, stress transfer requires either more than two covalent intermolecular crosslinks per molecule or many electrostatic or hydrogen bonds. Hydrogen bonding, however, is not the preferred way to confer stability on structural macromolecules: eventually the hydrogen bonds will break under continuous loading.

In addition, if these crosslinks or bonds are not strategically placed along the molecule (i.e., they will not transfer stress from one molecule to another), then the crosslinks are useless (see Figure 5.1). If the molecules are crosslinked into tight arrays the packing density can be higher as well as the effective stress transfer. Nature accomplishes the ability to transfer large loads by assembling biological macromolecules into organized structures that are then linked together into networks. For instance, collagen molecules form fibrils in tissues that in turn form fibers that are fused into bundles. Therefore, it is important to understand how biological macromolecules are self-assembled into network structures capable of storing energy and transferring loads. Finally, it is important to understand how stress is transferred between the extracellular matrix and cells in order to understand mechanochemical transduction.

FIGURE 5.1. Collagen crosslinking sites in tendon. Crosslinking sites on collagen occur in the helical and nonhelical ends as shown in the diagram. The dotted lines at the ends of the molecule represent the crosslinks that stabilize collagen molecules in tendon. Note the molecules are crosslinked end-to-end and the crosslinks connect rigid regions (light regions) of the molecule. Crosslinks that run from one end of the tendon to the other are required for stress transfer and energy storage.

The study of animal tissues is rather complex inasmuch as water molecules, ions, cells, macromolecules, tissues, and organs exist in equilibrium. From a structural point of view, biological tissues contain highly ordered arrays of macromolecules. One might wonder why biological structures need to be made up of highly ordered arrays of proteins, polysaccharides, and lipids. The reason is that individual polymer molecules cannot sustain the weight of gravity without rearranging. In the case of tissues, the molecules need to assemble into ordered structures and be crosslinked for the shape of the tissue to be maintained. In some cases, assemblies of macromolecules are purposely not crosslinked so that the shape can be changed quickly. For instance, cytoskeletal actin filaments are rapidly assembled and disassembled to allow for changes in cell shape. Noncovalent bonds hold these filaments together so that rapid shape changes can occur. In this example, covalent crosslinks would prevent rapid shape changes, however, actin filaments by themselves are rigid enough to maintain cell shape at any one instant. In contrast, collagen fibers in the skin must be crosslinked to form force-bearing units to prevent tearing when skin is stretched. This comparison is used to underscore the complexity of biological tissue structure and its relationship to physical properties. In some cases rigidity is sacrificed for structural flexibility. In other cases structural flexibility is sacrificed for permanence.

Biological macromolecules are largely found in animal tissues in the form of helices (α and collagen triple helix), β structures, amorphous or flexible chains, and combinations of these structures. It is the assembly of these units that makes up the crystalline arrays that are the noncellular components of tissues. It is amazing how the assembly of a few types of structures gives rise to many different tissue structures.

5.2 Theory of Assembly of Biological Macromoleucles

The assembly of biological macromolecules into fibrous and other supramolecular structures has been studied extensively since the early 1900s. Oosawa and Kasai (1962) pointed out that various biological macromolecules form intramolecular helices, that is, helices that are formed from hydrogen bonds within a single polymer chain; but no one considered the possibility of intermolecular helical structures that had hydrogen bonds between chains. The purpose of their paper was to present a simple theory of the helical aggregation of macromolecules and to compare theoretical predictions with experimental results. The theory developed suggested that an equilibrium distribution existed between monomers (single polymer chains) and linear and helical aggregates containing more than one polymer chain. When the concentration of macromolecules is increased, the helical aggregates begin to appear at the critical concentration determined by solvent conditions. Above the critical concentration very long helical aggregates coexist in equilibrium with a constant concentration of dispersed monomers. The authors used this theory to explain the equilibrium and kinetic features of the transformation of globular monomeric actin (G-actin) to fibrous aggregated actin (F-actin).

Another approach used for thinking about the mechanism of assembly of biological macromolecules involves considering it as a phase transition involving a solid phase (aggregate) in equilibrium with a solution phase containing single isolated polymer chains. Flory developed free energy calculations to understand the energy required to transfer molecules between the liquid and solid phases (see Silver, 1987). The free energy change required to mix solute molecules and solvent molecules is related to the change in the number of bonds formed or broken as well as the associated entropy change.

Specifically, he developed relationships for the free energy of mixing of solvent and solute (polymer) molecules with different solute axial ratios (length/width) and solute interaction parameters. In the absence of interactions between solute molecules, the free energy of the system lowers when rodlike molecules precipitate out of solution and form a separate solid phase. This is due to the fact that small water molecules must order themselves around large rodlike macromolecules in solution and therefore the system is most stable when the water molecules and rodlike molecules are separated in space into different phases, such as liquid and solid phases.

Flory's mathematical relationships for the free energy change that occurs when a solute is mixed with a solvent for the solution phase and the solid phase have been discussed as they apply to collagen assembly by Silver (1987). The free energy of mixing of solute and solvent per mole of molecules is equivalent to the change in chemical potential of a desired state (μ)

from that of the reference state (μ_1°). These relationships are given by Equations (5.1) through (5.4) where the subscripts 1 and 2 represent solvent and solute, respectively, and R is the gas constant, T is the absolute temperature, V is the volume fraction of solute, Z is the axial ratio of the solute (half length/half width), y is the disorientation index (the amount of disorientation of the solute), and μ_1 is the solute–solvent interaction parameter.

Isotropic or solvent phase $\left(\dfrac{\mu_1-\mu_1^0}{RT}\right)_l = \ln(1-V_2)+\left(1-\dfrac{1}{Z}\right)V_2+x_1V_2^2$ (5.1)

$$\left(\frac{\mu_2-\mu_2^0}{RT}\right)_l = \ln(V_2/Z)+(Z-1)V_2-\ln Z^2+x_1Z(1-V_2)^2 \qquad (5.2)$$

Anisotropic or solid phase $\left(\dfrac{\mu_1-\mu_1^0}{RT}\right)_s = \ln(1-V_2)+\dfrac{(y-1)}{Z}V_2+\dfrac{2}{y}+x_1V_2^2$

(5.3)

$$\left(\frac{\mu-\mu_2^0}{RT}\right)_s = \ln(V_2/Z)+(y-1)V_2+2-\ln y^2$$
$$+x_1Z(1-V_2)^2\,\Delta G = \#\,\text{moles}\,(\Delta\mu) \qquad (5.4)$$

What these equations tell us is that for molecules with large axial ratios (i.e., 100 or larger) the free energy is lowered (the lower the free energy the more stable a system is) for the solute by undergoing a phase transition from the liquid to the solid states; that is, the solute precipitates out. Increasing the solute concentration resulting in a precipitation of the macromolecular component further supports phase separation (See Figure 5.2). This observation suggests that polymer molecules that form α helices, collagen triple helices, or β structures in solution tend to spontaneously phase separate due to their long thin profiles and form a new solid phase. If the molecules exhibit attractive forces they are likely to self-assemble into higher-order structures. Below we discuss the use of Flory's equations to understand collagen self-assembly.

There are several self-assembling macromolecules that are of interest to us in this text. They include (1) collagen, the primary structural material found in the extracellular matrix; (2) actin, a component of the cell cytoskeleton that is involved in cell locomotion and in formation of the thin filaments of muscle; (3) microtubules, which are involved in cell mitosis, movement, and organelle movement; and finally (4) fibrinogen, which forms fibrin networks that minimize bleeding from cut vessels. Self-assembly is important in these systems because the function of these macromolecules can be modified via processes that increase the molecular axial ratio and hence decrease the solubility.

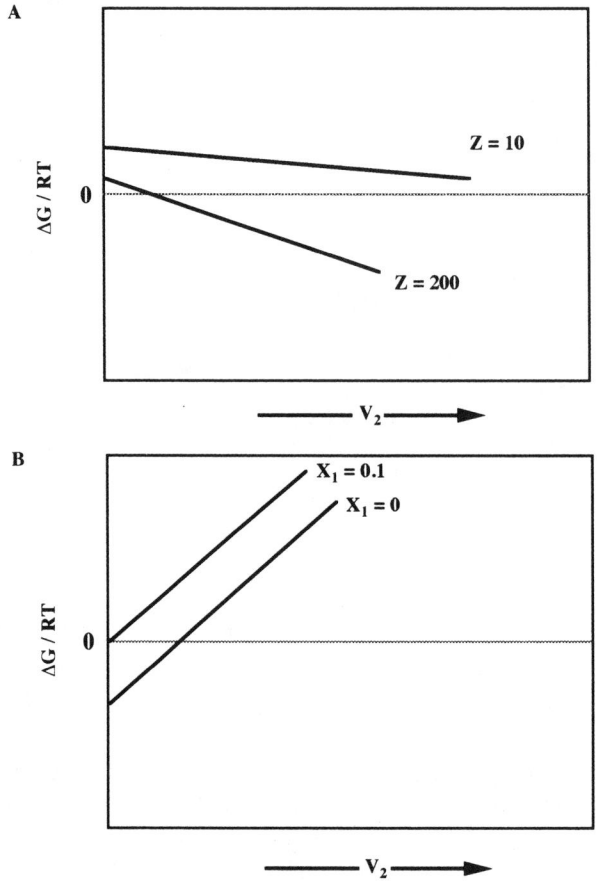

FIGURE 5.2. Free-energy change of mixing for rods and solvent molecules. Free energy change ($\Delta G/RT$) of (A) solid phase associated with transfer of a solute molecule (macromolecule) from the liquid to the solid state as a function of solute volume fraction (V_2) for low ($Z = 10$) and high ($Z = 200$) axial ratios; and (B) liquid phase as a function of solute volume fraction in the presence ($X_1 = 0.1$) and absence ($X_1 = 0$) of interactions between solute molecules. The diagrams show that separation of solute and solvent molecules occurs spontaneously for high axial ratios above a critical volume fraction and that the free energy of the solvent is raised by intermolecular interactions.

5.3 Methods for Studying Self-Assembly Processes

A number of methods have been developed for studying self-assembly processes including light scattering, ultracentrifugation, and electron microscopy. All of these methods have some associated problems when applied to studying self-assembly; however, applying several of these methods and then comparing the results has advanced our understanding of self-assembly processes.

5.3.1 Light Scattering

Phenomenologically, it was observed many years ago that when a solution of macromolecules self-assembled it became turbid and the resulting turbidity–time curve could be used to characterize this process. Typically the starting solution was transparent to light at wavelengths between 313 and 500 nm and once assembly proceeded the process was characterized by a sigmoidal (S-shaped) curve, containing lag, growth, and plateau phases as diagrammed in Figure 5.3. This type of curve was seen with collagen and it was hypothesized that during the lag phase nuclei assembled that later grew rapidly during the growth phase.

Silver and Birk (see Silver, 1987) have discussed the mathematics of analysis of the events that occur during the turbidimetric lag and growth phase. The turbidity per unit path length T, is equivalent to 2.303 times the absorbance (measured using a spectrophotometer) and is proportional to an optical constant H, times several factors: these include the molecular weight M, weight concentration of macromolecule c, and particle dissipation factor Q (Equation (5.5)). The particle dissipation factor is related to the largest dimension of the macromolecule; it is one if the macromolecule is small with respect to the light wavelength in solution and zero if the largest dimension is infinite. The optical constant is given by Equation (5.6) where n, dn/dc, No, and λ are the index of refraction of the solution, refrac-

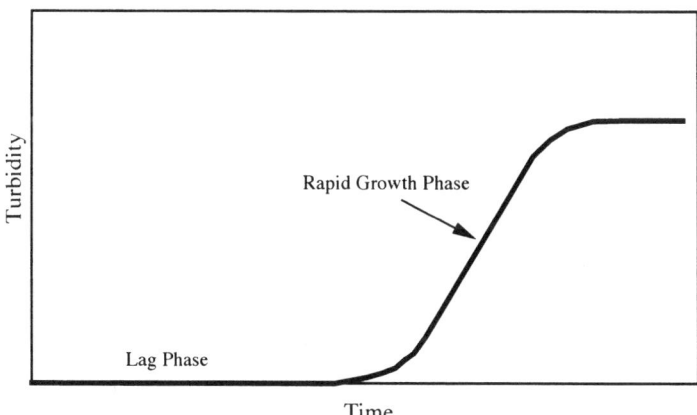

FIGURE 5.3. Turbidity–time curve illustrating collagen self-assembly. Turbidity–time curve illustrating lag phase, during which small linear and lateral aggregates form, and growth phase, during which unit fibers form that rapidly grow into fibers. The plateau is characteristic of termination of fibril growth.

tive index increment of the macromolecular component, Avogadro's number, and wavelength of light in solution, respectively.

$$T = HMcQ \tag{5.5}$$

$$H = \frac{32\pi^2 n^2 (dn/dc)^2}{3No\lambda^4} \tag{5.6}$$

Equation (5.5) is analogous to the equation for the scattered light intensity at an angle 90 degrees to the transmitted beam and Q is related to the particle scattering factor at a scattering angle of 90 degrees. Using this analogy, the increased turbidity reflects an increase in molecular weight. At the same time if the particle formed is getting longer then Q may cause the turbidity to decrease because of destructive interference. Therefore, it is evident from Equation (5.5) that increases in turbidity are a result of changes in size (M, molecular weight) and shape (Q, particle dissipation factor).

Q and H can be calculated for a rodlike assembly and the results depend on the light wavelength. At wavelengths of 300 and 650 nm, the turbidity per unit concentration (λ/c) is related to the mass per unit length (M/L) as indicated in Equation (5.7).

$$\begin{aligned} \lambda/c &= 6.57 \times 10^{-3} (M/L) & \lambda &= 300 \text{ nm} \\ \lambda/c &= 4.49 \times 10^{-4} (M/L) & \lambda &= 650 \text{ nm} \end{aligned} \tag{5.7}$$

As one can see from reviewing the mathematics above, the turbidity per unit concentration is proportional to the molecular weight per unit length. This implies that the turbidity lag phase can be characterized either by a constant molecular weight or a molecular weight per unit length that does not increase. The latter case occurs when self-assembly proceeds in a linear fashion and the length doubles every time the molecular weight does the same. This mathematical analysis of turbidity–time curves suggests that turbidity measurement is not the recommended way to follow linear growth of macromolecular assemblies. However, it is also concluded that rapid turbidity increase during the growth phase of turbidity–time curves is characteristic of systems that grow laterally.

Another manner to characterize self-assembly using light-scattering techniques is to measure the light intensity scattered at a fixed angle. Although the problems associated with measuring turbidity changes during self-assembly also affect measurement of the intensity of light scattered at an angle of 90 degrees, these problems can be averted by measurement of the scattered light intensity at very small scattering angles. Figure 5.4 illustrates that measurement of the scattered light intensity extrapolated to 0 degrees during self-assembly is more sensitive than that at a scattering angle of 90 degrees. Therefore, measurement of the light intensity at low scattering angles can be used to follow the early events associated with self-assembly.

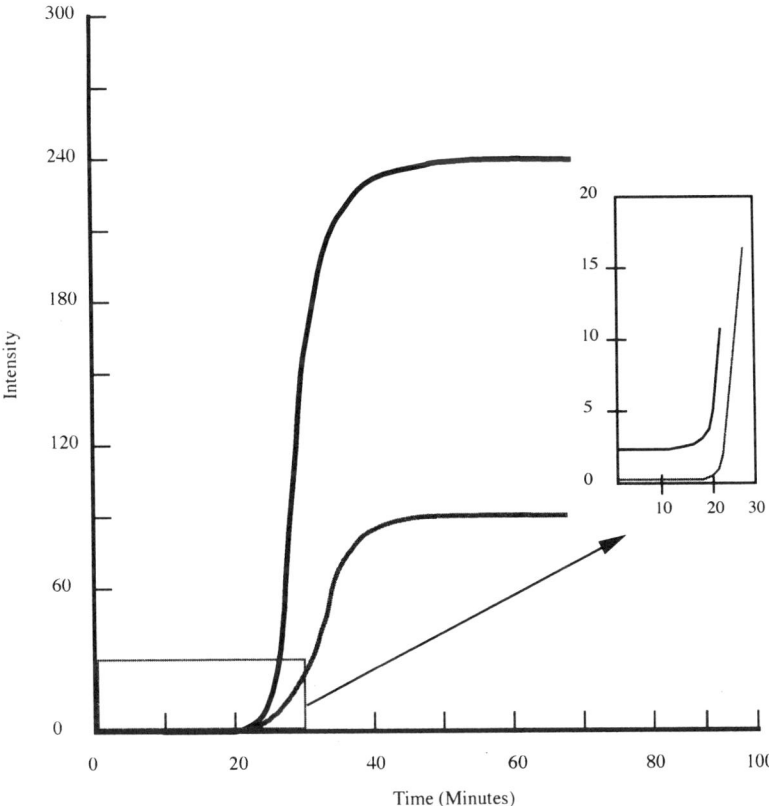

FIGURE 5.4. Scattering intensity during initiation of collagen self-assembly. Relative scattering intensity versus time at a scattering angle of 90° (lower curve) and extrapolated to 0° (upper curve) during the initiation of collagen self-assembly; changes are observed at a scattering angle of 0° before they are measured at 90°.

In addition, by measuring the intensity fluctuations at low scattering angles changes in the translational diffusion coefficient can be measured during the early phases of self-assembly as illustrated by Figure 5.5. The translational diffusion coefficient decreases as the ratio of the length to width increases as linear assembly occurs.

5.3.2 Equilibrium Ultracentrifugation

Another method for studying self-assembly has been to measure the concentration gradient as a function of position as a function of time during equilibrium centrifugation. The major criticism of this approach is that if the products that form during self-assembly are very large they can be sedimented out and removed from the sampling window. However, even with

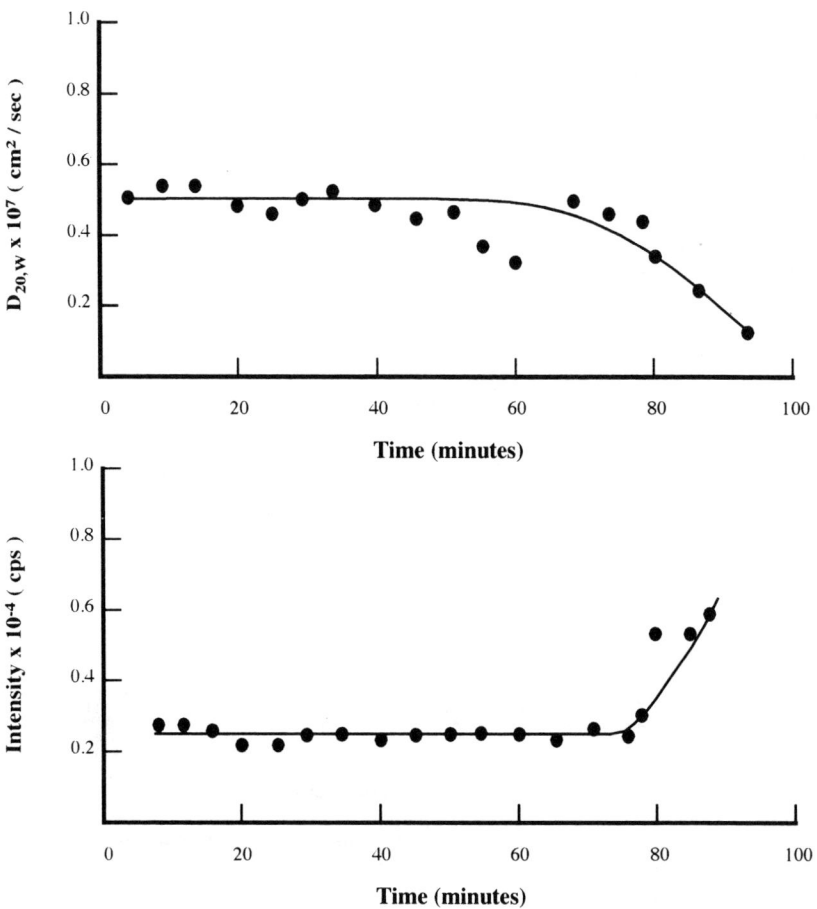

FIGURE 5.5. Measurement of physical properties during initiation of collagen self-assembly. Translation diffusion coefficient ($D_{20,w}$) (top) and intensity of scattered light at 90° (bottom) versus time for type I collagen. Note translational diffusion constant decreases, whereas intensity of scattered light remains initially unchanged.

these criticisms it is worth using this approach to observe small aggregates that form during early self-assembly steps.

5.3.3 Electron Microscopy

Although electron microscopy is always interpreted with care and caution, because of the possibility of artifacts, a picture is always of value in developing models of self-assembly products. In self-assembly experiments it is important to stop the process of self-assembly so that direct visualization can be achieved without modifying the process. This is always a concern; however, direct visualization has helped us understand the mechanism of assembly of most biological macromolecules.

5.4 Collagen Self-Assembly

Approximately 50 years ago it was first observed that purified collagen molecules in solution in vitro would spontaneously self-assemble at neutral pH at room temperature to form fibrils that appeared to be identical to those seen in vivo. Collagen self-assembly is characterized by an increase in the turbidity (Figure 5.3), increase in the scattered light (Figure 5.4), and a decrease in the translational diffusion coefficient (Figure 5.5). Self-assembly of collagen to form rigid gels was observed by Gross and coworkers (see Silver, 1987) and Jackson and Fessler (see Silver, 1987). In their experiments, collagen was solubilized and then heated to 37°C in a buffer containing a neutral salt solution. Under these conditions it was recognized that collagen molecules and aggregates of molecules would spontaneously self-assemble forming fibrils that had the characteristic 67 nm repeat distance when viewed in the electron microscope. Subsequent studies showed the addition of ions, alcohols, and other substances that affected both electrostatic interactions and hydrophobic bonds were able to modify the assembly of collagen. The ability to form fibers from self-assembled collagen fibrils has made it possible to study the properties of model systems that mimic the structure and properties of tendons and other tissues.

The mechanism by which collagen molecules self-assemble into fibrils has been a topic of intense research interest since the 1950s. Early studies evaluated the kinetics of the transition of collagen from the solution to the solid phase by raising the temperature over a wide range of pHs and ionic strengths. The results were interpreted to suggest that collagen self-assembly involved a phase transition with no change in molecular conformation. These workers suggested that it was controlled by the addition of molecules at the fibril surface at temperatures above 16°C, and limited by the rate of diffusion of collagen molecules at temperatures less than 16°C. Thermodynamic study results indicated that native collagen fibril formation was an endothermic process made thermodynamically favorable by the large increase in mobility of the water molecules. This occurred when water molecules and collagen formed separate phases. In contrast, more recent studies have concluded that water-mediated hydrogen bonding between polar residues promotes collagen assembly (see Silver et al., 2003).

Early electron microscopic studies suggested that linear growth of fibrils appeared to occur by addition of groups of collagen molecules to form a subfibril; lateral growth occurred by entwining these subfibrils. Interpretation of early studies of collagen self-assembly was confused because of the difficulty in obtaining solutions of collagen single molecules as the starting point for self-assembly. Many of the solutions contained aggregates of collagen molecules and therefore interpretation of the results was difficult.

There are several observations that help us understand the process of collagen self-assembly. The first suggests that under typical solution conditions used for preparing soluble collagen (i.e., low pH and low salt content), collagen molecules are in equilibrium with larger aggregates. The aggregate in equilibrium with single molecules has been estimated to be between 1.5 million and 5 million or between about 5 and 17 molecules. Other studies indicate that the propeptides on newly formed procollagen molecules appear to limit association, thereby limiting the size of the aggregate in equilibrium with single molecules to only 5 molecules.

Other study results suggest that self-assembly of type I collagen leads to the formation of characteristic units that range in length from 4-D staggered dimers (about 570 nm) to aggregates that are about 700 nm long (Figure 5.6). This corresponds to between 2 and 3 collagen molecules long. Taken together these values of length and width suggest that aggregates formed during the initial phases of self-assembly contain 5 to 17 molecules and are 2 to 3 collagen molecules long (Figure 5.6). Estimates of the diameter of the first aggregates formed are initially 1 to 2 nm and then 2 to 6 nm. These observations are consistent with previously published models which state that self-assembly involves an initial linear step that is followed by both linear and lateral growth steps that occur simultaneously; reported results are also consistent with the formation of a fibril subunit that contains about 5 molecules packed laterally and is two or three collagen molecules long, as diagrammed in Figure 5.7. These units appear to grow laterally by fusion because autocorrelation of diameter distributions of self-assembled type I collagen fibrils show a periodicity of about 4 nm.

Studies of self-assembly of collagen type I indicate that initiation involves the nonhelical end regions, and removal of these ends with enzymes prolongs initiation. Enzymes such as pronase, which cleave both the amino and carboxylic nonhelical ends, arrest self-assembly. Pepsin, which removes portions of the amino and carboxylic nonhelical ends, slows aggregation resulting in wide fibrils. Leucine aminopeptidase removes the amino terminal nonhelical end and leaves the carboxylic end intact causing a lengthened lag phase. Carboxypeptidase A and B remove the carboxyl nonhelical end leaving the amino end intact, decreasing lateral growth without affecting

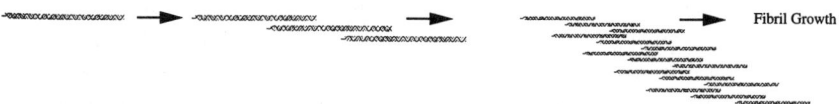

FIGURE 5.6. Collagen self-assembly. The diagram models the initiation of collagen self-assembly via formation of linear aggregates containing about three molecules that then laterally associate. The lateral assembly step may require a supramolecular twist, explaining why linear aggregation precedes lateral aggregation.

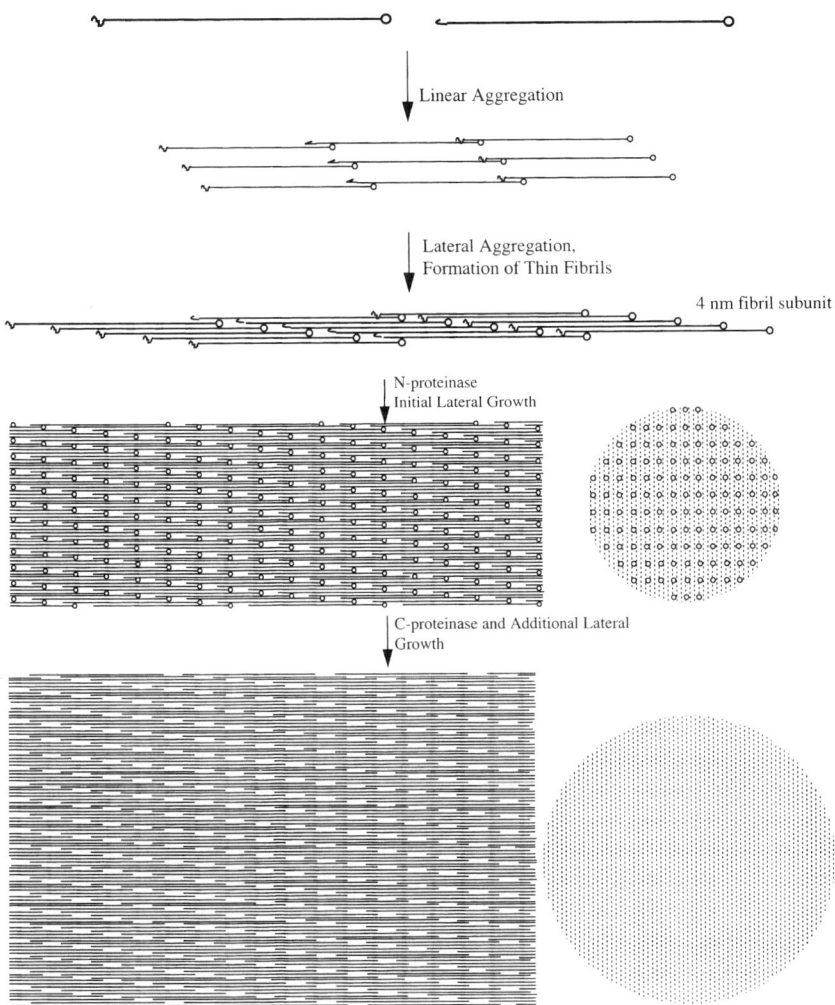

FIGURE 5.7. Diagram showing role of N- and C-propeptides in collagen self-assembly. The procollagen molecule is represented by a straight line with bent (N-propeptide) and circular (C-propeptide) regions. Initial linear and lateral aggregation is promoted by the presence of both the N- and C-propeptides. In the presence of both propeptides lateral assembly is limited and the fibrils are narrow. Removal of the N-propeptide results in lateral assembly of narrow fibrils; removal of the C-propeptide results in additional lateral growth of fibrils. As indicated in the diagram, the presence of the N- and C-propeptides physically interferes with fibril formation.

the lag phase. From these studies it is concluded that the amino terminal nonhelical end is involved in formation of the initial thin fibrillar unit and the carboxy terminal nonhelical end is involved in lateral fusion of thin fibrillar units. These conclusions parallel the conclusions made based on self-assembly of procollagen (see Figure 5.7) suggesting the effects of the propeptides and the nonhelical telopeptides on type I collagen self-assembly appear to be similar.

One of the drawbacks to studying self-assembly using type I collagen is that the fibrils that form are narrow (20 to 40 nm in diameter) and are not circular. Narrow diameters are likely caused by reactions that occur during self-assembly that favor linear as opposed to lateral growth. Factors such as glycine that block lysine-derived aldehydes during fibrillogenesis increase the width of collagen fibrils. Changes in ionic strength lead to changes in fibril diameter suggesting that lateral growth also involves electrostatic interactions. The nonuniformity and small diameters seen in collagen fibrils self-assembled from collagen molecules suggest that another factor must play a role in controlling fibril fusion and diameter.

Due to limitations associated with studying self-assembly using the collagen molecule, Prockop and coworkers developed a model of collagen self-assembly using selective cleavage of the procollagen I molecule using procollagen proteinases (see Silver et al., 2003). Solubility studies indicate that procollagen and pC-collagen (collagen containing the C-propeptide) are more soluble in solution than pN-collagen (collagen containing the N-propeptide) and collagen. Assembly of pN-collagen leads to wide D-periodic sheetlike structures. pC-collagen assembles to form tapelike and sheetlike structures (see Silver et al., 2003). Incubation of pC-collagen I with C-proteinase processes the pC-collagen to collagen and the collagen assembles into fibrils. Microscopy of the fibrils suggests that they are tightly packed and circular in outline. The average fibril diameter at 29°C is about 650 nm and at 37°C the average diameter is about 150 nm. These diameters are about a factor of ten larger than the diameters of fibrils formed from collagen. These observations suggest that addition of propeptides allows for formation of large diameter fibrils perhaps by promoting lateral over linear growth. For fibrils self-assembled from procollagen solutions, growth initially occurs at the pointed tip of the fibril; however, later growth occurs at both the pointed and blunt tips.

It is apparent that the ends of the molecule are involved in collagen self-assembly. These ends also overlap with helical parts of the molecule involved in crosslinking (Figure 5.1). Veis and George (see Silver et al., 2003) pointed out that the crosslinking and collagen cleavage sites have helix stabilities that are different than other parts of the molecule. They also reported that as the temperature is increased above 10°C, hydrogen bonds, which form the major triple-helix stabilizing force, become weaker. Weakening of hydrogen bonds within flexible sites on the collagen molecule may promote rotational freedom leading to interactions that support self-assembly.

5.4.1 Collagen Assembly in Developing Tendon

The ability of collagen molecules to assemble into crosslinked fibrils is an important requirement for the development of tissue strength. Although the process is under cellular control, the tendency for collagen molecules to form cross-striated fibrils is a property of the molecular sequence. Tendon is a multiunit hierarchical structure that contains collagen molecules, fibrils, fibril bundles, fascicles, and tendon units that run parallel to the geometrical axis (see Silver et al., 2003) as diagrammed in Figure 3.28. The fundamental structural element in tendon is type I collagen in the form of fibrils. Collagen is synthesized in the precursor form, procollagen, which contains nontriple helical extensions at both ends (Figure 2.23). The presence of amino (N) and carboxylic terminal (C) extensions on the collagen molecule have been shown to limit self-assembly of procollagen to about five molecules. Removal of the N- and C-propeptides by specific proteinases occurs prior to final fibril assembly. The C-propeptides are essential for both the initiation of procollagen molecular assembly from the constituent chains and lateral assembly of procollagen molecules (Figure 5.7). Procollagen molecular assembly in vivo initiates within intracellular vesicles. These vesicles are thought to move from regions within the Golgi apparatus to deep cytoplasmic recesses where they discharge their contents. Results of studies on embryonic skin suggest that the N-propeptides remain attached to fibrils 20 to 30 nm in diameter after collagen is assembled; however, after the N-propeptide is cleaved, fibril diameters appear to increase, suggesting that the N-propeptide is associated with initiation of fibrillogenesis. The C-propeptide is removed before further lateral fibril growth occurs (Figure 5.7).

Fibrillogenesis of type I collagen is specifically impaired in the skin of animals with a disease termed dermatosparaxis. In this disease the N-propeptide of the pro α 1(I) chain is not cleaved resulting in skin that is easily torn. Studies on dermatosparactic calf skin, in which 57% of the collagen molecules have intact N-propeptides, suggest that the presence of the N-propeptide on the fibril surfaces prevents tight packing of the collagen fibrils and results in skin fragility. Mass mapping of the dermatosparatic collagen fibrils shows that the N-propeptides are in a bent-back conformation that is within the overlap region. The finding of intact N-propeptides on fibril surfaces is also observed in cell cultures of skin fibroblasts from a patient with Ehlers Danlos syndrome type VII. Partial cleavage of the N-propeptide allows the N-propeptide to become incorporated within the body of the fibrils. This finding led to the proposal that the type I procollagen N-propeptides facilitate the fusion of small diameter fibrils (Figure 5.7).

The C-propeptide of fibril-forming collagens appears to regulate later steps in the assembly of procollagen into fibrils; it is removed from small-diameter fibrils during growth, possibly during fibril fusion. The C-propeptide has been observed in fibrils with diameters between 30 and 100 nm suggesting that it is involved in initiation and growth of fibrils (Figure 5.7). Procollagen and the intermediates pN-collagen (containing the N-propeptide) and pC-collagen (containing the C-propeptide) are present in developing tendon up to 18 days embryonic. Collagen oligomers isolated from developing chick tendons include 4-D staggered dimers of collagen molecules suggesting that this is a preferred molecular interaction for initiation of collagen fibrillogenesis in vivo. About 50% of the fibrils formed in 18-day-old chick embryos are bipolar (molecules run in both directions along the axis of the tendon) and the other half is unipolar. Analysis of the staining pattern of fibrils reveals the axial zone of molecular polarity to be highly localized.

During chick tendon development the structure and mechanical properties of tendon change rapidly. The morphology of embryonic development of collagen fibrils in chick tendon has been studied and characterized extensively. Two levels of structural organization of the collagen fibrils seem to occur during development of chick hind limb extensor tendons. Along the axis of the tendon, cytoplasmic processes of one or more axial tendon fibroblasts are observed to direct formation of groups of short collagen fibrils that appear to connect cells together (Figure 5.8). Also observed histologically during development is that groups of axial tendon cells are encircled by a second type of fibroblast that forms bundles of collagen fibers.

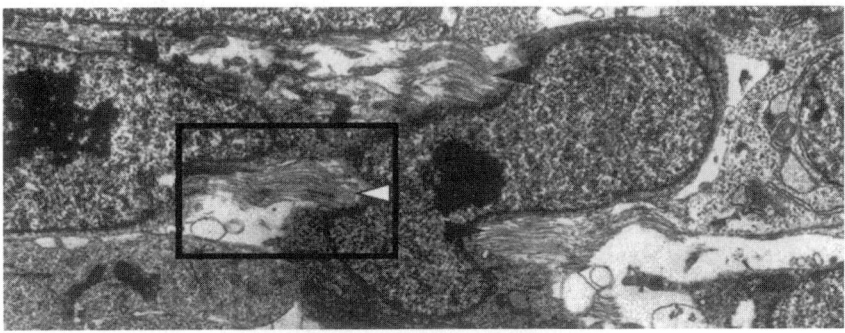

FIGURE 5.8. Directed cellular self-assembly of axial collagen fibrils during chick tendon development. Transmission electron micrograph showing collagen fibrils (see arrow in box) from a seven-day-old chick leg extensor tendon that appear to be connecting two fibroblasts during tendon development. Inset shows a high magnification view of the collagen fibrils that originate from invaginations in the cell membranes on either side of the fibril. The collagen fibrils shown are about 50 mm in diameter.

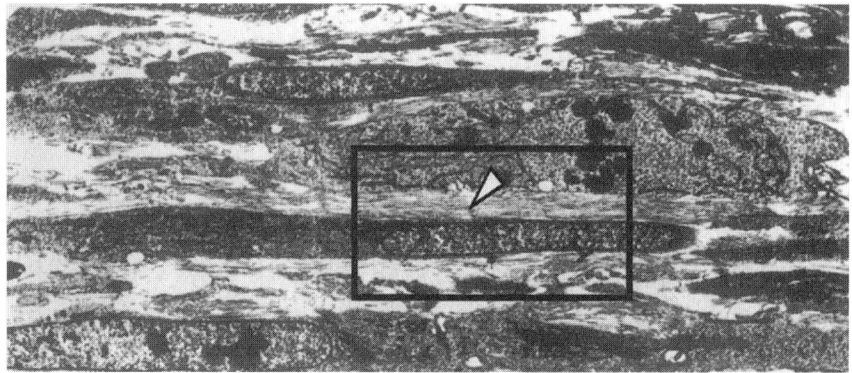

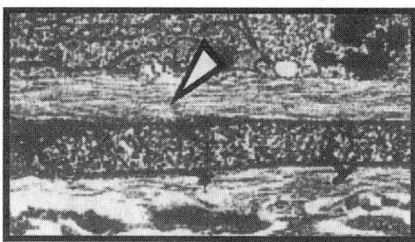

FIGURE 5.9. Lateral condensation of axial collagen-fibrils and alignment of tendon fibroblasts. Transmission electron micrograph showing collagen fibers from a ten-day-old chick leg extensor tendon. Note the fibrils (see arrow) and fibroblasts appear to be more highly aligned and densely packed compared to the same structures at day seven. Fibrils shown have diameters of about 50 nm. Insert shows a high magnification view of the relation between the collagen fibrils and the cell surfaces on either side. (Micrograph adapted from Silver et al., 2003.)

This type of cell encircles groups of collagen fibrils with a sheath that separates fascicles. Initially, axial tendon cells appear at both ends of growing fibrils (Figure 5.8). Once the fibrils begin to elongate they are then packed closely side to side (Figure 5.9). Later a planar crimp is introduced into collagen fibrils perhaps by the contraction of cells at the ends of the fibrils or by shear stresses introduced by tendon cells between layers of collagen fibrils (Figure 5.10). Results of recent modeling studies suggest that the molecule and fibril have many points of flexibility where crimp could develop (see Silver et al., 2003).

In cross-section, collagen fibers are made up of individual fibrils that appear to be released from invaginations in the cell membrane (Figure 5.9). Additional collagen diameter growth appears to occur by addition of materials that appear to originate inside the Golgi apparatus. Later during lateral growth, these invaginations in the cell membrane disappear causing lateral fusion of fibrils (Figures 5.11 and 5.12). Macroscopically this results in

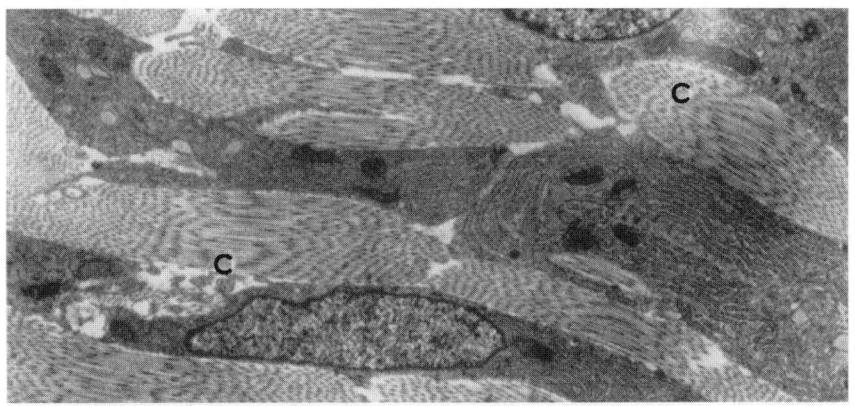

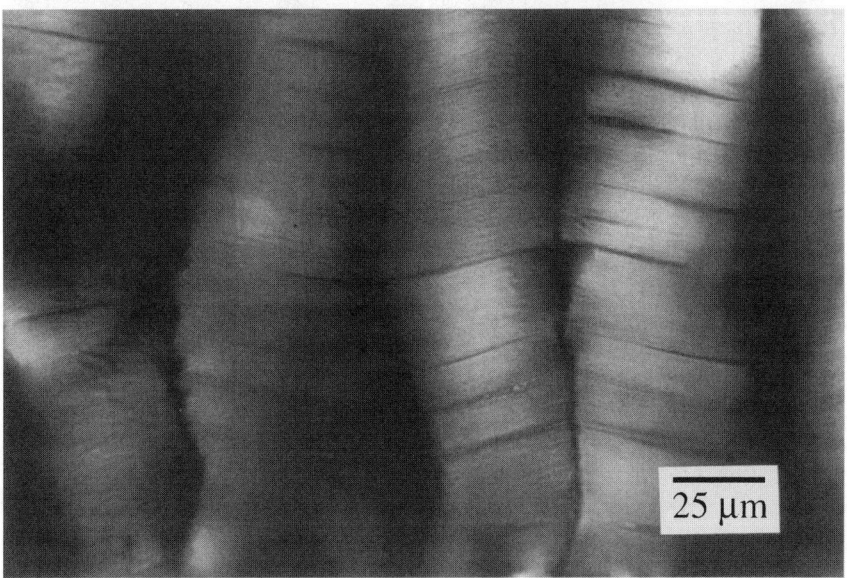

FIGURE 5.10. Formation of crimp in axial collagen fibrils during development of chick extensor tendon. (Top) Transmission electron micrograph showing collagen fibrils (C) from a 17-day-old chick leg extensor tendon. Note the fibrils appear to be going in and out of the plane of section consistent with the formation of a crimped planer zigzag pattern. Fibrils shown have diameters of about 100 nm. (See Silver et al., 2003) (Bottom) Polarized light micrograph illustrating the crimp pattern seen in a rabbit Achilles tendon.

FIGURE 5.11. Addition of axial collagen fibrils within invaginations in the cell membrane to a growing fibril. (Top) Transmission electron micrograph showing collagen fibril formation in invaginations within the cell membrane of a 14-day-old embryo. The arrows are placed in areas of the micrograph where collagen fibrils appear to be in the extracellular matrix and are in close proximity to the cell membrane. Insets A–D show the close relationship between cytoplasmic endoplasmic reticulum and collagen fibrils that appear to be in the extracellular matrix. The middle insets (C and D) show areas where collagen fibrils are within cellular membranes that appear to bud off and add to a growing fibril. The collagen fibril bundle (fiber) diameter marked by two Xs is 2 μm. (Bottom) High magnification view of insets shown in top micrograph. (See Silver et al., 2003).

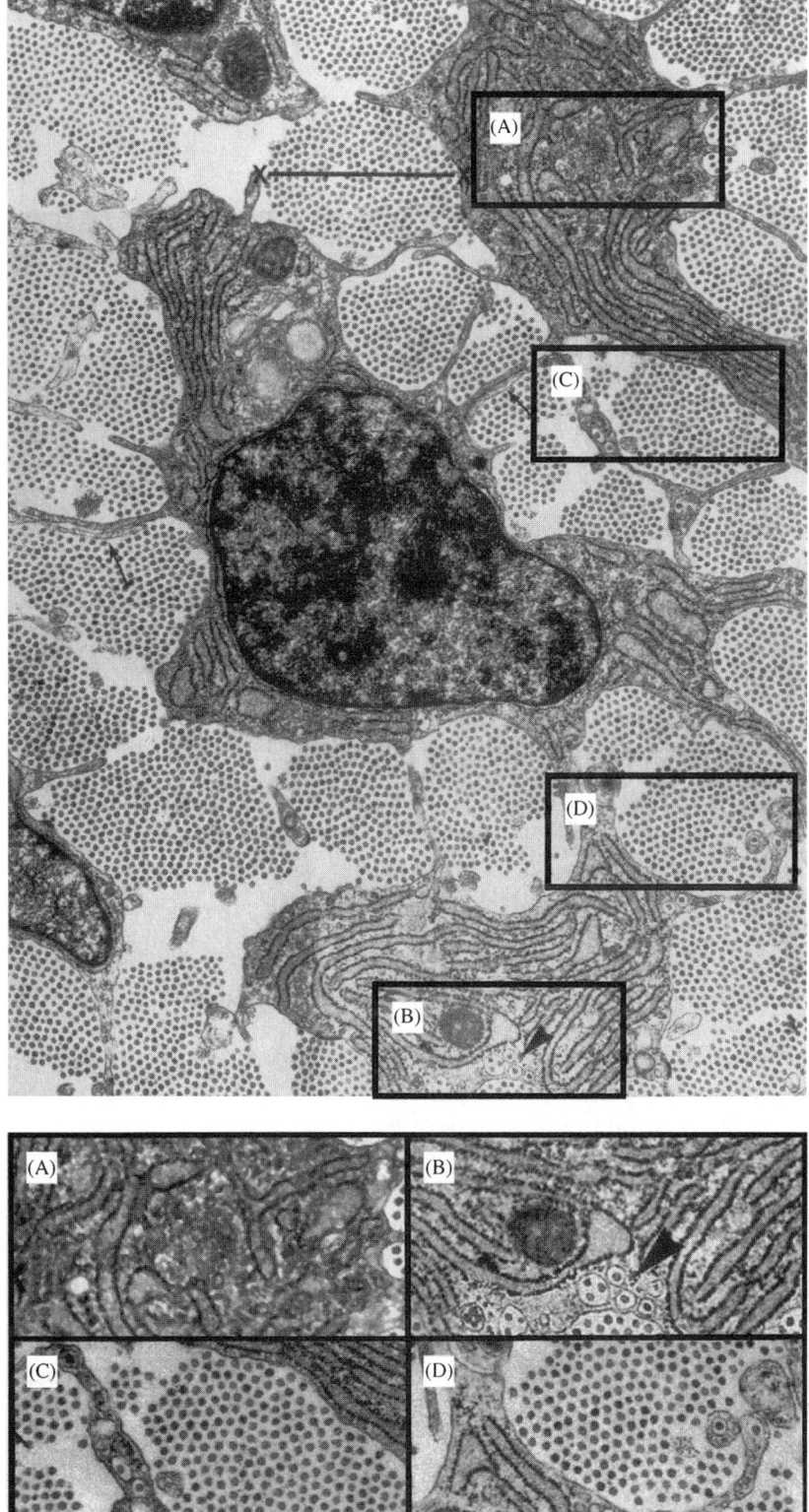

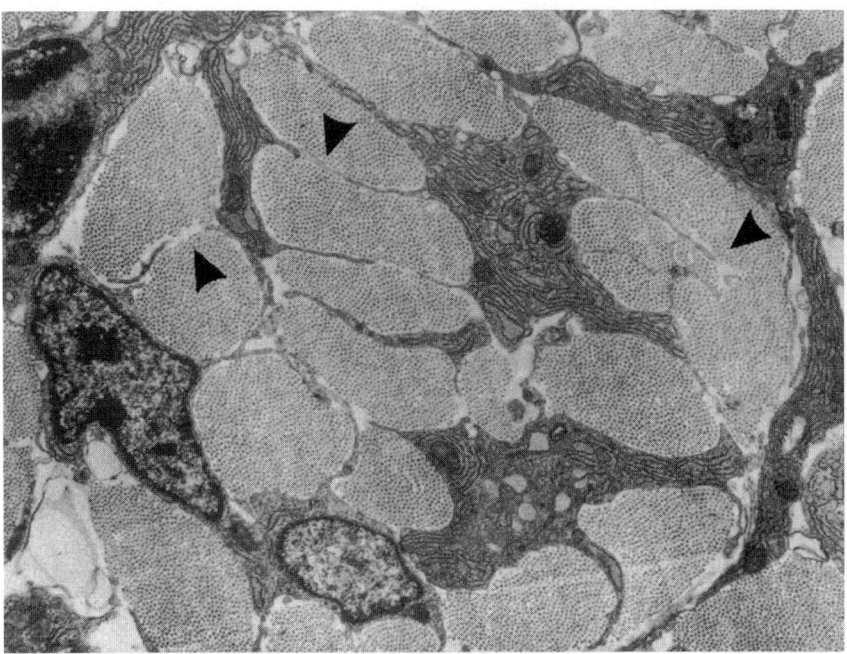

FIGURE 5.12. Lateral fusion of collagen fibrils during fascicle development of chick extensor tendon. Transmission electron micrograph showing the lateral fusion of collagen fibrils at day 17 of chick embryogenesis. Note that the demarcation between collagen fibrils (arrows) is less clear compared to the cross section shown at day 14 (Figure 5.11). Several fibrils appear to be in the process of fusion generating fibrils with irregular cross sections. The fibril bundle (fiber) diameter is still about 2 μm before fusion similar to that observed on day 14 (see Silver et al., 2003).

increases in fibril diameter and length that are necessary to bear large tensile loads.

Birk and coworkers have studied the manner in which collagen fibrils are assembled from fibril "segments" in developing chick tendon (see Silver et al., 2003). During development, fibril segments are assembled in extracytoplasmic channels defined by the fibroblast. In 14-day-old chick embryos, tendon fibril segments are deposited as units 10 to 30 μm in length. These segments can be isolated from tendon and studied by electron microscopy. Holmes and coworkers (see Silver et al., 2003) have shown that fibrils from 12-day-old chick embryos grow in length at constant diameter, and that end-to-end fusion requires the C terminal end of a unipolar fibril. By 18 days, embryonic fibril growth occurs at both fibril ends and is associated with increased diameter. Because fibril segments at 18 days cannot be isolated from developing tendon it is likely that fibril fusion and crosslinking occur simultaneously.

Fibril segments appear to be intermediates in the formation of mature fibrils and range in length from 7 to 15 μm in 14-day-old embryonic tendon. Between 14 and 17 days the bundles begin to branch and undergo rotation over several micrometers and the segments increase in length up to 106 μm. A rapid increase in length and diameter is seen between days 16 and 17 and is consistent with the rapid increase in tendon ultimate tensile strength. Although the fibril packing density of collagen does not change prior to and just after birth, the mean collagen fibril width increases and the cell volume fraction decreases. This is believed to be associated with growth by fibril fusion.

In mature tendon, collagen fibril bundles (fibers) have diameters between 1 and 300 μm and fibrils have diameters from 20 to over 280 nm. The presence of a crimp pattern in the collagen fibers has been established for rat tail tendon as well as for patellar tendon and anterior cruciate ligament; the specific geometry of the pattern, however, differs from tissue to tissue. It is not clear that the crimp morphology of tendon is actually present in tendons that are under normal resting muscular forces. However, the multilevel structural hierarchy in collagen is very important for transmitting loads in tendon.

5.5 Assembly of Cytoskeletal Components

Actin and tubulin are two important cellular components that are involved in cell shape and movement. Actin is present in all mammalian cells and is involved in cellular transport and phagocytosis (eating of extracellular materials), provides rigidity to cell membranes, and when bonded to tropomyosin and troponin, forms the thin filaments of muscle. Tubulin is the subunit from which microtubules are self-assembled. Microtubules are most commonly known for their role in cell division. The mechanisms of self-assembly of these macromolecules have been well studied and are important models of biological assembly processes. Below we examine each of these processes.

Actin exists in two states within the cytoplasm of the cell: in the monomeric form (G-actin) and in the self-assembled or fibrous form (F-actin). G-actin has been shown to exist as an oblate ellipsoid by electron microscopy whereas F-actin exists as a double-helical structure. G- and F-actin exist in cells in equilibrium with actin-binding proteins. These proteins are involved in the polymerization and depolymerization of F-actin.

5.5.1 Actin Self-Assembly

Actin is one of the most abundant cellular proteins and is present in all mammalian cells. F-actin is physically crosslinked to form the cellular cytoskeleton and its contraction allows cell deformation and movement during processes like phagocytosis. The G- to F-actin assembly and disassembly sequence in the cell cytoplasm allows shape changes during cell movement. Gelsolin is an actin-binding protein that inhibits assembly of F-actin by binding to G-actin. G-actin is globular in shape (see Figure 2.26) and consists of α helical regions connected by aligned regions containing β structures. The three-dimensional structure of G-actin includes large and small domains. In the cleft between the domains lies the ATP binding site and there is also a binding site for calcium.

In the presence of an actin-binding protein termed filamin, G-actin is polymerized into helical F-actin. Step one of polymerization involves formation of a G-actin–calcium complex. Step two involves replacement of the calcium with magnesium cation forming an altered G-actin; the altered form of G-actin binds an additional magnesium cation and then a dimer of two altered G-actin monomer forms. Binding of an additional altered G-actin monomer makes a trimer, the nucleus, with one pointed end and one barbed end. The addition of altered F-actin molecules to the nucleus causes elongation with the barbed end growing faster than the pointed end (see Figure 5.13).

G-actin is soluble in water and is transformed into F-actin in neutral salt solutions. Conversion of G-actin to F-actin is associated with hydrolysis of ATP leading to the formation of filaments that are up to several micrometers long and 7 to 10 nm wide. The process has been characterized as containing two steps, that is, nucleation and growth, and as is polymerization of other macromolecules, entropy driven. In muscle actin filaments tropomyosin binds to F-actin and mechanically stabilizes them.

5.5.2 Tubulin

Tubulin is the building block from which microtubules are formed. Microtubules are essential in many cellular processes including mitosis, cell shape changes, and internal organelle movement of cilia and flagella. Tubulin is made up of subunits that contain two components (thus they are dimers) termed α and β subunits, each of which has a molecular weight of 55,000 (Figure 5.14). Assembly of these subunits occurs longitudinally forming a protofilament and then 13 protofilaments close into a hollow cylinder, the basic structure of the microtubule; the microtubule has an outer diameter of 30 nm and an inner diameter of 14 nm.

Tubulin polymerization been followed turbidimetrically; the shape of the turbidity–time curve is very similar to that observed for collagen; that is, it is S-shaped containing lag, growth, and plateau phases. This has led to the interpretation that the lag phase involves linear growth of subunits into protofilaments, similar to those of collagen, which associate laterally into microtubules during the growth phase as diagrammed in Figure 5.15. During assembly the α subunit binds GTP and hydrolyzes GTP to GDP when the dimer adds to a microtubule. In addition, microtubule-associated

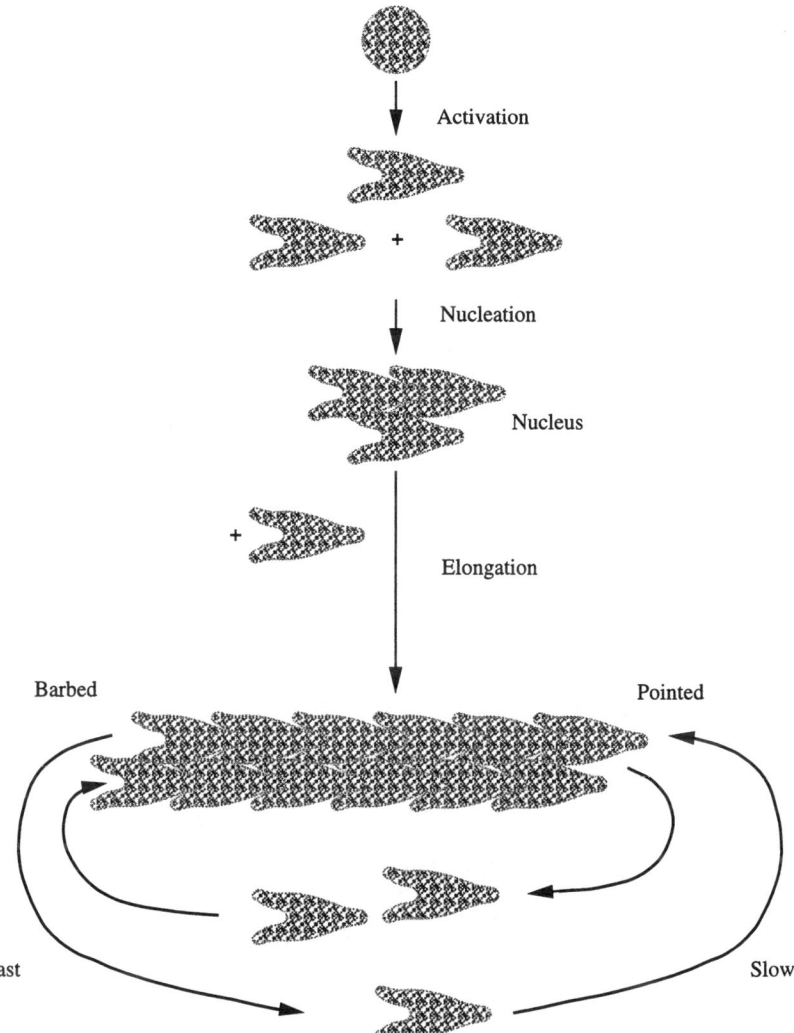

FIGURE 5.13. Assembly of actin filaments. The diagram shows the steps in the transition of actin monomer, G-actin, to actin filaments, F-actin. Monomers are activated by binding calcium and then exchanging calcium for magnesium, leading to nucleus formation. The nucleus consists of a trimer of G-actin with one pointed end and one barbed end. The addition of activated G-actin monomer to the nucleus causes elongation of the barbed end faster than the pointed end.

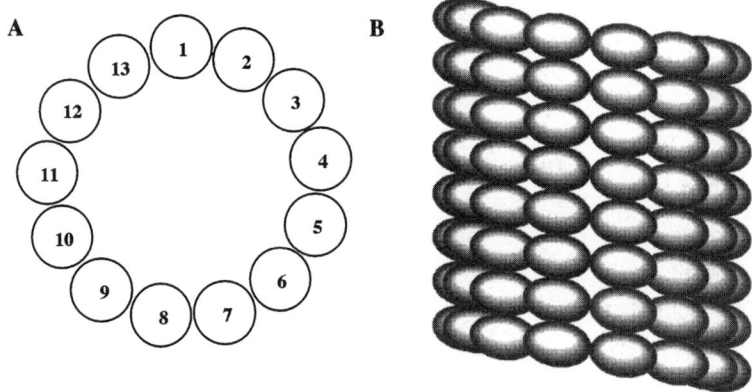

FIGURE 5.14. Schematic diagram of microtubular structure. A cross-section (A) and longitudinal section (B) of a microtubule are shown.

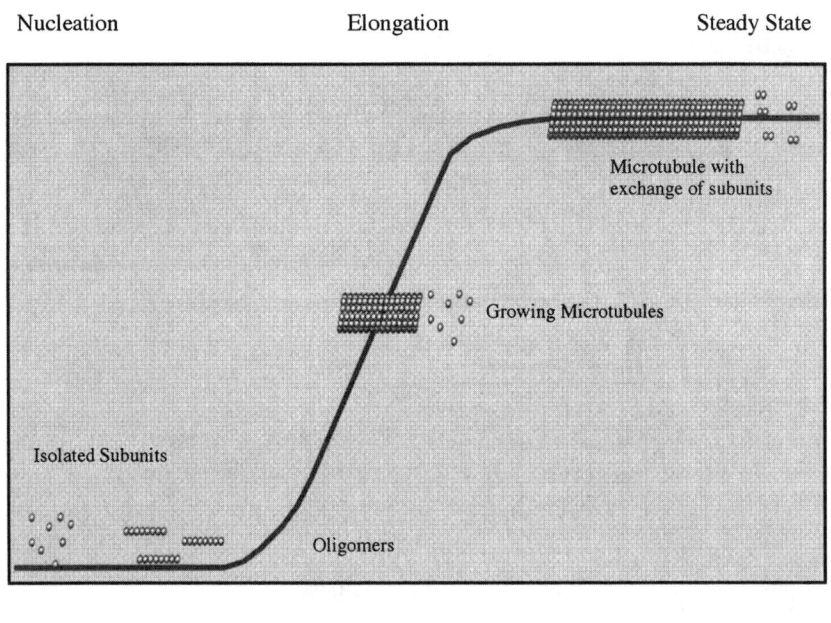

FIGURE 5.15. Turbidity–time curve for microtubule assembly. The diagram illustrates the turbidity–time changes that occur during microtubule assembly. Initially during the turbidity lag phase, tubulin monomers form rings of tubulin subunits that cause microtubule elongation during the growth phase.

proteins (MAPs) help to stabilize the structure and also help link the micro-tubules to other structures in the cell.

5.6 Actin–Myosin Interaction

The interaction of actin and myosin is key to the generation of contractile forces in skeletal muscles. Skeletal muscles are composed of thick and thin filaments; the thick filaments are composed primarily of myosin and thin filaments contain actin. The interaction between the two is a reversible self-assembly process that is followed by disassembly that separates actin from myosin. This separation initiates movement of myosin with respect to actin and results in shortening of the muscle, causing force generation. It is the assembly and disassembly process that we are interested in in this section.

The process involves binding of ATP to myosin. The myosin molecule has a length of over 16.5 nm and is approximately 6.5 nm wide and 4.0 nm deep at its thickest end (see Figure 2.27). The secondary structure contains several long α helices. The actomyosin complex is a very large macromole-cular complex. The starting point in the contractile cycle is taken as the point when myosin is tightly bound to actin in the absence of ATP. In this state it is believed that the space between the two parts of the head region is closed. The first step in the contractile cycle involves opening of the space between the components of the head when ATP binds which disrupts the actin-binding site on myosin (Figure 5.16). After ATP binding occurs and the space between the two head components narrows, hydrolysis occurs. This is followed by myosin starting to rebind to actin, which is followed by loss of γ-phosphate and initiation of the power stroke. The power stroke is completed as ADP is lost from the site between the two head components.

5.7 Fibrinogen

Fibrinogen is a protein that composes 3% of the weight of blood and is found in blood cells including platelets. It is involved in formation of fibrin networks that make up the noncellular component of blood clots and is converted into fibrin via the intrinsic and extrinsic clotting pathways. In either of these pathways prothrombin is converted to thrombin, which pro-teolytically removes fibrinopeptides A and B from the fibrinogen molecule. The fibrinogen molecule has a molecular weight of about 330,000; its struc-ture is diagrammed in Figure 2.30. The molecule consists of two terminal domains with molecular weights of about 67,200 with diameters of about 6.0 nm that are connected to a central domain by threadlike domains that are about 16 nm long. These domains have a coiled-coil structure and

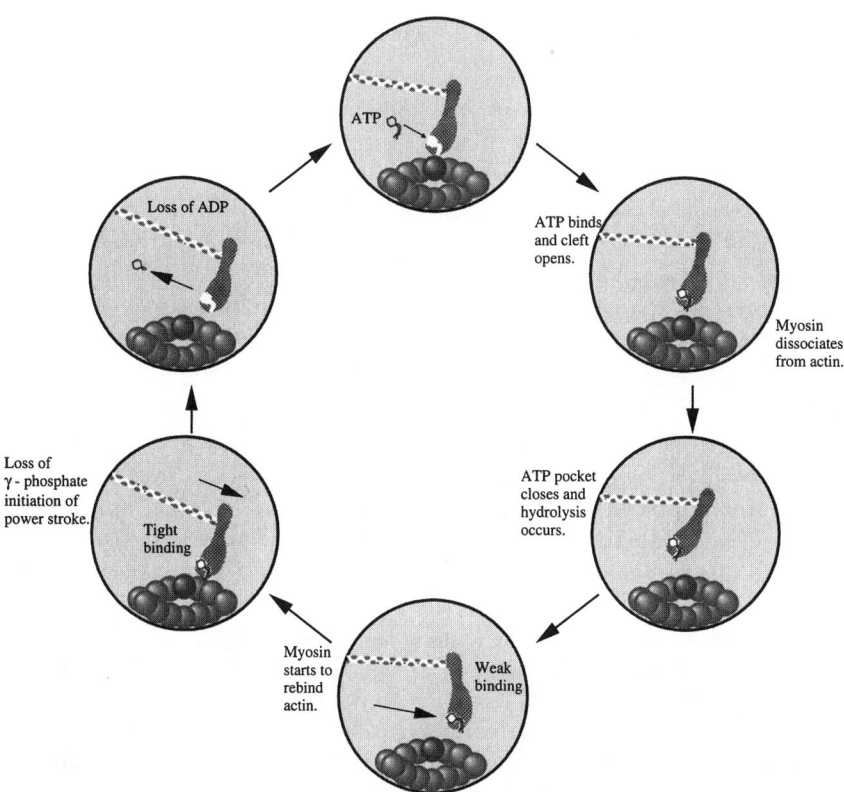

FIGURE 5.16. Summary of contraction cycle. Interaction between myosin head and F-actin involves the binding of ATP to a cleft in the myosin head (top) that associates myosin head from actin, causing hydrolysis of ATP into ADP. This is followed by binding of myosin head to actin, loss of γ-phosphate, and then loss of ADP and repeat of the cycle (see Silver and Christiansen, 1999).

contain α helices. The central domain has a molecular weight of about 32,600 and is about 5.0 nm in diameter. Removal of fibrinopeptides A and B (FPA and FPB) result in only a small decrease in molecular weight and allow assembly to proceed by the interaction of sites blocked by FPA and FPB. Historically it has been proposed that removal of FPA leads to linear polymerization whereas removal of FPA leads to lateral growth. This has led to the model of self-assembly that has initiation of assembly by removal of FPA and dimerization of two fibrinogen molecules, which is followed by removal of FPB and linear polymerization. Further assembly occurs by lateral packing of elements that are formed during the previous step as diagrammed in Figure 5.17.

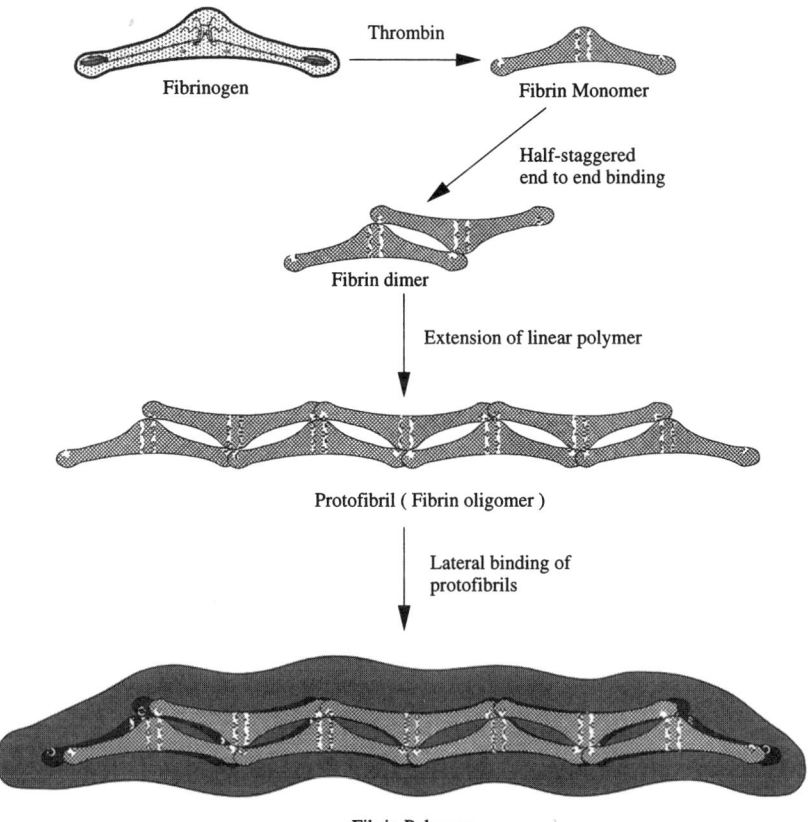

FIGURE 5.17. Conversion of fibrinogen to fibrin. The diagram illustrates the assembly of fibrin monomer after cleavage of fibrinopeptides A and B by thrombin, causing end-to-end binding of monomers followed by lateral growth.

5.8 Summary

The ability of biological macromolecules that are composed of helices including the α helix and the collagen triple helix to assemble into higher-order structures has some common elements. First, many of these processes are characterized by turbidimetric lag and growth phases suggesting that linear growth of a stable element or nucleus must occur prior to lateral growth. In many instances linear growth involves formation of dimers and trimers in an end-to-end or overlapped fashion possibly because aggregation in this manner favors formation of a new phase (solid) in equilibrium with the liquid phase. From another perspective linear aggregation decreases the probability for the molecules to tumble thereby stabilizing later lateral growth steps. Whether linear growth stabilizes a "nucleus" as in classical nucleation and growth kinetics is unclear; however, linear growth provides an element that has more surface area whereas lateral growth minimizes the surface area exposed for new growth. Maximization of the surface area would allow rapid growth of large numbers of macromolecules especially if each successive step in the assembly process used larger and larger building blocks. In addition, it would also provide the beginning of a twisted unit if growth involves formation of a superhelix.

Turbidity increases during the growth phase reflect the rapid lateral growth of large units. This favors interpretation of biological assembly as multistep processes because formation of large structures can only occur rapidly if high molecular weight units grow in diameter by fusion. The ultimate consequence of self-assembly either at the cellular or extracellular levels is the formation of structures that can transmit or resist large forces.

A practical example of the need for biological structures to grow rapidly over a small time frame is observed when one characterizes the strength of tissues just prior to and after birth. In the prenatal period, the extracellular matrix is rather weak and the collagen fibrils appear discontinuous, at least in tendons. Just after birth, associated with the beginning of movement, collagen fibril diameters increase, as does the mechanical strength. The only way that the fibrils could grow rapidly in length and in width appears to involve fibril fusion. However, as we show in Chapter 6, the mechanical properties of at least collagen are not only dependent on collagen fibril diameters but they are also dependent on how the molecules are crosslinked. In the absence of continuous networks of crosslinked molecules the ability of collagen fibers and other assemblies of macromolecules is limited.

Suggested Reading

Berg RA, Birk DE, Silver FH. Physical characterization of type I procollagen in solution: Evidence that the propeptides limit self-assembly, *Int J Biol Macromol.* 1986;8:177–182.

Birk DE, Silver FH. Collagen fibrillogenesis *in vitro*: Comparison of types I, II and III, *Arch Biochem Biophys.* 1984;235:178–185.

Cassel JM, Mandelkern L, Roberts DE. The kinetics of the heat precipitation of collagen, *J Am Leather Chem Assoc.* 1962;51:556–577.

Cooper A. Thermodynamic studies of the assembly *in vitro* of native collagen fibrils, *Biochem J.* 1970;118:355–365.

Gale M, Pollanen MS, Markiewicz P, Goh MC. Sequential assembly of collagen revealed by atomic force microscopy, *Biophys J.* 1995;68:2124–2128.

Gaskin F, Cantor CR, Shelanski MI. Turbidimetric studies of the *in Vitro* assembly and disassembly of porcine neurotubules, *J Molec Biol.* 1974;89:737–758.

Kadler KE, Holmes DF, Trotter JA, Chapman JA. Collagen fibril formation, *Biochem J.* 1996;316:1–11.

Korn ED, Carlier M-F, Pantaloni D. Actin polymerization and ATP hydrolysis, *Science*. 1987;238:638–644.

Oosawa F, Kasai M. A theory of linear and helical aggregations of macromolecules, *J Mol Biol*. 1962;4:10.

Rayment I, Holden HM. The three dimensional structure of a molecular motor, *Trends Biochem Sci*. 1994;19:129–134.

Silver FH. Self-assembly of connective tissue macromolecules. In: *Biological Materials: Structure, Mechanical Properties, and Modeling of Soft Tissues*. New York: NYU Press, 1987; Chapter 5, 150–153.

Silver FH. *Biomaterials, Medical Devices and Tissue Engineering: An Integrated Approach*. London: Chapman & Hall; 1994; Chapter 1.

Silver FH, Freeman JW, Seehra GP. Collagen self-assembly and the development of tendon mechanical properties. *J Biomech*. 2003;36:1529.

Veis A, George A. *Fundamentals of Interstitial Collagen Self-Assembly. Extracellular Matrix Assembly*. New York: Academic Press; 1994;15–45.

Ward NP, Hulmes DJS, Chapman JA. Collagen self-assembly in vitro: Electron microscopy of initial aggregates formed during the lag phase, *J Mol Biol*. 1986;190:107–112.

Yuan L, Veis A. The self-assembly of collagen molecules, *Biopolymers*. 1972;12: 1437–1444.

6
Mechanical Properties of Biological Macromolecules

6.1 Introduction

To a first approximation the mechanical properties of self-assembled arrays of biological macromolecules reflect the behavior of the individual macromolecules. The greater the rotational freedom that a molecule has, based on the allowed areas of the conformational plot, the more compliant or flexible is the material. In mechanical terms we usually speak of modulus or stiffness. Amorphous polymer chains, similar to those that characterize parts of the elastin molecule, are easily deformed like a rubber band. Macromolecules in helical or extended conformation have higher moduli and therefore are more difficult to stretch. The collagen triple helix and the β structure are more extended in space and therefore are more rigid than the α helix. Of course the viscosity and the stiffness also depend on the shape factor as discussed in Chapter 4. The more elongated the molecule and the higher the shape factor, the stiffer is the molecule. In this chapter we concentrate on determination of the mechanical properties of isolated macromolecules. In doing so we assume that biological tissues act ideally and do not exhibit viscoelasticity, and do not show any time dependence. As we discuss in Chapter 7, all biological macromolecules exhibit time dependence and therefore the treatment of the mechanical properties is more complex. We use this approach to attempt to simplify the analysis for the reader who has little knowledge of mechanical properties.

Tissues are composites of macromolecules, water, ions, and minerals, and therefore their mechanical properties fall somewhere between those of random coil polymers and those of ceramics. Table 6.1 lists the static physical properties of cells, soft and hard tissues, metals, polymers, ceramics, and composite materials. The properties listed in Table 6.1 for biological materials are wide ranging and suggest that differences in the structure of the constituent macromolecules, which are primarily proteins, found in tissues give rise to the large variations in strength (how much stress is required to break a tissue) and modulus (how much stress is required to stretch a tissue). Because most proteins are composed of random chain structures, α

TABLE 6.1. Mechanical properties of tissue and synthetic materials

Material	Ultimate tensile strength (MPa)	Modulus (MPa)
Soft tissue		
Arterial wall	0.5–1.72	1.0
Hyaline cartilage	1.3–18	0.4–19
Skin	2.5–16	6–40
Tendon/ligament	30–300	65–2500
Hard tissue (bone)		
Cortical	30–211	16–20 GPa
Cancellous	51–93	4.6–15 GPa
Polymers		
Synthetic rubber	10–12	4
Glassy	25–100	1.6–2.6 GPa
Crystalline	22–40	0.015–1 GPa
Metal alloys		
Steel	480–655	193 GPa
Cobalt	655–1400	195 GPa
Platinum	152–485	147 GPa
Titanium	550–860	100–105 GPa
Ceramics		
Oxides	90–380 GPa	160–4000 GPa
Hydroxyapatite	600	19 GPa
Composites		
Fibers	0.09–4.5 GPa	62–577 GPa
Matrixes	41–106	0.3–3.1
Aortic cell	—	90×10^{-6a}

[a] Data from Sato et al., 1990.
Source: Silver and Christiansen, 1999.

helices, β structures, and collagen triple helices it is important to understand the mechanical behavior of each of these structural elements. In this chapter we focus on introducing important aspects of measuring mechanical properties of different proteins that are known to be composed of a few common structural features.

Simple continuum mechanical treatment of the behavior of uniform materials (isotropic) that are tested by pulling the ends of a long thin specimen in tension (uniaxial tension) leads to the following relationship between the shear modulus G, and E, the tensile modulus, where v is Poisson's ratio.

$$G = E/2(1+v) \qquad (6.1)$$

In this case the relationship between stress and strain is given by Equation (6.2) for a material that exhibits a linear relationship between stress σ and strain ε.

$$\sigma = E\varepsilon \qquad (6.2)$$

As discussed further in Chapter 7, linearity between stress and strain for an assembly of macromolecules is indicative of inducing a conformational change either by uncoiling an amorphous polymer such as elastin or stretching a helical macromolecule such as collagen. Typically there are no covalent bonds broken during during this type of deformation.

The following analyses are oversimplified in the case of most tissues because most tissues show directional dependence of the mechanical properties (they have different ultimate tensile stresses and moduli when pulled in different directions). For this reason the information shown in this chapter is an oversimplified picture of the mechanical behavior. Therefore the values of ultimate tensile strength and moduli presented are estimates that depend on how the experimental results were obtained.

6.2 Mechanical Properties of Model Polypeptides

The four protein conformations that provide mechanical stability to cells, tissues, and organs include the random coil or amorphous structure that characterizes a part of the structure of elastin, the α helix, which is represented by the keratin molecule, the collagen triple helix, and the β structure of silk. In humans the β structure is found only in short sequences connecting parts of other structures such as the α helix, but serves as an example of the relationship between protein structure and properties. The ultimate tensile strength and modulus of each structure differs as discussed below.

Elastin, which is the major component of elastic fibers, has the most molecular flexibility. Macroscopically this translates into fibers that undergo large deformations up to several times the original specimen length. A typical stress–strain curve for elastic fibers is shown as the initial part of the stress–strain curve of aorta (Figure 6.1). The slope (tangent) of this curve at high strains is the modulus and has been reported to be about 4 MPa when it is corrected for the elastin content. However, the modulus is dependent on the strain-rate suggesting that the deformation mechanism may be more complicated than just tension-induced uncoiling of the elastin chain during loading.

In contrast to elastin, the generalized stress–strain curve for collagenous tissues is S-shaped. This general S-shape has been misinterpreted as indicating that all tissues containing collagen exhibit the same deformation mechanism. As discussed below, the deformation mechanisms for tendon and vessel wall are quite different and as a result close inspection of the stress–strain curves indicates that the magnitude of the stress after correction for the collagen content of each tissue is an order of magnitude different. Therefore, at this point we only comment on the relative differences between the stress–strain behaviors of elastin (Figure 6.1) and collagen (Figure 6.2). The major differences include the higher extensibility of elastin and the much lower stress required to stretch elastic tissue to failure. These differences reflect the different biological functions of these

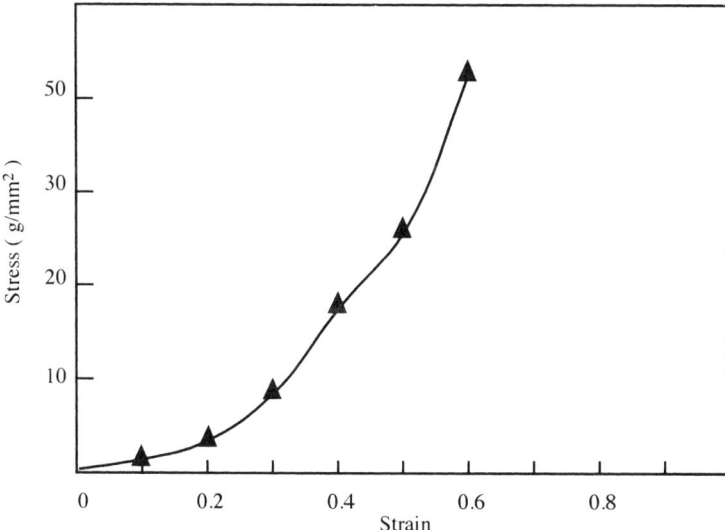

FIGURE 6.1. Stress–strain curve for aorta. Tensile stress–strain curve for human thoracic aorta in the circumferential direction obtained at a strain rate of 50% per minute. At strains less than 0.2, the elastic fibers dominate the behavior, whereas above 0.2, alignment of collagen fibers occurs. (Adapted from Silver, 1987.)

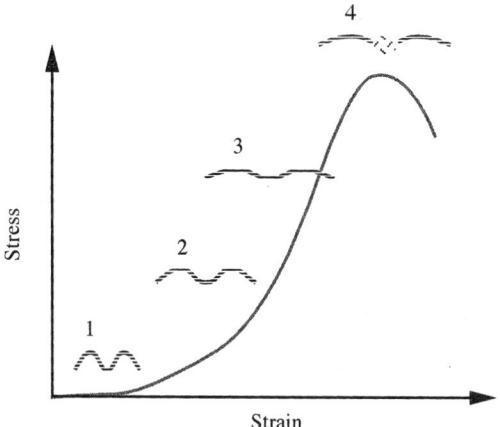

FIGURE 6.2. Generalized uniaxial stress–strain curve for connective tissue primarily containing collagen. The stress–strain curve for unmineralized connective tissue consists of two nonlinear regions and a linear region. The initial nonlinear region (region 1) involves uncrimping or geometrical alignment, or both, of the collagen fibers or expression of fluid. The linear region (regions 2 and 3) is a result of stretching and sliding of collagen fibers that fail by defibrillation in the final nonlinear region (region 4).

macromolecular components; elastin and elastic tissue help transfer the stress to the neighboring collagen fibers whereas collagen fibers store and transmit energy to the musculoskeleton and dissipate energy applied to external tissues such as skin. The modulus for collagen from the generalized stress–strain curve for collagenous tissue after correction for the collagen content is between 20 MPa and about 8 GPa, and the strain at failure is between about 10% and about 100%. This illustrates how the extended collagen conformation is stiffer and stronger than the compact amorphous chain conformation of elastin.

The next example of the mechanical properties of a standard polypeptide structure is that of silk, which is composed of extended β sheets. The stress–strain curve for drag silk of the spider (Figure 6.3) shows an ultimate tensile strength (UTS) of about 800 MPa, a modulus of about 7.0 GPa, and a strain at failure of about 30%. The value of the UTS and modulus appears to be higher than that of collagen; as we show in Chapter 7 after correction for orientation effects and other complications the values for collagen are

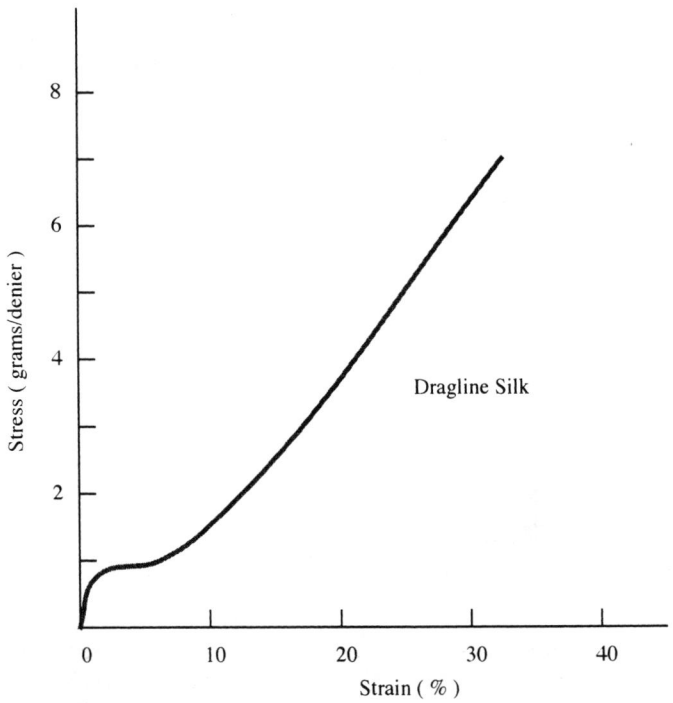

FIGURE 6.3. Stress–strain curve for silk. The ultimate strength of dragline silk is about 800 MPa, with a modulus of about 7 GPa and strain at failure of 30%. Note that these values are higher than those for collagen, reflecting the higher axial rise per residue along the molecule.

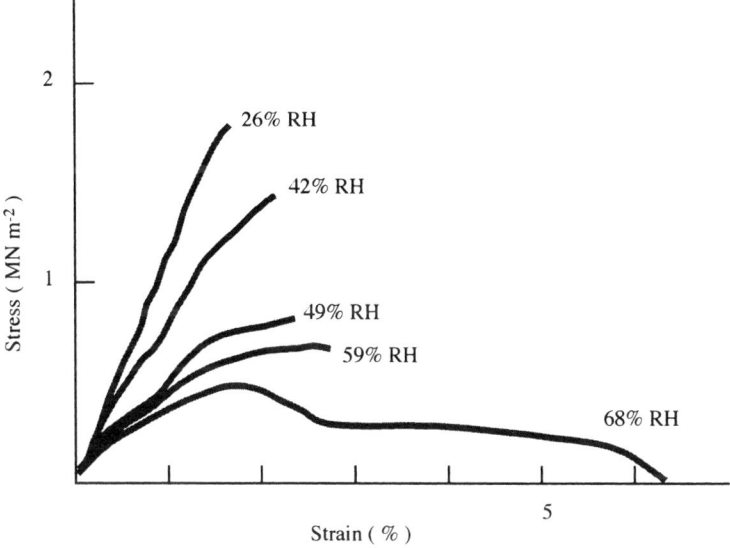

FIGURE 6.4. Stress–strain curve for stratum corneum. At low moisture content, the stress–strain properties of stratum corneum are almost linear, with a UTS of about 1.8 GPa and a modulus of about 120 MPa. Note that the properties are lower than those of collagen and silk.

about the same as silk. However, the higher values for silk compared to amorphous elastin supports the theory that the UTS and modulus are directly related to the average displacement of the amino acids along the backbone of a macromolecule or the shape factor. In silk the value of the axial displacement along the chain per amino acid is about 0.36 nm, which is many times greater than that of an amorphous polymer chain. This means that macromolecules with high shape factors are stiff and strong. But of course this requires that the molecules be crosslinked into a stress-bearing network.

Our last example of the mechanical properties of a protein is that of keratin found in the top layer of skin. The stratum corneum in skin is almost exclusively made up of different keratins that have an α-helical structure. The helices do not run continuously along the molecule so the structure is not ideal. However, the stress–strain characteristics are shown in Figure 6.4 and demonstrate that at low moisture content the stress–strain curve for keratins in skin is approximately linear with a UTS of about 1.8 GPa and a modulus of about 120 MPa. These values are between the values reported for elastin and silk, which is consistent with the axial rise per amino acid being 0.15 nm for the α helix. Thus the α helix with an intermediate value of the axial rise per amino acid residue has an intermediate value of the modulus. This reinforces the concept that stiff strong macromolecules have high axial ratios and shape factors.

6.3 Mechanical Properties of Collagenous Tissues

Well, as you might guess, the mechanical behavior of a tissue is much more complicated than just determination of the axial ratio and shape factor of the constituent macromolecules. As we discuss for tissues with collagen acting as the major structural component, the axial ratio is just one factor contributing to the behavior. The other factors that contribute to dictating the mechanical properties of collagenous tissues found in ECMs is the orientation of the collagen fibers, the relationship between collagen fibers and other components of the tissues (i.e., elastic fibers, proteoglycans, and smooth muscle), and the extent of crosslinking. In addition, tissues containing layers, such as the aorta, must be considered as composites of elements with different orientations. With this in mind let us begin to look at these factors individually.

6.3.1 Mechanical Properties of Oriented Collagen Networks

Much of our understanding of the mechanical behavior of collagen has come from studying the mechanical properties of oriented collagen networks, including tendons. Although the macroscopic structure of tendons varies somewhat from location to location, the basic structural elements, including aligned collagen fibrils and fibers are found throughout these tissues as discussed in Chapter 3. Tendons are multicomponent cablelike elements that cyclically store elastic energy and then transmit this energy to bones in the musculoskeleton. Tendons are loaded along or close to the tissue axis under normal physiological conditions, therefore uniaxial mechanical testing gives a reasonable approximation of the mechanical properties in vivo. However, as discussed previously, all ECMs in vivo are under an internal tensile mechanical load so that tendons actually operate in vivo in a prestressed state. It is difficult to predict the exact prestress present in any tissue, because that is likely to vary with location and age, so the exact range of positions on the stress–strain curve over which the tissue operates is debatable.

The stress–strain curve for tendon (Figure 6.5) is normally divided into three regions: (1) a low modulus toe region, (2) a linear region, and (3) a yield and failure region. The toe region is characterized by a very low modulus and has been attributed to the straightening of the planar crimp within collagen fibers of tendon. It is also due in part to the geometrical alignment of the collagen fibrils with the tensile axis, because other tissues

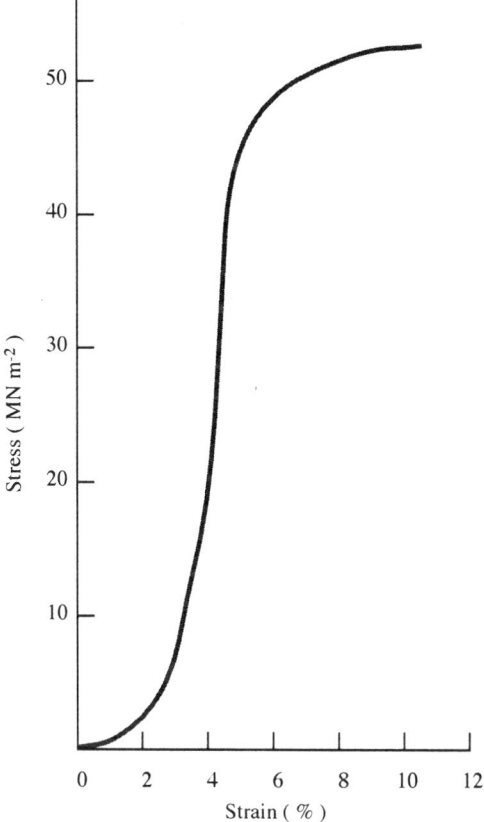

FIGURE 6.5. Typical stress–strain curve for tendon. The diagram illustrates the stress–strain curve for an isolated collagen fiber from tendon. Note that collagen fibers from tendon fail at UTS values above 50 MPa and at strains between 10 and 20%. The slope of the linear portion of the curves at high strains is 2 GPa.

without the planar crimp exhibit a toe region. Values of UTS, strain, and modulus are given for tendon and other soft ECMs in Table 6.2. For tendon the UTS and modulus vary between 50 and 150 MPa and 100 and 2500 MPa, respectively, although these values have not been corrected for the differences in collagen content. These differences are due to variations in collagen content, structure, and the strain rate used in testing. The strain-rate dependence of the properties of tendon suggests that an alternative approach is needed to evaluate the mechanical properties of ECMs.

Dura mater (the collagenous covering over the brain) and pericardium (collagenous layer covering the heart) are made up of collagenous layers, each with an aligned collagen network. In the case of dura mater, the networks are aligned parallel and perpendicular to the midline of the brain. In

TABLE 6.2. Mechanical properties of soft tissues

Tissue	Maximum strength (MPa)	Maximum strain (%)	Elastic modulus (MPa)
Arterial wall	0.24–1.72	40–53	1.0
Tendons/ligaments	50–150	5–50	100–2500
Skin	2.5–15	50–200	10–40
Hyaline cartilage	1–18	10–120	0.4–19

Source: Adapted from Silver et al., 1992.

pericardium there are three layers of networks that can be superimposed by rotation through a 120° angle. When tested along the collagen fiber direction, dura mater and pericardium behave similar to tendon. This behavior is also seen in the superficial layer of articular cartilage as discussed in Chapter 7.

6.3.2 Mechanical Properties of Alignable Collagen Networks

Articular cartilage is a complex of several layers containing collagen fibers with different orientations (see Chapter 3). In the superficial zone the collagen fibers are oriented parallel to the surface whereas in the intermediate layers the fibrils are at angles to the vertical direction. Mechanical tests conducted along the fiber direction suggest that the mechanical properties of the superficial layer of cartilage approach those of tendon when the collagen content is used to normalize the height of the stress–strain curves. The mechanical properties of articular cartilage are listed in Table 6.2. Although the mechanical properties of articular cartilage appear to be dependent on the behavior of the collagen fibers, the high content of proteoglycans appears to also affect the properties. This is in contrast to other tissues such as skin and vessel wall where the behavior appears to be affected by the presence of other fibrous networks.

6.3.3 Mechanical Properties of Alignable Composite Networks

Some ECMs such as skin and vessel wall exhibit behaviors that appear to be similar to that of tendon, but the magnitudes of the actual stresses appear too low after the values have been corrected by dividing by the collagen contents. This is due to the contribution of other networks besides collagen to the behavior. The mechanical behavior of skin and vessel wall to a first approximation is similar to that of tendon and cartilage. However, after correction for the collagen content the UTS and moduli are still below that of tendon and cartilage. In the case of skin the value is only somewhat smaller than that of tendon (see Figure 6.6 and Table 6.3); however, for vessel wall

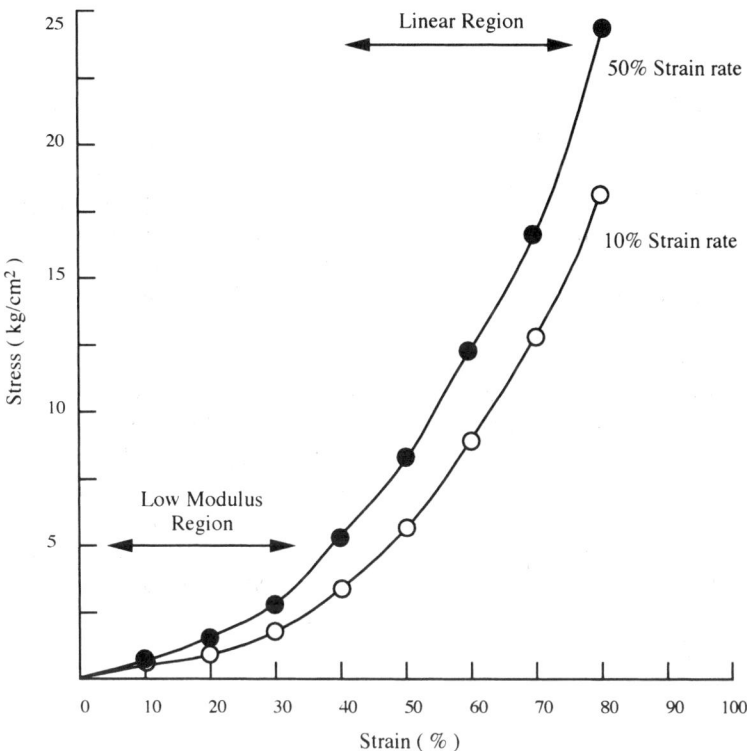

FIGURE 6.6. Stress–strain curve for skin. Stress–strain curves for wet back skin from rat at strain rates of 10 and 50% per minute. The low modulus region involves the alignment of collagen fibers along the stress direction that are directly stretched in the linear region. Disintegration of fibrils and failure occurs at the end of the linear region. (Adapted from Silver, 1987.)

the values are extremely small when compared to the properties of tendon (see Figure 6.1 and Table 6.2). In addition, note the strong strain-rate dependence of the stress–strain curve. At higher strain rates the material appears stiffer. This complicates the interpretation of the stress–strain properties of tissues as is discussed in Chapter 7.

TABLE 6.3. Properties of skin

Tissue	Failure strain	Failure stress (MPa)	Tangent modulus (MPa)
Abdominal skin (human)	0.8–1.00	4–14	1–24
Back skin (rat)	0.9–1.0	5–12.5	6–44

Source: Adapted from Silver, 1987.

6.3.4 *Mechanical Properties of Hard Tissue*

Mechanical properties of unmineralized ECMs depend on the collagen content, orientation, strain rate, loading direction, degree of crosslinking, and presence of other stress-bearing networks including cells that actively contract. The mechanical properties of mineralized tissues are even more complex because of issues concerning preservation, cutting and machining, density, mineral content, fat content, water content, and specimen orientation. Because bone is a composite structure containing lamellar and osteonic components, the properties depend to a high degree on the specimen orientation.

In an attempt to simplify the discussion, we ignore the fact that the modulus and properties of bone are dependent on the testing direction and mineral content. A typical stress–strain curve for cortical bone is illustrated in Figure 6.7. Mineralized ECMs show a much higher modulus and UTS, and the strain at failure is markedly decreased. In the same manner that increased crosslinking increases the UTS of unmineralized tissue, mineral deposition acts as a crosslink and improves the UTS and the modulus of bone. The UTS for cortical bone varies from 100 to 300 MPa, the modulus varies from several to more than 20 GPa, and the strain at failure falls to only 1 to 2%.

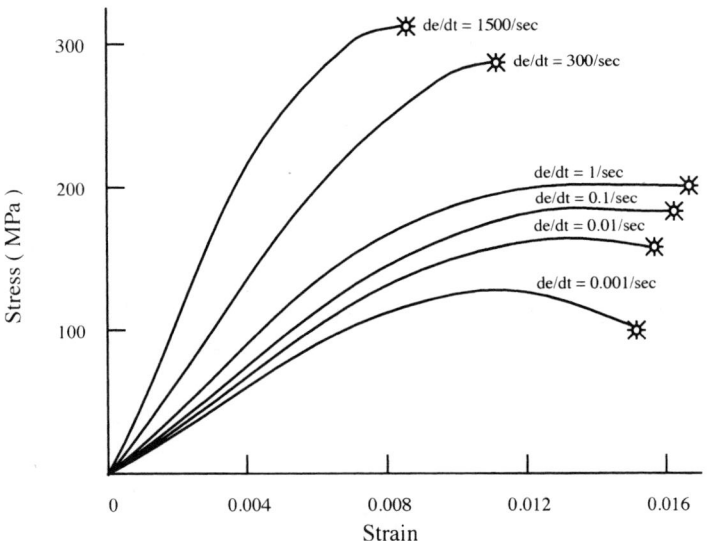

FIGURE 6.7. Strain-rate dependence of stress–strain curve for bone. The diagram illustrates dependence of the stress–strain curve on the strain rate. Note that at higher strain rates the curve moves to the left and becomes more elastic. (see Silver and Christiansen, 1999).

Mechanical properties of cancellous (spongy) bone are dependent on the bone density and porosity, and therefore the strength and stiffness of spongy bone is much lower than that of cortical or osteonic bone. The axial compressive strength is related to the square of the bone density.

6.4 Cellular Contribution

Due to the fact that cells contribute to the active stress in ECMs, as is discussed relating to the active tension in skin and other ECMs, it is important to consider how this may occur. Endothelial cells are normally hexagonal, but under shear stress, they change morphology to an elongated shape with the long axis aligned with the flow direction. Accompanying this shape change is a change in intracellular ion flux, gene and protein expression, and cytoskeletal structure. Cells contain F-actin, microfilaments, and microtubules that could contribute to the active tension observed in ECMs. During shear-induced structural changes to the endothelial cell, prominent changes occur in the microfilaments containing F-actin. Microfilaments are grouped together with myosin and other actin-binding proteins to form stress fibers, 20 to 50 nm in diameter. The elastic modulus of an endothelial cell has been estimated to be about one-thousandth of a MPa. Stress fibers have been estimated have moduli somewhat higher but much less than the numbers that were described for even amorphous proteins such as elastin. This leads us to question how such low modulus materials could provide enough tension to contract the much stiffer collagen fibers. The only way that can be theoretically possible for cells to be responsible for generating the active tension in collagen fibers in ECMs is if they can pull directly on the collagen fibers and each cell is in parallel with many other cells. Thus the nature of the interactions between cells and collagen fibers in ECMs is beyond the scope of this book and the mechanism behind active force generation is likely to be a challenging problem.

6.5 Summary

Transduction of external mechanical forces by ECMs is a challenging problem. Tissues that passively transfer energy from the environment to cells can do it efficiently if they contain macromolecules in an extended conformation such as the collagen triple helix. Although the triple helix must be somewhat rigid in order to transfer the stress between muscle and bone it also must be flexible enough to store elastic energy during locomotion. The larger the shape factor the more stress is required to stretch a helical conformation of a protein as discussed in Chapter 4. In turn this energy can be transferred to cell membrane components. How strong stiff materials such as collagen are able to transfer energy to cells without disrupting cell structure is still unclear, as is the mechanism by which low modulus cells increase the tension in collagen fibers. What is clear is that nature has designed the structural material outside cells to transduce mechanical energy through changes in helical conformation and that these changes can be controlled through specific interactions at the cell surface as we show in Chapter 9.

Suggested Reading

Cowin SC. Mechanics of materials. In: *Bone Mechanics*. Cowin SC, ed. Boca Raton, FL: CRC Press; 1989; Chapter 2, 15–42.

Sato M, Theret DP, Wheeler LT, Ohshima N, Nerem RM. Application of the micropipet technique to the measurement of cultured porcine aortic endothelial cell viscoelastic properties. *J Biomech Eng.* 1990;112:263.

Silver FH. Mechanical properties of connective tissue. In: *Biological Materials: Structure, Mechanical Properties, and Modeling of Soft Tissues*, New York: NYU Press; 1987; Chapter 6.

Silver FH, Christiansen DL. Mechanical properties of tissues. In: *Biomaterials Science and Biocompatibility*, New York: Springer; 1999; Chapter 7.

Silver FH, Kato YP, Ohno M, Wasserman AJ. Analysis of mammalian connective tissue: Relationship between hierarchical structures and mechanical properties. *J Long-Term Effects Med Implants.* 1992;2:165.

7
Viscoelastic Mechanical Properties of Tissues

7.1 Introduction

Unfortunately all biological materials are viscoelastic and therefore all the properties reported for tissue or components of tissues in Chapter 6 are only approximations. Many workers respond to this by suggesting that we should ignore the time dependence, but this leads to an error as high as 100% or more in determining the stress that can be supported without failure occurring in a tissue. In addition, this leads to designing implants that have different moduli than normal tissue. Modulus mismatch at the tissue–implant interface leads to stress concentrations on the host tissue and cellular hyperplasia or necrosis. Neither of these alternatives is acceptable for promoting long-term compatibility of an implant. As we show below, if we assume that the elastic contribution is not strain-rate dependent, which is only true for collagen networks and not true for smooth muscle or elastic tissue, the strain-rate dependence amounts to $d\varepsilon/dt$. This means that assuming the viscosity is constant (which is not true), the change in the viscous part of the viscoelastic behavior is proportional to the ratio of the strain rates, which for biological tissues can be higher than an order of magnitude. What this means, if you are having a hard time following the logic, is that the time-dependent part of the viscoelastic behavior can outweigh the time-independent part. Luckily, extracellular matrix is thixotropic; for example, it shear thins (the viscosity decreases with increasing strain rate) and therefore the time-dependent contribution is not as large as it would be if the viscosity of tissues did not decrease with increased shear rate.

Because all tissues are viscoelastic this means that their mechanical properties are time dependent and their behavior is characterized both by properties of elastic solids and those of viscous liquids. The classic method to characterize a viscoelastic material is to observe the decay of the stress required to hold a sample at a fixed strain (stress relaxation) or by the increasing strain required to hold a sample at a fixed stress (creep) as diagrammed in Figure 7.1 and explained further in Figure 7.2. Viscoelastic materials undergo processes that both store (elastic) and dissipate (viscous)

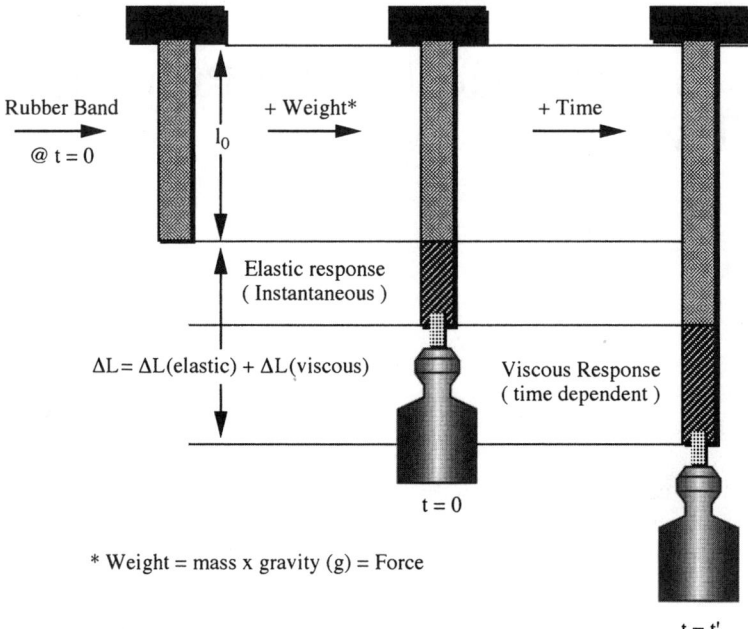

FIGURE 7.1. Definitions of elasticity and viscoelasticity. If a weight is placed on a rubber band at time zero and the initial length is l_o, then instantaneously the weight causes a length increase that is followed by a continual slow length increase until the rubber band reaches some new equilibrium length $l_o + \Delta L$. The elastic response is the length increase that occurs instantaneously, and the viscous response is the time-dependent length component.

energy. The elastic processes that store energy during deformation appear to involve conformational changes at the macromolecular level whereas the viscous processes appear to involve sliding of structural units by each other during deformation. In the case of polymeric materials such as rubber the elastic part of viscoelastic behavior involves free energy changes associated with a loss of entropy that arises as the number of chain conformations is limited by stretching. Viscous behavior of polymeric materials is associated with plastic flow of polymeric chains during mechanical deformation. Similar processes occur for macromolecules that make up tissues.

These processes are fairly well understood for synthetic polymers and are beginning to be understood for natural polymers. In order to understand mechanochemical transduction processes we need to be able to model how changes in macromolecular chain conformation trigger the chemical signals that alter gene expression and tissue physiology. The purpose of this chapter is to attempt to relate changes in the time-independent mechanical properties of biological macromolecules (elastic properties) to the conforma-

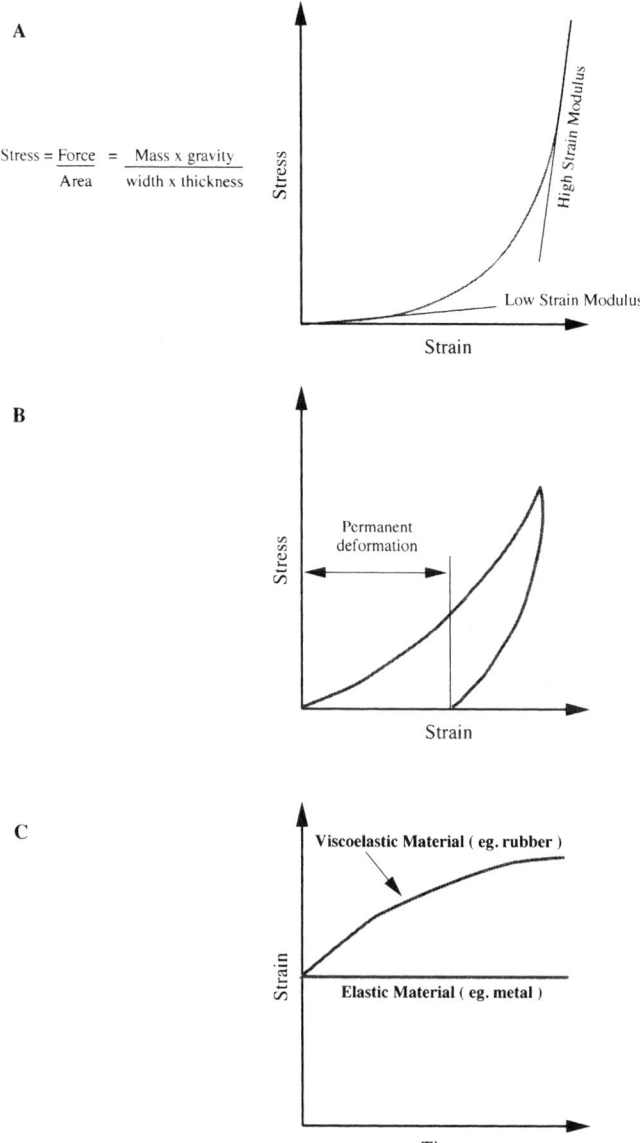

FIGURE 7.2. Definitions of stress, strain, and modulus. Stress is defined as force per unit area, and strain is the change in length divided by the original length. When stress is plotted versus strain, then the slope is the modulus (A). When the load is removed, any strain remaining is called permanent or plastic deformation (B). When elastic materials are loaded, they are characterized by a constant strain as a function of time, whereas viscoelastic materials have strains that increase with time (C).

tional changes that occur at the molecular level. Conformational changes that are induced by mechanical loads applied to biological macromolecules are stimulators of mechanochemical transduction processes that involve the cell membrane. These processes are described in Chapter 9.

7.2 Viscoelastic Behavior

To fully understand the behavior of biological materials we need to address the issue of viscoelasticity. When a weight is placed on viscoelastic material, there is an instantaneous elastic response and a time-dependent viscous response (see Figure 7.1). For polymers the elastic response reflects the change in macromoleular conformation, which is usually time independent if no bonds are broken. The viscous response is the flow of macromolecules by each other similar to what happens during the flow of fluids in a tube. Fluid flow is a time-dependent process. Polymers exhibit viscoelastic behavior because they have both a time-independent response and a time-dependent response.

Other definitions that are required to understand the following discussion of mechanical properties of biological macromolecules include stress, strain, plastic, Hookean, and linear. When a stress (force per unit cross-sectional area) is applied to a material a resulting strain (change in length divided by the original length) occurs as is shown in Figure 7.2. The curve that reflects the loading of most biological materials is shown in Figure 7.2(A). Notice it is nonlinear and can be fit by at least two straight lines or a series of terms. If the relationship between stress and strain during loading and unloading is the same then the material is referred to as Hookean; that is, it follows Hooke's law (stress is proportional to the strain times a constant). If the unloading curve for the material is different from the loading curve, and there is some permanent deformation in the unloaded material, the material has undergone plastic deformation (see Figure 7.2(B)). When a viscoleastic material is placed under a fixed load, the strain is time dependent (see Figure 7.2(C)).

7.3 Molecular Basis of Elastic and Viscous Properties

The molecular basis of elastic and viscous properties is a subject of intense research interest. Although there is evidence to suggest that the time-independent (elastic) properties of collagen involve reversible axial stretching of the collagen triple helix, there are relatively few other studies to suggest that this mechanism may apply to other biological macromolecules. In addition, the available literature suggests that the time-dependent (viscous) properties of biological tissues reflect molecular and fibrillar slippage of collagen. Because most of the work reported in the literature separating the elastic and viscous behavior of biological macromolecules involves collagen, we use this information to illustrate how to separately determine the elastic and viscous behaviors of biological macromolecules.

7.4 Experimental Determination of Elastic and Viscous Mechanical Properties

The experimental method used to separate the elastic and viscous components of the mechanical behavior of biological tissues has been developed for ECMs. The testing procedure is simple; a sample tissue is loaded in tension and then pulled through a series of strain increments (Figure 7.3). For example, a piece of tendon is carefully removed from an animal and then the sample is stretched in the grips of a tensile testing device. If one prepares tendon fibers from a rat tail, which easily can be teased from the tail tendon, and stabilizes the ends of the sample by gluing them to either sandpaper or pieces of a tongue depressor, they can then be inserted into

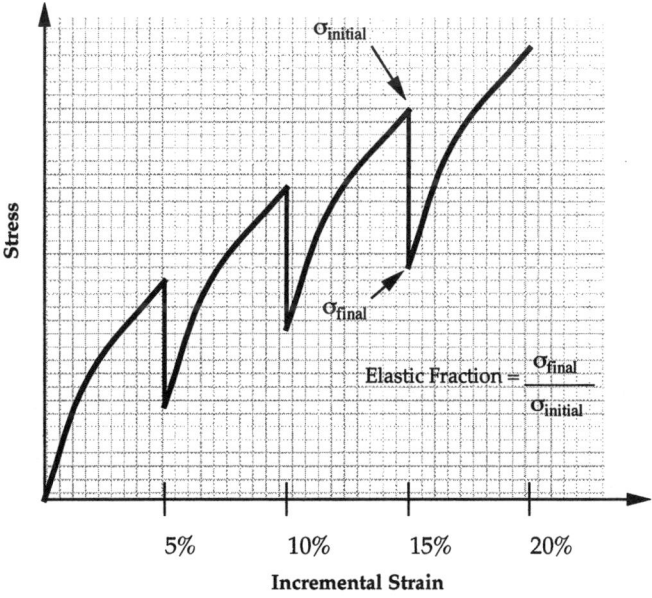

FIGURE 7.3. Determination of elastic and viscous components. Incremental stress–strain curve constructed by stretching a specimen in strain increments of 2 to 5% and allowing the specimen to relax to an equilibrium stress before an additional strain increment is added. The elastic fraction is defined as the equilibrium stress divided by the initial stress. (Adapted from Silver, 1987.)

the grips of the tensile testing device. The sample is now held in place and the mechanical properties can be tested without the specimen slipping when the crosshead is lowered to stretch the sample.

A strain with increment of 2% is then applied axially to the tendon and the motion of the crosshead is stopped at that point. The initial (total) stress is then recorded as well as the stress after a period of time when it (the stress) no longer decreases with increasing time. This final stress value is the time-independent stress termed the elastic stress. The stress lost to viscous slippage (viscous stress) is the difference between the initial stress and the final stress. The elastic stress is just the stress at equilibrium and is plotted versus strain to get an elastic stress–strain curve. The elastic stress, as pointed out in Chapter 6, is the stress stored at the molecular level as a change in conformation of a helical or extended macromolecule. In theory, the slope of the elastic stress–strain curve is proportional to the molecular stiffness of the molecule being stretched.

The viscous stress at any strain is related to the axial ratio and shape factor (see Chapter 3) because the resistance to flow or movement is related to friction generated during the sliding of two molecules past each other. Therefore, in theory the length of the element that is dissipating energy by frictional movement can be obtained from the viscous stress at any strain. Again this should be related to the shape factor v for the macromolecule. Remember the shape factor for rodlike molecules is much higher than for spheres and therefore rodlike molecules can dissipate much more energy through viscous slippage than is dissipated by spherical molecules. Therefore we can dissipate much more energy using macromolecules with high shape factors. Energy dissipation occurs primarily in muscle, which is composed of actin and myosin filaments with high shape factors.

7.4.1 Determination of Elastic and Viscous Stress–Strain Curves for Tendon

Elastic and viscous stress–strain curves can be experimentally determined from incremental stress–strain curves measured on samples of different tendons. Typical elastic and viscous stress–strain curves for rat tail and turkey tendons are shown in Figures 7.4 and 7.5. For both types of tendons the curves at high strains are approximately linear. As we discuss in Chapter 8, the elastic modulus can be calculated for collagen, because most of the tendon is composed of collagen and water, by dividing the elastic slope by the collagen content of tendon. When this is done the value of the elastic modulus of collagen in tendon is somewhere between 7 and 9 GPA.

From the viscous stress–strain curve using Equations (4.1), (4.2), and (8.2) we can calculate the collagen fibril length. The collagen fibril lengths in tendon range from about 20 μm for during tendon development to in excess

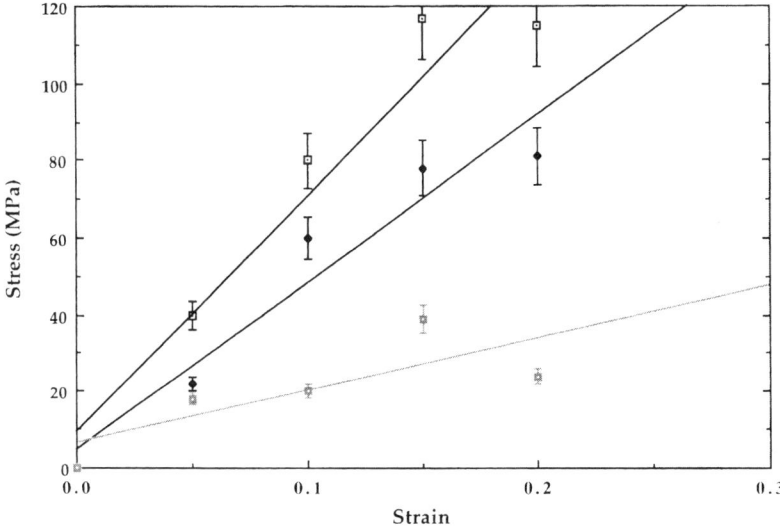

FIGURE 7.4. Total, elastic, and viscous stress–strain curves for collagen fibers from rat tail tendon. The total stress–strain curve (open boxes) was obtained by collecting all the initial, instantaneous, force measurements at increasing time intervals and then dividing by the initial cross-sectional area. The elastic stress–strain curve (closed diamonds) was obtained by collecting all the force measurements at equilibrium and then dividing by the initial cross-sectional area. The viscous component curve (closed squares) was obtained as the difference between the total and the elastic stresses. Error bars represent one standard deviation of the mean.

of 1 mm for mature tendons (Figure 7.6). The fibril length calculated from the viscous stress correlates with that measured by electron microscopy as shown in Figure 7.6.

7.4.2 Determination of the Elastic and Viscous Stress–Strain Curves for Model Collagen Fiber Systems

Analysis of incremental stress–strain curves for tendon is simplified by analyzing the stress–strain curves for model systems consisting of self-assembled type I collagen fibers crosslinked to different degrees. Incremental stress–strain curves measured on fibers made from purified type I collagen obtained from rat tail tendon are similar to incremental stress–strain curves obtained from tests conducted on tendons (see Figures 7.7 and 7.8). For uncrosslinked collagen fibers (Figure 7.7), both the elastic and viscous stress–strain curves are approximately linear as for tendon, however, the viscous stress–strain curve is above the elastic one. This is in contrast to tendon where the elastic curve is above the viscous curve. In addition, the

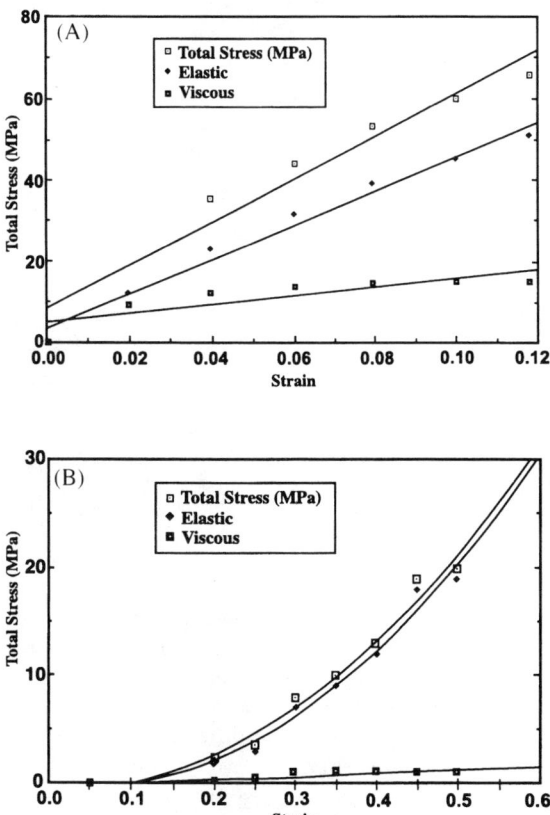

FIGURE 7.5. Representative total elastic and viscous stress–strain curves for unmineralized and mineralized avian tendons. The curves show the total, elastic, and viscous stress–strain relations for gastrocnemius tendon segments proximal to the bifurcation point, B, in Figure 3.29 for animals after (A) and prior to (B), the onset of mineralization. Note the different scales in A and B and the increased slope of the elastic stress–strain curve and decreased strain to failure for mineralized (A) compared to unmineralized (B) tendons.

magnitude of the stress is much higher for tendon than for uncrosslinked collagen fibers. Comparison of incremental stress–strain curves for uncrosslinked collagen fibers and tendons suggests that in the absence of crosslinks the elastic stress is lower than the viscous stress and the magnitude of the stress that can be borne by self-assembled collagen fibers is much lower than that for tendon.

When the uncrosslinked self-assembled collagen fibers tested above were crosslinked by removal of water (dehydration) the resulting incremental stress–strain curves approached those observed for tendons (see Figure

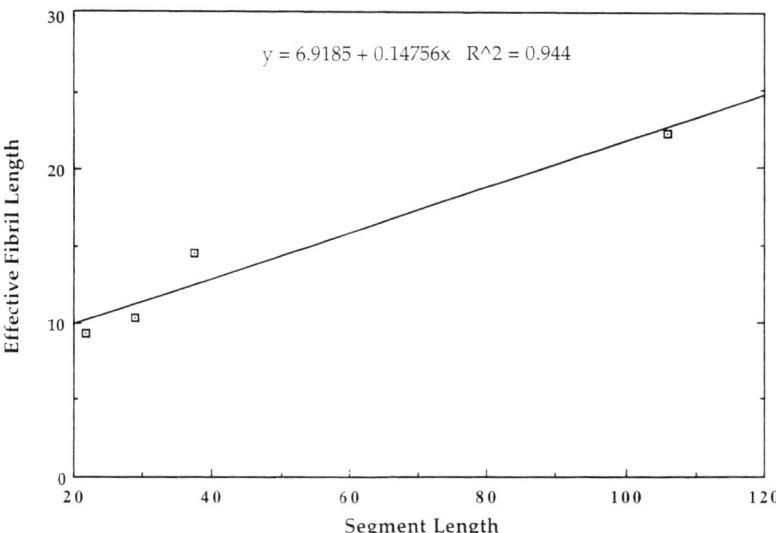

$$y = 6.9185 + 0.14756x \quad R^2 = 0.944$$

FIGURE 7.6. Effective mechanical fibril length versus fibril segment length. Plot of effective fibril length in µm determined from viscous stress–strain curves for rat tail tendon and self-assembled collagen fibers versus fibril segment length. The correlation coefficient (R^2) for the line shown is 0.944 (see Silver et al., 2003).

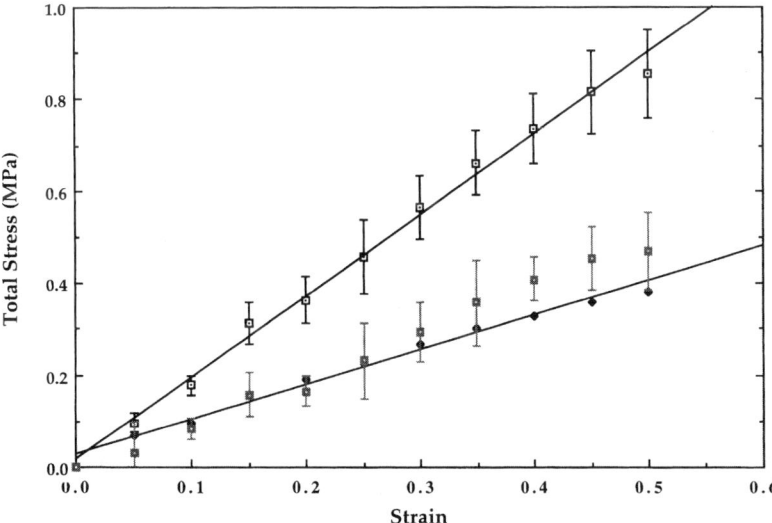

FIGURE 7.7. Total, elastic, and viscous stress–strain curves for uncrosslinked self-assembled type I collagen fibers. Total (open squares), elastic (filled diamonds), and viscous (filled squares) stress–strain curves for self-assembled uncrosslinked collagen fibers obtained from incremental stress–strain measurements at a strain rate of 10%/min. The fibers were tested immediately after manufacture and were not aged at room temperature. Error bars represent one standard deviation of the mean value for total and viscous stress components. Standard deviations for the elastic stress components are similar to those shown for the total stress but are omitted to present a clearer plot. The straight line for the elastic stress–strain curve closely overlaps the line for the viscous stress–strain curve. Note that the viscous stress–strain curve is above the elastic curve suggesting that viscous sliding is the predominant energy absorbing mechanism for uncrosslinked collagen fibers.

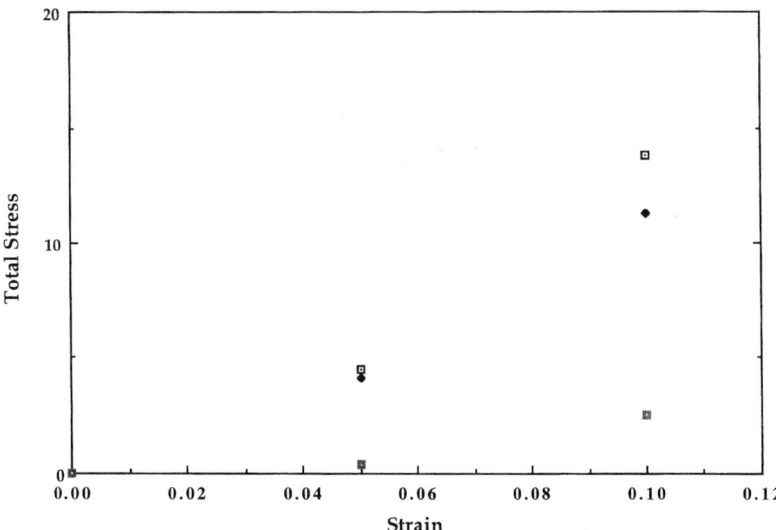

FIGURE 7.8. Total, elastic, and viscous stress–strain curves for crosslinked self-assembled collagen fibers. Total (open squares), elastic (filled diamonds) and viscous (filled squares) stress–strain curves for self-assembled collagen fibers obtained from incremental stress–strain measurements at a strain rate of 10%/min. The fibers were tested after aging for three months at room temperature and pressure. Note that the viscous stress–strain curve is below the elastic curve suggesting that elastic deformation is the predominant energy-absorbing mechanism for crosslinked collagen fibers.

7.8). These results suggest that the formation of crosslinks between collagen molecules in self-assembled collagen fibers significantly increases the stress that the fibers are able to bear elastically. This indictates that crosslinking is an important aspect of mechanotransduction because in the absence of crosslinks, forces and stress levels that are transduced by cells are dramatically reduced. In addition, as pointed out in Chapter 9, internal stresses are already present in the collagen fibers of the ECM, and thus the balance between external loading and internal loading must influence mechanotransduction and not just the absolute value of either of these forces and stresses.

The mechanical properties of collagen fibers including ultimate tensile strength (UTS) and elastic modulus are both highly dependent on collagen fibril length (Figure 7.9). This suggests that although the collagen molecule has high rigidity based on the semirigid structure of the triple helix, collagen molecules must be connected into longer structural units to maximize the benefit of their molecular properties. This can only be effectively achieved by the formation of chemical crosslinks between molecules in the fibril.

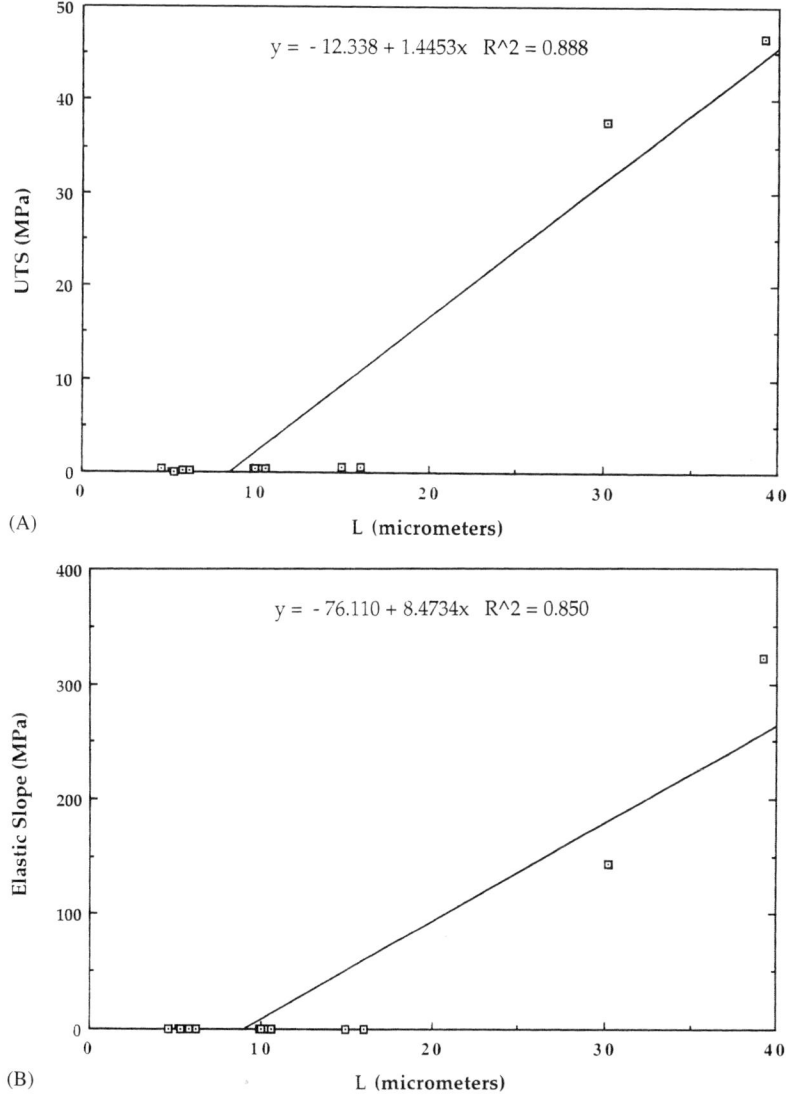

FIGURE 7.9. Relationship between mechanical properties and fibril length (L) for self-assembled collagen fibers. Plot of UTS (A) and elastic slope (B) versus L in μm for self-assembled type I collagen fibers stretched in tension at strain rate of 50%/min. Points with fibril lengths less than 20 μm are for uncrosslinked self-assembled type I collagen fibers and the points above 20 μm are for crosslinked fibers. The correlation coefficient for the best fit line is given by R^2.

7.4.3 Determination of Elastic and Viscous Stress–Strain Curves for Skin

Incremental stress–strain curves have been determined for skin and puri-fied human dermis and are essentially the same. These curves are more complicated than the behavior of tendon in that the curves cannot be approximated by a single straight line as can be seen in Figure 7.10. In con-trast the elastic and viscous stress–strain curves can be approximated using two straight lines. Again the elastic stress–strain curve is above the viscous curve suggesting that energy storage is enhanced over energy dissipation.

The passive mechanical properties of skin have been shown to reflect the behavior of collagen and elastic networks that make up the majority of the macromolecular components of the dermis. Elastic fibers contribute to the initial low modulus portion of the behavior providing recovery to the collagen networks in skin, whereas the collagen fibers prevent premature mechanical failure of skin and bear the loads at high strains. Collagen fibers have been modeled as a biaxial-oriented wavy fiber network, which aligns with the load direction during elastic fiber deformation. When skin is stretched in uniaxial tension, the collagen fibers become aligned with the

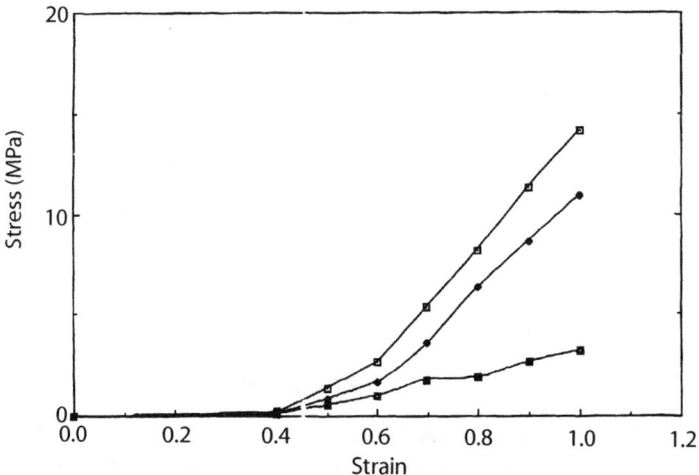

FIGURE 7.10. Stress–strain curve for skin. Total stress–strain curves (open boxes) were obtained by collecting all the initial, instantaneous, force measurements at increasing time intervals and then dividing by the initial cross-sectional area and multiplying by 1.0 + the strain. The elastic stress–strain curve (closed diamonds) was obtained by collecting all the force measurements at equilibrium and then dividing by the initial cross-sectional area and multiplying by 1.0 + the strain. The viscous component curve (closed squares) was obtained from the difference between the total and the elastic stresses.

load axis and the skin on stretching takes on a tendonlike morphology. At high strains the elastic modulus of skin is about 4 GPa after correction for the collagen and water content.

Analysis of elastic stress–strain curves for skin and dermis suggest that the elastic and collagen fibril networks are in parallel and can be fit using two Voigt elements (a Voigt element is an elastic spring in parallel with a viscous dashpot as discussed in Chapter 8) in parallel. The elastic moduli of the springs in the two Voigt elements are consistent with the elastic moduli of elastic fibers and collagen fibrils as discussed in Chapter 8. This suggests that at low degrees of deformation the collagen fibrils remain wavy and offer little resistance to deformation, as previously hypothesized, and that the slope of the elastic stress–strain curve reflects stretching of first elastic and then collagen fibers. The slope of the elastic stress–strain curve of collagen has been reported to be strain-rate independent supporting a model of collagen fibers as almost "ideal" springlike elements. At high strain rates the slope of the viscous stress–strain curve decreases, which is consistent with reports that skin shear thins and is thixotropic. Additionally, both the elastic and viscous properties of skin have been reported to change during aging.

7.4.4 Determination of Elastic and Viscous Properties of Vessel Wall

Elastic and viscous stress–strain curves have been measured for human aorta as well as other arteries. The curves are all similar in that the stress is much lower than that for skin (see Figure 7.11). The lower stress values are consistent with a different network structure of vessel wall compared to skin, which is reflected by the smooth muscle content of aortic tissue. Smooth muscle is absent from skin. The curves for different vessels have similar shapes, however, on careful review the curves have much lower values for the high strain moduli. This relates to the differences in the structure of the media from each of these vessels and potential crosslinking differences.

7.4.5 Determination of Elastic and Viscous Properties of Cartilage

Incremental stress–strain curves for "normal" human articular cartilage are shown in Figure 7.12. There is no normal cartilage in most adults because osteoarthritis begins early after adulthood starts due to wear and tear injury to cartilage. This cartilage was was removed from areas of joints that appeared "normal" in adults undergoing total knee replacement. The tensile stress–strain curve of the normal cartilage appears almost linear after an initial low modulus region. Elastic moduli for articular cartilage

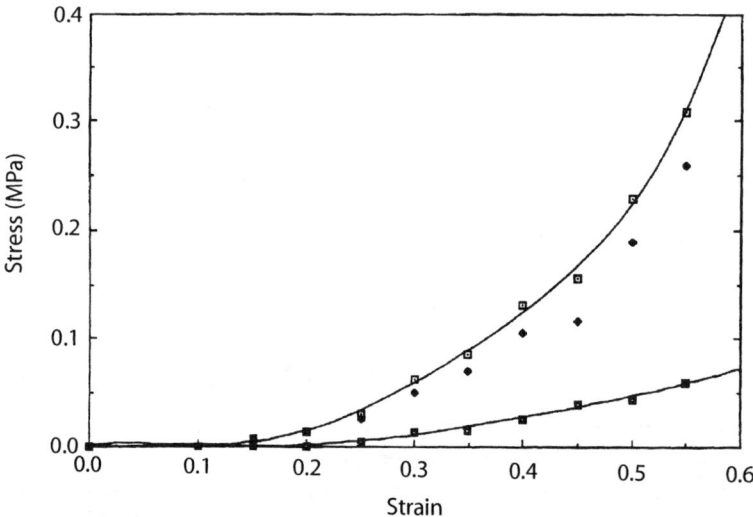

FIGURE 7.11. True stress–strain curve for aorta. Total, elastic, and viscous stress–strain curves for aorta. The total stress–strain curves (open boxes, top) were obtained by collecting all the initial, instantaneous, force measurements at increasing time intervals and then dividing by the initial cross-sectional areas and multiplying by 1.0 + the strain. The elastic stress–strain curves (closed diamonds, middle) were obtained by collecting all the force measurements at equilibrium and then dividing by the initial cross-sectional areas and multiplying by 1.0 + the strain. The viscous component curves (closed squares, bottom) were obtained as the difference between the total and the elastic stresses. Error bars represent one standard deviation of the mean.

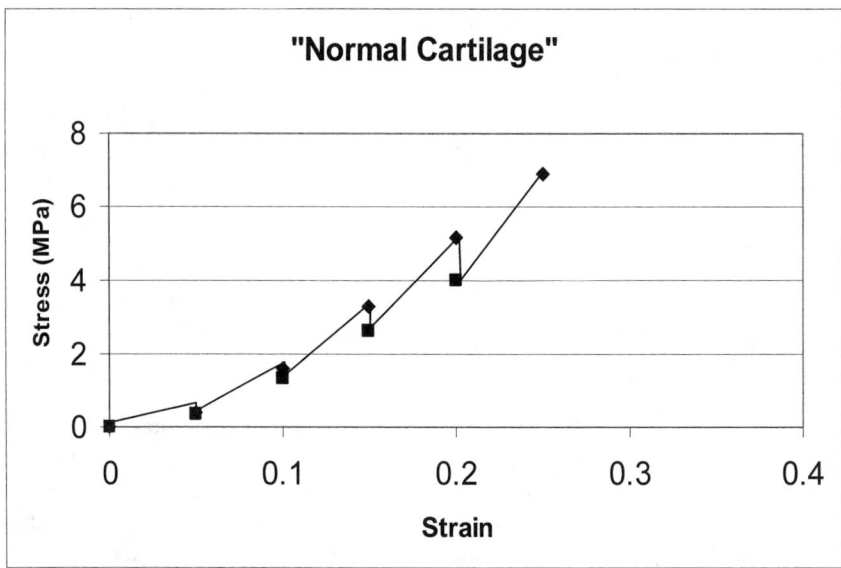

FIGURE 7.12. Elastic and viscous stress–strain curves for "normal" articular cartilage. Elastic **(top)** and viscous **(bottom)** stress–strain curves were obtained by plotting the equilibrium (elastic) and the total-equilibrium (viscous) stresses for visibly "normal" cartilage.

samples taken along the collagen fibril direction approach 8 GPa, approximately equal to the value found for tendon. For cartilage samples shown in Figure 7.12, the elastic modulus is about 2 GPa, due a lack of orientation of the collagen fibers along the tensile testing direction.

7.4.6 Determination of Elastic and Viscous Properties of Mineralized Tendon

Elastic and viscous stress–strain curves for unmineralized and mineralized turkey tendons are plotted in Figure 7.5. In general, the elastic stress–strain curves for tendons with low mineral content (0.029 weight fraction of mineral) are lower than those that are seen for mineralized tendons (mineral content about 0.3).

At low mineral content, the elastic stress–strain curve for turkey tendons has a very long, low modulus region and, as the mineral content increases, the low modulus region is replaced with an almost linear relationship between elastic stress and strain. The slope of the elastic stress–strain curve increases with mineral content and approaches 8 GPa.

Viscous stress–strain curves for turkey tendons at low mineral contents (0.029 or less) increase very slowly with strain and appear to reach a plateau value (see Figure 7.5). Note the low ratio of the viscous to elastic stress at low strains suggesting that unmineralized tendons are very efficient at storing energy. At high mineral contents (about 0.3), the slopes of the viscous stress–strain curves increase rapidly; the slope appears to plateau at high strains. Mineralized tendon appears to be able to bear much higher loads but is less efficient at storing energy.

7.4.7 Effects of Strain Rate and Cyclic Loading

Determination of elastic and viscous stress–strain behavior for different ECMs indicates that although the generalized stress–strain curve generated during a continuous constant rate-of-strain experiment has a similar shape, the behavior of each tissue is slightly different. Whereas the elastic behavior of tendon is approximately linear due to direct loading of primarily collagen fibers, the behavior of cartilage, skin, and vessel wall is somewhat more complex. The elastic behavior of tendon has been reported to be strain-rate independent making the elastic modulus of collagen in tendon between 7 and 8 GPa at any strain rate.

In cartilage, collagen and PGs in the interfibrillar matrix are the predominant macromolecules. Although the elastic modulus of cartilage is below that of tendon, the high content of PGs appears to contribute to an increase in the elastic modulus at high strain rates unlike the behavior of tendon. In skin, the low strain elastic modulus due to loading of elastic tissue is strain-rate dependent but the high strain elastic modulus of collagen is not. This suggests that the elastic and collagen fibers are parallel networks and are not connected in series. If connected in series the strain-rate dependence of the elastic behavior would occur at both low and high strains. However, the elastic modulus of collagen in skin is slightly lower than that of tendon. This has been postulated to reflect the difference in crosslink chemistry between skin and tendon.

The cyclic dependence of the stress–strain curve has been examined in the literature since the 1950s. If a tendon is cycled through more than ten loading and unloading cycles the stress–strain curve continues to change due to changes in the material properties of the tendon. If, however, the elastic stress–strain curve (up to strains of 20%) is measured for skin during several cycles it is observed that the changes are very small between cycles. In contrast most of the change associated with cyclic behavior of tendon is due to changes in the viscous stress–strain curve, which can be large. If the strain is kept below about 20% for skin, some of the cyclic dependence of the viscous stress–strain curve is found to be reversible. It has been hypothesized that the reversible changes in the cyclic viscous stress–strain behavior of skin are due to water loss during stretching and the irreversible changes are due to irreversible fibrillar slippage. These results suggest that preconditioning of ECM samples prior to mechanical testing only affects the viscous behavior of the tissue and not the elastic behavior, further suggesting that preconditioning may not be necessary because the removal of the tissue from the body will change the starting point for the stress–strain curve.

7.5 Internal Loads in ECMs and the Net Load and Stress

We have presented information on the elastic and viscous stress–strain behaviors for a variety of different ECMs in preparation for relating changes in external loading and mechanochemical transduction processes. In order to determine the exact external loading in each tissue that stimulates mechanochemical transduction processes we must take into account the balance between passive loading incorporated into the collagen network in the tissue and active loading applied externally. Inasmuch as the passive load is different for each tissue and is also a function of age (the tension in tissues decreases with age), the net load experienced at the cellular level is difficult to calculate.

In vivo measurements suggest, at least in vessel wall and perhaps tendon, that under normal conditions ECMs operate in the beginning of the high strain region of the stress–strain curve, therefore we can use the stress obtained in this manner to estimate the unloaded stress on a tissue. The unloaded elastic stress can then be corrected for the applied external elastic loading and the cellular stress can be estimated knowing the tissue morphology and the number of cells per unit area. Determination of the net cellular elastic force is important to establish a relationship between external and internal loading and extent of up- and down-regulation of mechanochemical transduction.

7.6 Summary

In order to determine the net stresses and forces acting at the cellular level we must first understand the physics of how forces are transferred externally and internally in ECMs. Each ECM has a different morphology and network structure by which stresses and loads are transferred to and from cellular components. In tendons the collagen network is loaded in tension during locomotion. At the cellular level the external tension is superimposed on the internal tension that is observed when a tendon springs back on cutting. The elastic component of the viscoelastic response represents direct stretching of the collagen triple helix, which as discussed in Chapter 9 causes direct stretching of integrin receptors that are present on the cell membranes. The elastic response, because it is applied directly to specific receptor sites on individual cells is likely to cause a conformational change in the collagen triple helix as it unfolds. This conformational change can in turn cause conformational and other energy changes in the cell membrane as discussed in Chapter 9. The viscous response and the sliding of neighboring collagen fibrils by each other is likely to cause direct stretching of cell membranes or membrane components, and cellular junctions through creation of nonspecific shear effects. As discussed in Chapter 9 there are several mechanical mechanisms believed to trigger activation of mechanochemical transduction pathways within the cell.

The determination of forces and stresses acting at the cellular level in other ECMs is more complicated. In other ECMs, stresses and forces are transferred to cells through collagen, elastic fiber, and smooth muscle networks in a more complex fashion than occurs in tendon. In order to calculate the stresses and forces that occur at the cellular level we must gain new insight into how the components are connected and how the cells are attached. Until this is accomplished we can only guess at the cellular loads in each of these tissues.

Suggested Reading

Cowin SC. Mechanics of materials. In: Cowin SC, ed. *Bone Mechanics*, Boca Raton, FL: CRC Press; 1989; Chapter 2, 15–42.

Landis WJ, Silver FH. The structure and function of normally mineralizing avian tendons. *Comp Biochem Physiol Part A*. 2002;133:1135.

Parry DAD. The molecular and fibrillar structure of collagen and its relationship to mechanical properties of connective tissue. *Biophys Chem*. 1988;29:195.

Saskai N, Odijima S. Elongation mechanism of collagen fibrils and force-strain relationships of tendon at each level of structural hierarchy, *J Biomech*. 1996;9:1131.

Silver FH. Mechanical properties of connective tissue. In: *Biological Materials:
Structure, Mechanical Properties, and Modeling of Soft Tissues*. New York: NYU
Press; 1987; Chapter 6.

Silver FH, Christiansen DL. Mechanical properties of tissues. In: *Biomaterials
Science and Biocompatibility*. New York: Springer; 1999; Chapter 7.

Silver FH, Freeman JW, Seehra GP. Collagen self-assembly and the development of
tendon mechanical properties. *J Biomech*. 2003;36:1529.

Silver FH, Horvath I, Foran DJ. Viscoelasticity of the vessel wall: The role of collagen and elastic fibers. *Crit Rev BME*. 2001;29:279.

Silver FH, Kato YP, Ohno M, Wasserman AJ. Analysis of mammalian connective
tissue: Relationship between hierarchical structures and mechanical properties,
J Long-Term Effects Med Implants. 1992;2:165.

8
Models of Mechanical Properties of ECMs

8.1 Introduction

We have discussed the mechanical properties of macromolecules and tissues and introduced the concept of viscoelasticity and the complexity that is introduced by this type of behavior. Although the components of ECMs are limited to cells, ions, water, collagen and elastic fibers, proteoglycans, and smooth muscle, the variation in arrangement of these components leads to wide variation in the mechanical properties of these tissues. It is important to understand the physical basis for the mechanical behavior of tissues. This is a complex task because of the time dependence as well as variations by which the components are attached. Below we attempt to address the physical basis of the mechanical behavior.

Models of mechanical behavior of tissues have been difficult to develop primarily because of the time dependence of the viscoelasticity. Analysis of viscoelastic behavior of even simple polymers at strains greater than a few percent is not accurate. In addition, most tissues undergo strains larger than a few percent, which makes the analysis require an understanding of the elongation behavior. In this chapter we focus on using modeling techniques to analyze the physical basis for determination of the tensile behavior of ECMs found in connective tissue.

8.2 Modeling Techniques

Much of the modeling of the behavior of ECM has involved the use of curve fitting to predict the observed experimental mechanical properties. The approach has been useful in developing mathematical expressions relating stress, strain, and moduli. Without these models experimental data would have to be handled as a series of data points. Some of the approaches used for mathematical equation development include use of beam models, geometrical models for straightening bent fibers, models relating stress and strain raised to a power, and the strain energy density function as previously described in detail (see Silver, 1987 and Silver et al., 1989 for reviews).

Another approach that has physical merit is to model the behavior of viscoelastic materials as a series of springs (elastic elements) and dashpots (viscous elements) either in series or parallel (see Figure 8.1). If the spring and dashpot are in series, which is described as a Maxwell mechanical element, the stress in the element is constant and independent of the time and the strain increases with time.

Although the use of springs and dashpots was initially used to represent the behavior of tissues in a phenomenological manner (i.e., there was no physical significance to the spring and dashpot), it is now becoming evident that for collagenous tissues the spring can be modeled to represent stretching of the collagen triple helix and fibril whereas the dashpot represents the sliding of collagen fibrils past each other. This has been further supported by the observations that purified collagen fibers to a first approximation show no strain-rate dependence of the elastic stress–strain behavior whereas the viscous behavior is strain-rate dependent. This observation appears not to be true for elastic fibers, smooth muscle, and proteoglycan rich ECMs. The viscous behavior of collagen solutions and molecules is not Newtonian; however, the thixotropy (shear thinning) observed in collagenous tissues appears to be associated with the PGs and hyaluronan. Both high molecular weight hyaluronan and PGs are well known to shear thin at high strain rates due to reversible breakage of the hydrogen bonds between sugar rings in the polysaccharide backbone. This suggests that the sliding behavior of collagen fibrils is influenced by the PGs that are present on the surface and between fibrils.

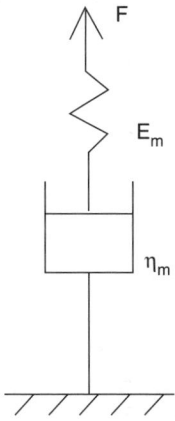

FIGURE 8.1. Diagram showing Maxwell mechanical model of viscoelastic behavior of connective tissues. In this model an elastic element (spring) with a stiffness E_m is in series with a viscous element (dashpot) with viscosity η_m. This model is used to represent time dependent relaxation of stress in a specimen bold of fixed length.

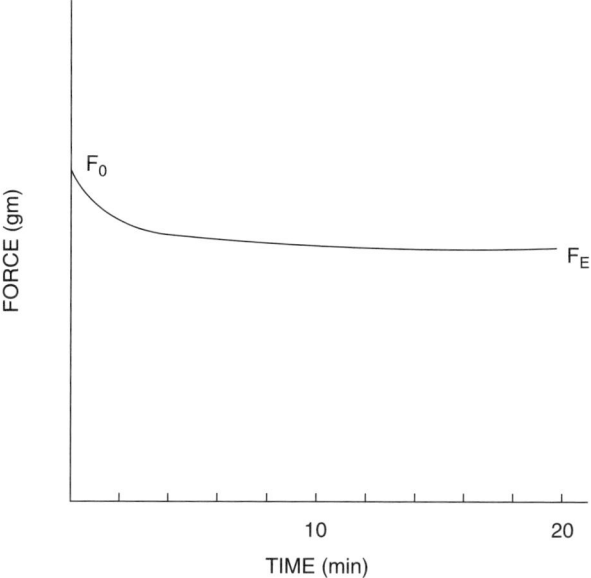

FIGURE 8.2. Force relaxation of tendon under an applied load: when a connective tissue such as tendon is kept at a fixed length (strain), the force (stress) is observed to decrease with time. The elastic component of the force is the force that remains at equilibrium F_n and is time independent. The elastic fraction is the force stored that is recovered when the load is removed.

The problem with the Maxwell model shown in Figure 8.1 is that the stress predicted by the model decays to zero as shown in Equation (8.1),

$$\sigma = \sigma_0 \exp(-t/\tau) \tag{8.1}$$

where σ_o is the initial stress, t is the time, and τ is the decay time, which is found from the ratio of the viscosity of the viscous element divided by the spring constant. The Maxwell model deformed to a fixed extension results ultimately in the stress decaying to a value of zero. All ECMs found in connective tissue transmit loads and maintain equilibrium values of internal stresses, therefore the stress does not decay to zero in these tissues as shown in Figure 8.2. In fact as shown in Figure 8.3 stress at equilibrium for ECMs tested decays to no less than 50% of the initial stress. Therefore, the Maxwell model does not adequately model the stress-bearing behavior of ECMs.

Because we are modeling the stress–strain behavior of tissues in which all tissues are under internal tension conditions where the stress is a function of time, then this element is not particularly useful for even superficial modeling of the stress–strain behavior of ECMs. The other element is a spring in parallel with a dashpot, termed the Voigt or Kelvin element. In

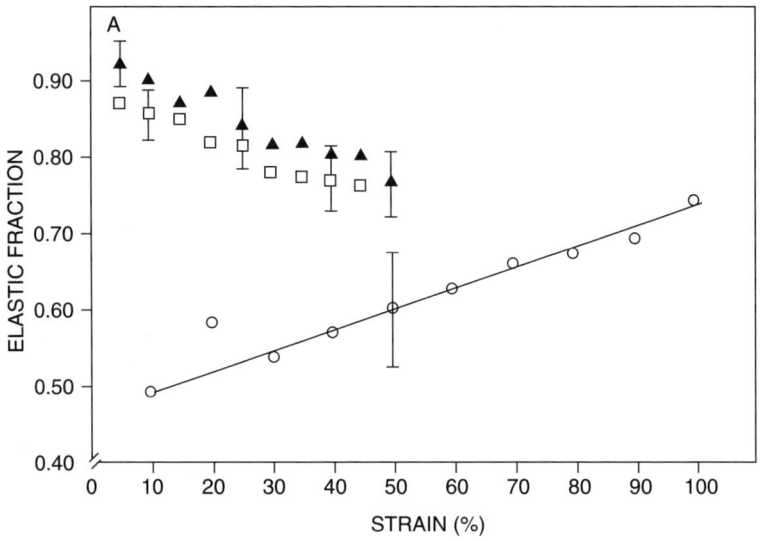

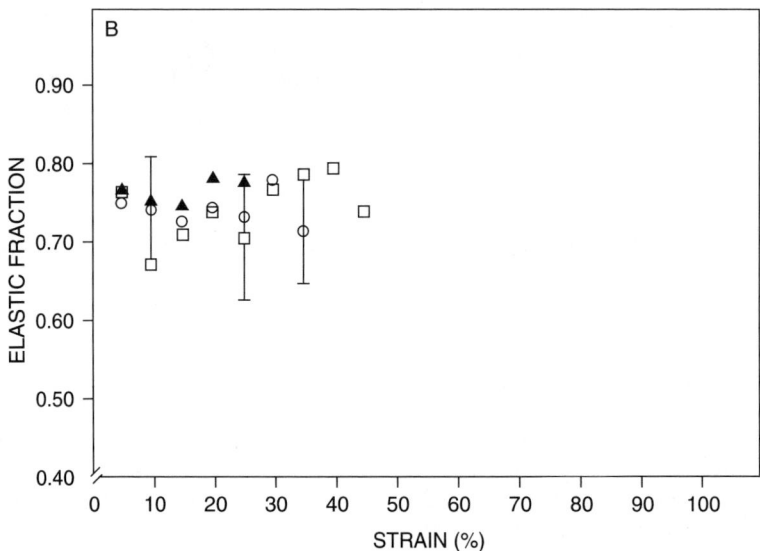

FIGURE 8.3. Strain dependence of elastic fraction of connective tissues: using a sequential incremental loading program, the elastic fraction as a function of strain (%) for: (A) aorta tested in the circumferential (▲) and axial directions (□), and skin (○), (B) pericardium (□), psoas major tendon (▲), and dura mater (○). (From Dunn and Silver, 1983.)

this case the total stress (σt) is the sum of the stress in the spring (σe) and the stress in the dashpot (σv) as given by Equation (8.2).

$$\sigma t = \sigma e + \sigma v \tag{8.2}$$

The stress in the elastic element is assumed to be linear and equal to the product of the elastic modulus (E) and the strain (ε), and the stress in the viscous element is given by the product of the viscosity (η) and the strain rate ($d\varepsilon/dt$) as given by Equation (8.3).

$$\sigma t = E\varepsilon + \eta(d\varepsilon/dt) \tag{8.3}$$

If the elastic modulus is obtained from the slope of the elastic stress–strain curve, then we can evaluate the first term on the right-hand side in Equation (8.3) from experimental data elastic stress–strain curves. The second term on the right-hand side in Equation (8.3) can be evaluated from the product of the strain rate, which is set in a constant strain-rate experiment, and the viscosity. As we discussed in Chapter 3, the viscosity of a macromolecule is related to the shape factor v, therefore we can evaluate the second term on the right-hand side of Equation (8.3) from the product of the shape factor and the strain rate.

In the simplest case the stress required to strain a viscoelastic material to a particular strain is the sum of an elastic term and a viscous term. This is somewhat more complicated for most tissues but this thought process can be used to understand the behavior of tendon after the crimp is straightened. Below we use this approach to model the behavior of tendon.

8.3 Mechanical Modeling of Aligned Connective Tissue

The mechanical behavior of ECMs has already been classified based on the directional alignment of the collagen fibers with the loading axis in tension. The behavior is more complex as described below for skin, but we wait until Section 8.4 to get any more complex in our analyses. To a first approximation the behavior of an oriented network of aligned ECMs depends on the collagen fiber orientation. However, this applies to ECMs rich in collagen as we discuss in Section 8.4. To a first approximation the behavior of tendon reflects the stretching of the collagen fibers and we do not have to worry about the deformation of other components. We do not discuss the behavior of ligament in this section because it does not totally follow the definition of an aligned collagen network, inasmuch as elastic tissue is present in significant amounts. This is in contrast to other published work where the behavior of tendon and ligament are considered to be the same.

Tendon contains dense ECM composed primarily of aligned collagen fibers. The modeling of this tissue is simpler than other ECMs because the analysis requires considering the mechanism of collagen fibril deformation, which has been studied in great depth. On a molecular basis the initial part

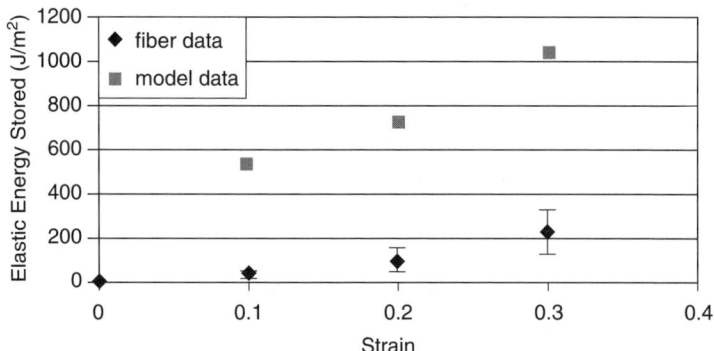

FIGURE 8.4. Elastic energy storage versus strain curves for experimental unmineralized self-assembled type I collagen fibers and for the microfibrillar model. Experimental tests were conducted at 22°C and a strain rate of 10%/min on unmineralized collagen fibers. This plot assumes that every macroscopic strain of 10% in the collagen fiber results in a molecular strain of 1%. Data from the model and the tested fibers have the same general trend, a low slope region followed by a high slope region. There is an increase in elastic energy stored in both the model and the fibers with increasing strain.

of the stress–strain behavior reflects stretching of the collagen triple helix, sliding of collagen molecules by each other, and then sliding of collagen fibrils by each other. The strain at the level of the triple helix is about 10% of the macroscopic strain (Mosler et al., 1985). This suggests that if we multiply the elastic strain by 0.10 to get the molecular strain and then correct the slope of the elastic stress–strain curve for tendon by dividing by the collagen content (about 0.5), we can calculate an elastic modulus for the collagen molecule in tendon. The values of the elastic modulus of the collagen molecule in tendon has been determined for rat tail tendon and mineralized turkey tendons and are listed in Table 8.1. These values fall between a value of 4.2 and 7.69 GPa.

Experimental models of the stress–strain properties of tendon have come from analysis of the mechanical properties of self-assembled collagen fibers with different extents of crosslinking. The values of the elastic modulus of self-assembled type I collagen fibers are also given in Table 8.1. These data suggest that the presence of crosslinks between collagen fibers are important for allowing stress to be transferred between collagen molecules in the fibril. In the absence of crosslinking the transfer of stress is very limited. These data suggest that stress in ECMs is borne primarily by the direct stretching of collagen triple helices and the transfer of stress by shear between collagen molecules and fibrils is small. This means that even in the presence of proteoglycans, the transfer of stress between uncrosslinked collagen molecules and fibrils is probably small.

TABLE 8.1. Elastic moduli for collagen in connective tissue

Molecule	Tissue	Elastic Modulus (GPa)
Type I Collagen	Self-Assembled	6.51
Type I	Rat Tail Tendon	7.69
Type I	Turkey Tendon	4.20 (no mineral)
Type I	Turkey Tendon	7.22 (mineral 0.245)
Types I and III	Skin	4.4
Type II	Articular Cartilage	7.0 (surface parallel)
Type II	Articular Cartilage	2.21 (surface perpendicular)
Type II	Articular Cartilage	4.91 (whole parallel)
Type II	Articular Cartilage	1.52 (whole perpendicular)
Type II	Articular Cartilage	0.092 (arthritic-whole perpendicular)
Elastin	Skin	0.040
Elastin	Vessel Wall	0.013

Another aspect of the study involves the strain-rate behavior of the elastic behavior of tendon. The strain-rate dependence of the elastic behavior of collagen fibers at different strain rates dose not appear to be statistically dependent on strain rate. This is consistent with a mechanism of collagen deformation where there are no hydrogen or other bonds broken during deformation but only stretched. Results of another molecular modeling study suggest that the energy required to stretch the triple helix along its molecular axis is to a first approximation equivalent to the area under the stress–strain curve suggesting that elastic energy appears to be stored by uncoiling of the flexible regions found within the triple helices (see Figure 8.4). Based on molecular modeling it appears that the increase in steric energy stored during stretching involves separating pairs of charged amino acid residues.

From the stress value at a particular strain found on the viscous stress–strain curve (see the second term found in Equation (8.3)) we can calculate the viscosity from the stress if we know the strain rate. Because

TABLE 8.2. Collagen fibril lengths based on mechanical measurements

Tissue		Fibril Length (mm)
Rat Tail Tendon		0.860
Self-Assembled Collagen Fibers		0.0373
Turkey Tendon (no mineral)		0.108
Turkey tendon (0.245 mineral content)		0.575
Human Skin		0.0548
Articular Cartilage	(surface parallel)	1.265
	(surface perpendicular)	0.688
	(whole parallel)	0.932
	(whole perpendicular)	0.696
	Osteoarthritic (whole perpendicular)	0.164

the strain rate is set during the tensile deformation of the sample, the viscosity can then be calculated at a particular strain. The shape factor v is related to the viscosity divided by the volume fraction of macromolecules, therefore it can be calculated using Equation (8.4), where σ_v is the viscous stress, η_o is the viscosity of water, $d\varepsilon/dt$ is the strain rate, and ϕ is the volume fraction of polymer.

$$v = [\sigma_v - \eta_0]/[d\varepsilon/dt(\phi)] \qquad (8.4)$$

Because collagen fibrils are thin long elements, their shape factors can be estimated from the equations used for prolate ellipsoids (see Figure 4.1). For prolate ellipsoids the shape factor is equal to a constant times the ratio of the major semiaxis length a, divided by the minor semiaxis length b, raised to power of 1.81 (Equation (8.5)).

$$V = k(a/b)^{1.81} \qquad (8.5)$$

We can calculate k for collagen knowing that a is 150 nm and b is 0.75 nm. Knowing the collagen fibril diameters for tendon determined from electron microscopy, we can calculate a fibril length by substituting the diameter into Equation (8.4) and solving for b, the fibril length. This has been done for rat tail tendon, turkey tendon, and self-assembled type I collagen fibers. The results of these calculations suggest that the fibril length varies from 0.037 mm to 0.860 mm as listed in Table 8.2.

These results suggest that the mechanical behavior of tendon in tension can to a first approximation be modeled as a spring and a dashpot in parallel with the stretching of the spring physically representing the stretching of the collagen triple helix and the crosslinks that hold the molecules together in a continuous stress-bearing network. The viscous element represents irreversible sliding of the collagen fibrils past each other during tensile deformation. Please note there are additional considerations including molecular sliding and axial tilting that occur during tensile deformation of tendon; these mechanisms are beyond the scope of this text and as a result the model presented is somewhat of an oversimplification. However, this model gives us a good idea of the starting point to generate mathematical expressions that can be used to represent the mechanical behavior of ECMs. The task only becomes more complicated when we consider the behavior of nonaligned (irregular) connective tissue.

8.3.1 Mechanical Models of Mineralized Tendon

Mechanical models of mineralized tendons follow from the analyses done for tendon. The elastic moduli of mineralized turkey tendons have been calculated from the experimental elastic incremental stress–strain curves. For mineralized tendons the stress–strain curves are linear at mineral content of about 0.3 and the elastic modulus is about 8 GPa and increases with

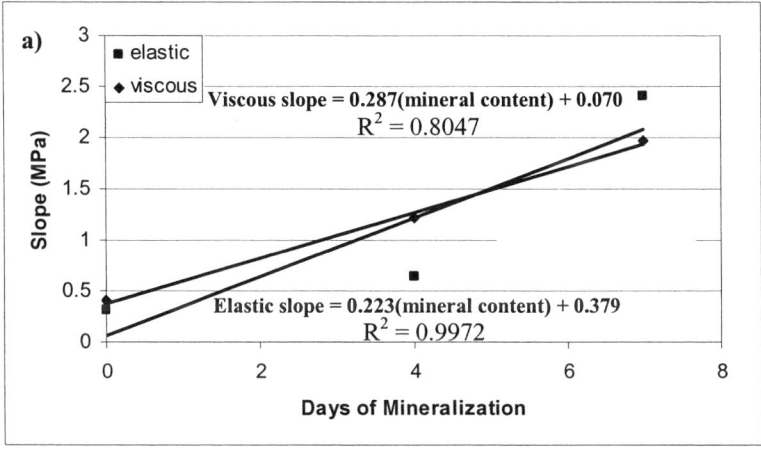

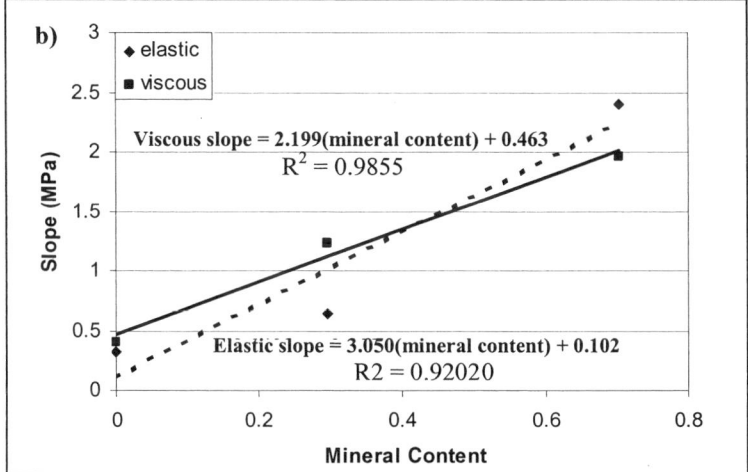

FIGURE 8.5. Plots of elastic modulus versus mineral content (b) and days of mineralization (a) from incremental stress–strain tests performed on mineralized self-assembled type I collagen fibers. Slopes were obtained from the straight portions of the elastic and viscous stress–strain curves.

mineral content (See Figure 8.5). Values of fibril lengths calculated for mineralized turkey tendon are 0.108 to 0.575 mm.

Molecular modeling results suggest that the increased elastic modulus of mineralized turkey tendon compared to unmineralized tendon is due to the effect of calcium and phosphate ions increasing the resistance of the flexible regions in collagen to axial deformation. These ions appear to act as bridges between collagen α chains and collagen molecules.

8.4 Mechanical Models of Orientable Connective Tissue

Unfortunately, the behavior of tendon gives one the misconception that physical modeling of ECMs is rather straightforward. The problem only gets more complicated when one considers skin. Because skin contains an almost biaxial-oriented collagen network, which orients with the load direction when it is stretched along the direction of one of the collagen fiber principal directions, then in theory it should be fit by a spring in parallel with a dashpot similar to the model used for tendon. Unfortunately, as shown in Figure 7.10, the elastic and viscous stress–strain curves for skin cannot be fit by single straight lines. If we concentrate on the elastic behavior, two straight lines can be used to approximate the elastic stress–strain curve: one line for the low strain region and one for the high strain region. This is consistent with the literature, which suggests that the low strain behavior reflects the behavior of the elastic fiber network and the high strain behavior reflects the behavior of the collagen fibers. This makes modeling more difficult because we now have a spring and dashpot in parallel (Voigt element) for both the elastic and collagen networks. The Voigt elements representing the elastic and collagen fibers must also be in parallel because loss of elastic fiber content associated with aging shifts the stress–strain behavior to the left as if the elastic fiber contribution were removed. Therefore, skin can be represented by two Voigt elements connected in parallel. The collagen fibers in skin must be wavy until the elastic fibers transfer the load to the collagen fibers so that the collagen network does bear loads at low strains.

Using this approach we can analyze the elastic fiber contribution by analyzing the low strain behavior of the stress–strain curve and collagen fiber contribution, the high strain behavior. The elastic and viscous stress–strain curves for skin have been analyzed using incremental stress–strain curves. The elastic modulus for elastic fibers after correction for the elastin content was found to be between 0.013 and 0.04 GPa as listed in Table 8.1, and the elastic modulus for the collagen fibers was found to be 4.4 MPa. The low strain behavior of skin was reported to be strain-rate dependent whereas the high strain behavior was not. This suggests that elastic fiber stretching may involve a change in the structure of elastin or fibrillins whereas that of collagen does not. Fiber length calculated from the viscous stress was 0.055 mm for skin as listed in Table 8.2.

8.5 Mechanical Models of Composite ECMs

Although the structure of connective tissue has classically been considered as aligned (regular) and unaligned (irregular), from a mechanical perspective tissues such as cartilage are composites of layers, regular (aligned) and irregular (unaligned) ECMs. This makes modeling of cartilage complex. The analysis of cartilage mechanical behavior is only possible knowing the behavior of the surface zone (aligned collagen network) and the deeper zones (alignable collagen network). Incremental elastic and viscous stress–strain curves of articular cartilage have been reconstructed in the literature. The elastic modulus parallel to the surface along the collagen fibrils is 7.0 GPa and that perpendicular to the surface is 2.21 GPa as shown in Table 8.2. Values of the elastic modulus for normal intact cartilage are somewhere between 2 and 7 GPa suggesting that the behavior can be modeled as a composite of layers of collagen fibers with different collagen fibril orientations.

Collagen fibril lengths have been determined for the superficial zone (aligned collagen fibers) and the intermediate zone (unaligned collagen fibers) from the slopes of the incremental viscous stress–strain curves. Collagen fibril lengths of between 0.696 mm to 1.265 mm are reported in Table 8.2.

Each layer of cartilage can be considered as a spring and dashpot in parallel and with a characteristic elastic modulus and collagen fibril length. The elastic modulus goes from about 8 GPa when the collagen fibrils and fibers are aligned with the tensile direction to about 2 GPa when the fibrils and fibers are perpendicular to the tensile direction. The Voigt elements for each layer are then connected in parallel to represent the behavior of the whole tissue.

8.6 Mechanical Models of Vessel Walls

We have seen that the mechanical models for irregular connective tissue are quite complex. The behavior is made even more confusing in tissues containing smooth muscle tissue including vessel walls. The frustrating part is that the addition of smooth muscle not only adds another component to the model, but it also adds another orientation to consider. To make a long story even longer, analysis of incremental stress–strain curves of vessel walls suggests that the low and high strain behaviors cannot be simply analyzed like skin. Although the low strain behavior is to a first approximation a reflection of the behavior of elastic tissue, the high strain behavior reflects the behavior of collagen and smooth muscle in series (see Figure 7.10 and compare it with Figure 7.11 for skin). The maximum elastic stress is far too low in vessel wall compared to skin to be explained by loading of the collagen fibers. From this it was concluded that the behavior could be modeled as a Voigt element representing the behavior of the elastic fibers, connected in parallel to two Voigt elements in series (representing the behavior of the collagen and smooth muscle fibers). The values of elastic modulus of collagen in vessel wall is lower than 1 GPa whereas that for elastic tissue is about 0.0130 GPa as listed in Table 8.2. The low value of the elastic modulus of collagen in vessel wall suggests that collagen and smooth muscle are in series in this tissue.

8.7 Summary

We have made much progress in the last decade using springs and dashpots to model the physical behavior of simple ECMs such as tendon, however, we are still struggling with developing models of more complex ECMs such as vessel wall. In vessel wall the series connection between collagen and smooth muscle is likely the important link between external loading and up-regulation of mechanochemical transduction. However, we still can only dream of formulating physical models that will help us understand these complex biological processes. Although the Voigt model consisting of a spring in parallel with a dashpot appears to represent a very simple model for the behavior of the collagen fibril, it is less than adequate for representing the behavior of irregular connective tissue with unaligned collagen fibers. More complex assemblies of Voigt elements can be used to approach the behavior of more complex ECMs.

We know from studies of hypertensive animals that blockage of fluid flow in the arterial system not only increases blood pressure but leads to vessel dilation and increases in wall thickness. There appears to be a direct relationship between external (increased blood pressure) mechanical stimulation and up-regulation of mechanochemical transduction processes by increasing the tensile loads that are placed on collagen fibers. This increase in external mechanical stimulation is then directly transferred to smooth muscle cells within the vessel wall. Increased tensile forces lead to increased activation of MAPK pathways as discussed in Chapter 9. We now have the beginning information that details how external mechanical loading influences tissue growth and development.

Suggested Reading

Mosler E, Folkhard W, Knorzer E, Nemetschek-Gansler H, Nemetschek Th, Koch MH. Stress-induced molecular arrangement in tendon collagen. J. Mol. Biol 1985;182:589–596.

Silver FH. Mechanical properties of connective tissue. In: *Biological Materials: Structure, Mechanical Properties, and Modeling of Soft Tissues*. New York: NYU Press; Chapter 7. 1987.

Silver FH, Christiansen DL, Buntin CM. Mechanical properties of aorta: A review. *Crit Rev Biomed Eng* 1989;17:323.

9
Mechanochemical Sensing and Transduction

9.1 Introduction

Gravity plays a central role in vertebrate development and evolution. Gravitational forces acting on mammalian tissues cause the net muscle forces required for locomotion to be higher on Earth than on a body subjected to a microgravitational field. As body mass increases during development, the musculoskeleton must be able to adapt by increasing the size of its functional units. Thus mechanical forces required to do the work (mechanical energy) of locomotion must be sensed by cells and converted into chemical energy (synthesis of new tissue; see Silver et al., 2003 for background reading).

Extracellular matrices are multicomponent tissues that transduce internal and external mechanical signals into changes in tissue structure and function through a process termed mechanochemical transduction. Under the influence of an external gravitational field, both mineralized and unmineralized vertebrate tissues exhibit internal tensile forces that serve to preserve a synthetic phenotype in the resident cell population. Application of additional external forces alters the balance between the external gravitational force and internal forces acting on resident cells leading to changes in the expression of genes and production of protein that ultimately may alter the exact structure and function of the ECM. Changes in the equilibrium between internal and external forces acting on ECMs and changes in mechanochemical transduction processes at the cellular level appear to be important mechanisms by which mammals adjust their needs to store, transmit, and dissipate energy that is required during development and for bodily movements.

Mechanosensing is postulated to involve many different cellular and extracellular components. Mechanical forces cause direct stretching of protein–cell surface integrin binding sites that occur on all eukaryotic cells. Stress-induced conformational changes in the extracellular matrix may alter integrin structure and lead to activation of several secondary messenger pathways within the cell. Activation of these pathways leads to altered regulation of genes that synthesize and catabolize extracellular matrix proteins as well as to alterations in cell division. Another aspect by which mechanical signals are transduced involves deformation of gap junctions containing calcium-sensitive stretch receptors. Once activated, these channels trigger secondary messenger activation through pathways similar to those involved in integrin-dependent activation and allow communications between cells with similar and different phenotypes. Another aspect of mechanosensing is through activation of ion channels in the cell membrane. Mechanical forces have been shown to alter cell membrane ion channel permeability associated with Ca^{+2} and other ion fluxes. In addition, the application of mechanical forces to cells leads to the activation of growth factor and hormone receptors even in the absence of ligand binding. These are some of the mechanisms that have evolved in vertebrates by which cells respond to changes in external forces that lead to changes in tissue structure and function.

9.2 How Is Gravity Sensed by Cells?

Gravity plays a central role in vertebrate development and evolution. As body mass increases during development, increased muscular forces are required to propel large vertebrates; this requires that the musculoskeleton be able to adapt by increasing the size of its functional units. This argument suggests that inherent to biological systems subjected to gravitational forces is the need to adapt to the level of external force through anabolic and catabolic processes. Thus mechanical forces required to do the work (mechanical energy) of locomotion must be sensed by cells and converted into chemical energy (synthesis of new tissue). How this conversion is accomplished is an important question that is useful for understanding problems of medical interest and also for considering how nature has evolved efficient systems to store mechanical energy.

In addition, mechanical forces also act as selective factors and developmental cues in both vertebrates and invertebrates through mechanochemical transduction processes. Thus the processes for mechanosensing and mechanochemical transduction have evolved over several billion years partly as a result of living in a gravitational field. When astronauts are exposed to microgravity, they experience muscle atrophy and bone resorption as a result of the decreased force required for locomotion or other bodily movements. This demonstrates the need to understand how mechanosensing and mechanochemical transduction work in order to better understand the effects of space travel on the human physiology. The questions that are addressed in this chapter include, how is mechanical loading sensed by cells and how is mechanical energy transduced into chemical energy.

Mechanosensing is used to describe the process by which cells sense mechanical forces. Mechanochemical transduction is the phrase that is used to try to describe the biological processes by which external forces such as gravity influence the biochemical and genetic responses of cells and tissues. Specifically, these responses include stimulation of cell proliferation or apoptosis (death) and synthesis or catabolism of components of the extracellular matrix. These processes cause either increases in chemical energy (conversion of amino acids or other small molecules into macromolecules) or decreases in chemical energy (depolymerization of macromolecules).

9.2.1 Gravity and Cellular Responses

Extracellular matrices (ECM) are the primary structural materials found in connective tissue in vertebrates that serve to maintain tissue shape (skin), aid in locomotion (bone), transmit and absorb mechanical loads (tendon and ligament), prevent premature mechanical failure (tendon, ligament, skin, and blood vessel wall), partition cells and tissues into functional units (fascia), act as scaffolds that define tissue and organ architecture (organ parenchyma), act as storage devices for elastic energy (tendon and blood vessel wall), and as the substrate for cell adhesion, growth, and differentiation of a variety of cell types.

The ability of cells and tissues found in ECM to faciliate locomotion in a gravitational field dictates that the cells be able to adapt their responses to changing energy requirements. What this means is that information in the form of external mechanical forces must somehow set in motion a series of biological steps that lead to increased muscle mass and to associated musculoskeletal tissues. Increased tissue mass is required to amplify the amount of work that can be done by the organism in response to growth or environmental demand. This is equivalent to conversion of mechanical energy into chemical energy in the form of macromolecular components of the cell. This process leads to a decrease in tissue mass when the organism is subjected to microgravity or under disuse conditions. Thus, changes in external mechanical loading must trigger cellular processes that can generate additional energy required to do work in response to elevated environmental demand and that under disuse conditions can stimulate catabolism of unnecessary musculoskeletal components that are no longer required for locomotion.

Mechanochemcal transduction is thought to involve several different macromolecular components and processes (see Figures 9.1 through 9.4). One process involves direct stretching of protein–cell surface integrin binding sites that occur on all eukaryotic cells (integrin-dependent mechanisms). Stress-induced conformational changes in ECM may alter integrin structure and lead to activation of several secondary messenger pathways within the cell. Activation of these pathways leads to altered regulation of genes that synthesize and catabolize extracellular matrix proteins as well as

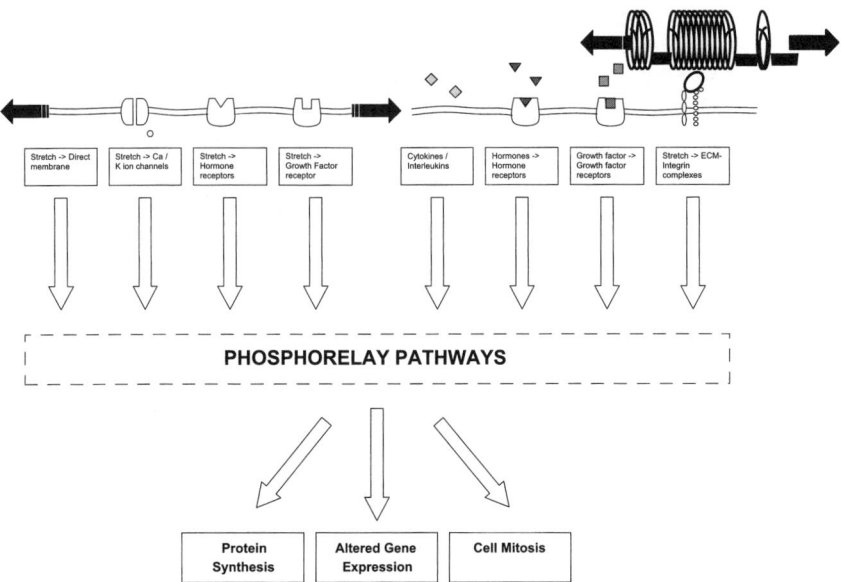

FIGURE 9.1. Relationship between the phosphorelay system and cell behavior. This diagram shows the relationship between stimuli external to the cell and the resulting activation of phosphorelay pathways. Phosphorelay pathways can be activated by direct membrane stretch, which is caused by osmotic, hydrostatic, electromechanical, and hydrodynamic effects; ion channel stretching; stretching of intercellular junctions (not shown); stretching of growth factor and hormone receptors; stretching of cell surface integrins; the presence of extracellular cytokines; and by combinations of these effects. In some cases, membrane stretching can activate growth factor receptors even in the absence of growth factors (ligand that normally activates the receptor). Any of these effectors can potentially activate the phosphorelay pathways shown in Figure 9.3.

to alterations in cell division. The second process by which mechanochemical transduction occurs involves deformation of gap junctions containing calcium-sensitive stretch receptors. Once activated, these channels trigger secondary messenger activation through pathways similar to those involved in integrin-dependent activation and allow cell-to-cell communications between cells with similar and different phenotypes. Another process by which mechanochemical transduction occurs is through the activation of ion channels in the cell membrane. Mechanical forces have been shown to alter cell membrane ion channel permeability associated with Ca^{+2} and other ion fluxes. The application of mechanical forces to cells leads to the activation of growth factor and hormone receptors even in the absence of ligand binding as recently reviewed.

Integrins are a family of cellular transmembrane receptors that provide physical and biochemical bridges between components of the ECM (see

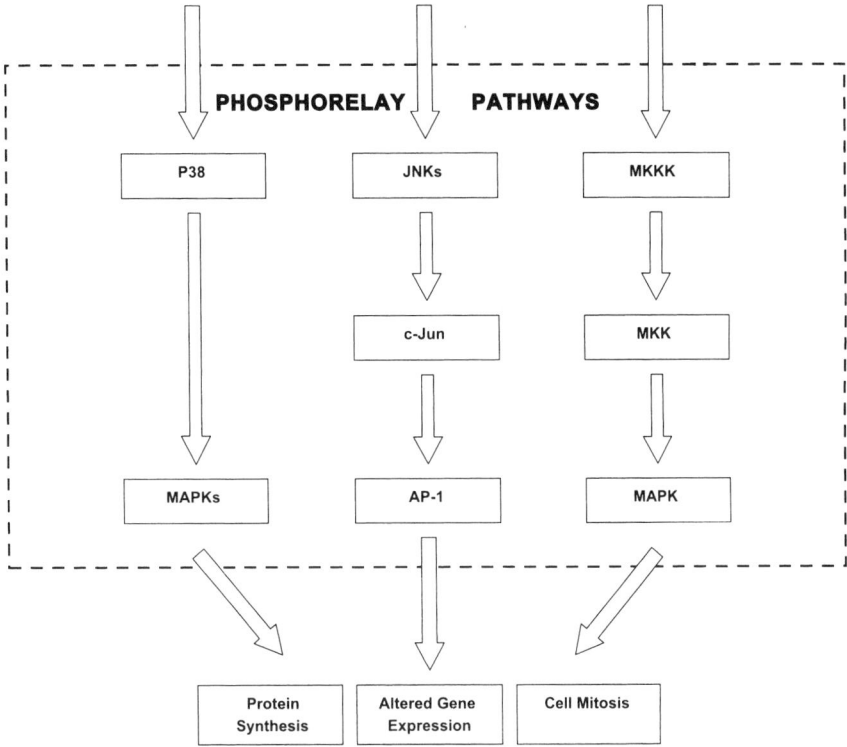

FIGURE 9.2. Diagram of activation of the phosphorelay system by physical forces. Mitogen-activated protein kinases (MAPKs) are part of a phosphorelay system, which are activated through cell membrane and cytoskeletal stretching such as diagrammed in Figure 9.1. Tension applied to the extracellular matrix (ECM) occurs as a result of mechanical, osmotic, hydrostatic, fluid, and electromechanical effects, leading to stretching of the cell membrane. Cell mitosis, gene expression, and protein synthesis occur by generation of secondary messengers that activate the pathways shown. These pathways of the phosphorelay system include the extracellular signal-regulated kinase (ERK-1/2) pathway, which is part of the MAPK kinase kinase (MKKK) pathway (right), the c-Jun kinase (JNK) (JNK1, JNK2 and JNK3) pathway (middle), and the p38 pathway (left). The ERK-1/2 pathway is stimulated by growth factors, cytokines, and G protein and can lead to increased cell mitosis. The JNK pathway is stimulated by growth factors and environmental stresses, altering gene expression and controlling programmed cell death. The p38 pathway regulates protein expression of many cytokines such as interleukin-1, which has been implicated in modulating the response to mechanical loading in a number of tissues such as cartilage.

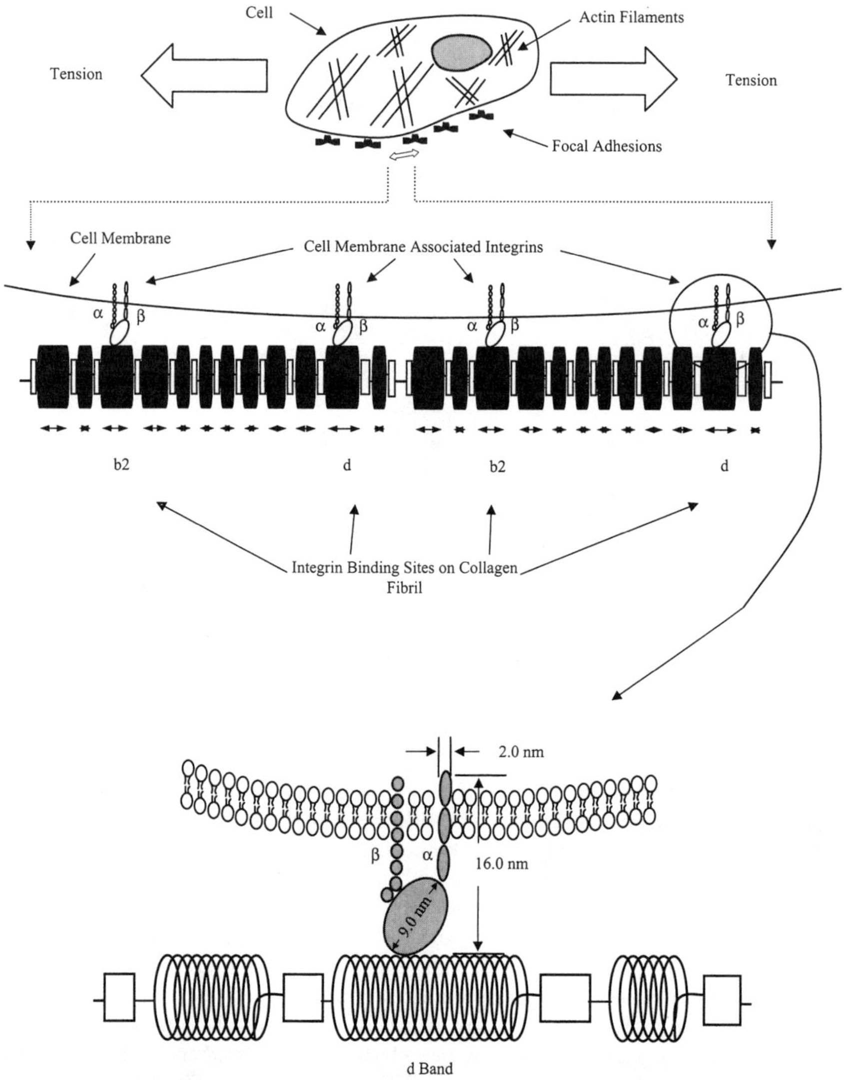

FIGURE 9.3. Integrin-ECM interactions and possible modes of energy transfer from collagen fibrils in the ECM to the cell cytoskeleton. This diagram illustrates how a cell binds to a collagen fibril at the b2 and d bands via integrin α- and β-chains at focal adhesion complexes and the influence of tensile forces. Tensile forces applied to collagen fibrils found in ECM lead to stretching of flexible domains found within the positive staining bands, identified by electron microscopic studies. The collagen fibril is represented by a series of springs (black cylinders) and rigid connectors (white rectangles) that transfer tensile loads and elastic energy to the cell cytoskeleton (actin filaments) via stretch-induced conformational changes in integrin α- and β-chains, shown connecting the cell membrane and collagen fibrils of the ECM. An expanded view of the d band (bottom) illustrates how energy transfer could occur between collagen fibrils and cell surface integrins when the springlike flexible region is stretched, altering the conformation of the integrin head region. Specific integrin binding sites on type I collagen fibrils are found in the b2 and d bands.

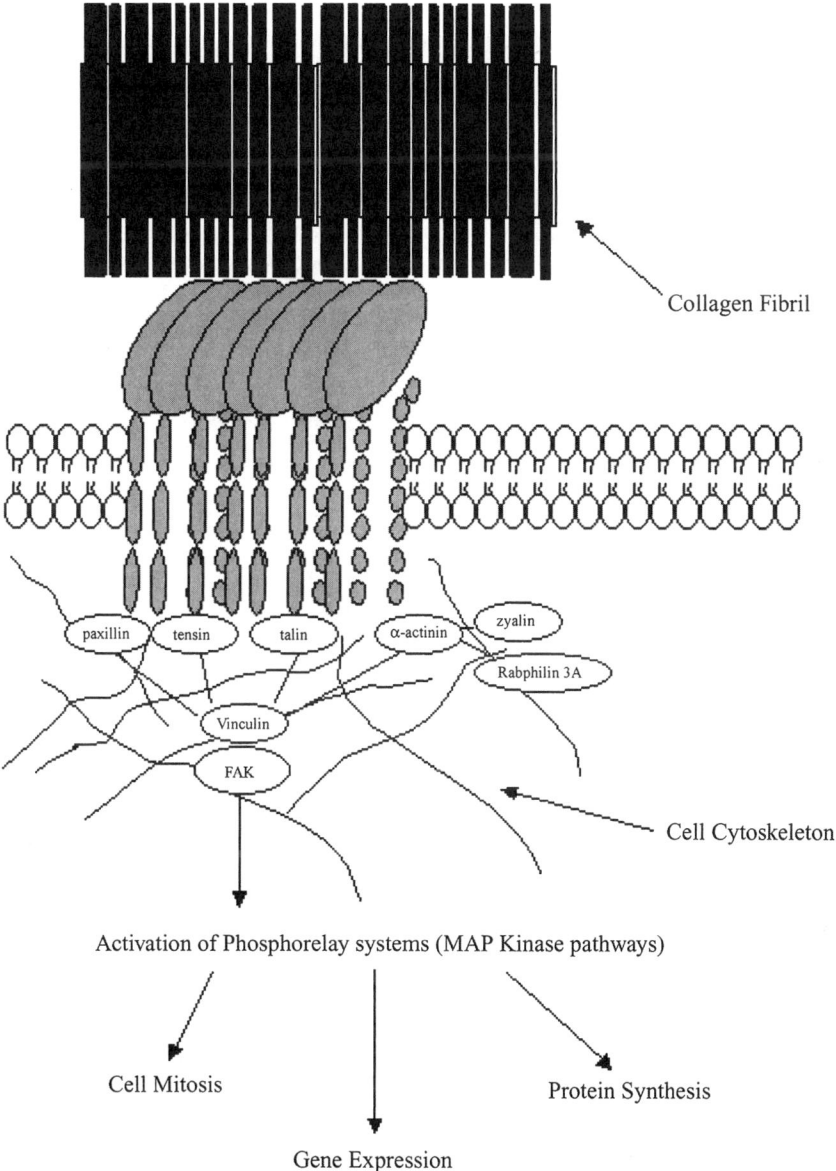

FIGURE 9.4. Integrin-mediated activation of mechanochemical transduction. This diagram illustrates how tension applied through collagen fibrils (top) at the ECM-cell interface leads to synthesis of new tissue via activation of MAPK phosphorelay systems. Energy transfer to the attached integrin subunits from the collagen fibril of the ECM putatively leads to physical changes in the cell cytoskeleton (actin filaments and other cytoskeletal molecules shown) that affect activation of focal adhesion kinase (FAK) and stimulation of secondary messengers. Secondary messengers activate several different pathways that are part of the phosphorelay system. This process leads to activation of MAPK, up-regulation of protein synthesis and cell mitotic activity, and changes in gene expression.

Figures 9.1, 9.3) and the intracellular physiological environment. Structurally, integrins contain a head domain that includes a calcium-dependent ligand-binding portion, and two tails, or legs, that anchor the receptor in the cell membrane and to the cell cytoskeleton. Mechanochemical transduction pathways that influence ECM gene expression appear to involve both integrin-dependent and integrin-independent processes. Binding of integrins to ECM molecules including collagen fibrils (Figure 9.4), laminin, and fibronectin results in the formation of cytoplasmic multiprotein assemblies composed of cytoskeletal and signaling molecules that transfer energy stored in the ECM to cytoskeletal elements within the cell. Integrin–ligand interactions in conjunction with intracellular regulators, including tyrosine kinases and phosphatases, control formation of the protein assemblies and their resultant activities within the cell cytoplasm. Integrin-mediated cellular responses include activation of signal transduction, force-induced cytoskeletal rearrangement, and coregulation of growth factor and perhaps hormone activity.

From an engineering perspective, under normal conditions cells appear to transduce increases in the net external force acting on a tissue into internal cellular tension and other intracellular responses. For instance, increased external loading of the arms and legs during exercise results in an increase in muscle mass as well as an increased skin surface area required to cover the expanded muscle tissue. This response is promoted in the presence of a gravitational field that acts as a resistance to motion. For this reason, the ability for mammals to undergo movement reflects the evolution of multicellular organisms on Earth. This evolution involves the development of transduction mechanisms by which external stimuli such as gravity alter cellular function.

9.3 Intracellular Signal Transduction Mechanisms

During evolution, mechanical as well as other stimuli such as light, pH, and temperature are critical environmental parameters that are sensed by cells. It has been postulated that intracellular signal transduction pathways were originally developed by cells to respond to those basic stimuli; this response has been preserved and further developed during evolution (see Liu et al., 1999 for background reading). This means that cells appear to respond to a variety of environmental stimuli by activating the same transduction pathways. This concept suggests that transduction pathways which regulate cell differentiation and dedifferentiation (cancerous transformation) may also regulate how cells respond to mechanical forces. Ultimately, this means mechanical stimuli might activate the same signaling pathways that are activated by hormones and growth factors; however, the exact activation mechanisms may be different. The signaling events initiated by mechanical as well as other stimuli include generation of secondary messengers (molecules found within the cell cytoplasm that act to activate other messengers that move from the cytoplasm to the cell nucleus), changes in the phosphorylation status of intracellular proteins (proteins within the secondary messenger pathways are activated and deactivated by adding or removing a phosphate group), amplification through enzymatic cascades (signals amplified by going through a series of steps), and transmission via a complicated network of signaling molecules (changes in cell nuclear events occurring only after a series of molecules have been activated; Figure 9.3).

The phosphorylation and dephosphorylation of enzymes within the cell, many of which are found attached or in close proximity to the cell membrane, are some of the ways that secondary messenger signaling pathways are activated. Protein phosphorylation is also involved in control of cell proliferation and differentiation with phosphorylation being controlled genetically by the synthesis and regulation of molecules termed protein kinases (PKs; e.g., see focal adhesion kinase (FAK) in Figure 9.4). When activated, PKs transduce external forces into chemical and physical changes within the cell. Effects of PK activation are regulated through the activation of secondary messengers that transmit signals within a cell.

PKs are enzymes that covalently attach phosphate to the amino acid side chains of serine, threonine, or tyrosine of specific proteins inside cells. Such phosphorylation of proteins can control their enzymatic activity, their interaction with other proteins and molecules, their location within the cell, and their propensity for degradation by proteases. An example of a protein kinase is focal adhesion kinase which is phosphorylated after integrin binding to a substrate (see Figure 9.4); phosphorylation of FAK is integrin mediated and leads to activation of an intracellular signaling pathway termed MAP kinase. Mitogen-activated protein kinases (MAPKs) constitute a family of kinases that regulate cellular activities including gene expression, mitosis, movement, metabolism, and programmed cell death (see Cobb and Goldsmith, 1995 and Johnson and Lapadat, 2002 for background reading; Figure 9.2). Substrates that are phosphorylated include other protein kinases, transcription factors, and cytoskeletal proteins. Protein phosphatases remove the phosphates that were transferred to the protein substrate by the MAPK.

MAPKs are activated through several different mechanisms, one of which involves activation of tyrosine kinase receptors, which are transmembrane proteins that transduce signals controlling cell growth, survival, motility, and differentiation. These receptors are activated by binding of a ligand (usually a growth factor); the receptors form dimers and then undergo autophosphorylation, resulting in the creation of docking sites that bind to signal transduction molecules downstream (FAK in Figure 9.4 creates a docking site in conjunction with cell cytoskeletal components). Autophosphorylation of growth factor receptors occurs when the receptor encounters a growth factor such as epidermal growth factor (EGF), platelet-derived growth factor (PDGF), or fibroblast growth factor (FGF) resulting in downstream signaling. Growth factor receptor phosphorylation in the presence of integrins suggests that integrins can induce clustering of growth factor receptors and downstream signaling even in the absence of the growth factor. Other growth factors such as insulinlike growth factor-1 (IGF-1) and transforming growth factor-β (TGF-β) have been implicated in regulation of integrin expression by different cell types.

MAPKs are part of a phosphorelay system composed of three kinases that are sequentially activated and that are regulated by phosphorylation. MAPKs act as substrates for MAPK kinases (MAPKKs) that activate MAPKs by phosphorylation, which increases MAPKs' activity in catalyzing phosphorylation of substrates for MAPKs. MAPK phosphatases reverse the phosphorylation and deactivate MAPK. MAPK kinase kinases (MAPKKKs) are the third component of the phosphorelay system and selectively activate MAPKK by phosphorylation. Subfamilies of MAPKs in multicellular organisms include three families named the extracellular-regulated kinases, ERK1 and ERK2; the c-Jun NH2-terminal kinases, JNK 1, JNK 2, and JNK 3; and the four p38 enzymes, p38α, p38β, p38γ, and p38δ (see Johnson and Lapadat, 2002 for a review; Figure 9.2).

ERK1 and ERK2 are other components of the MAPK phosphorelay system that are widely expressed; they are involved in regulation by meiosis, mitosis, and postmitotic function in differentiated cells. Many different stimuli activate ERK1 and ERK2 pathways including growth factors, cytokines, ligands for heterotrimeric guanine nucleotide-binding protein (G protein)-coupled receptors, and carcinogens. ERKs1 and 2 are activated through a three-kinase phosphorelay module that includes MAPKKK c-Raf1, B-Raf, or A-Raf, which can be activated by proto-oncogene RAS (Johnson and Lapadat, 2002). Mutations that convert RAS to an activated oncogene activate ERK1 and ERK2 pathways, which lead to increased proliferation of cells. Mechanical stresses and strains have been shown to activate the ERK1/2 pathways as noted below in the discussion of specific cell types.

JNK1, JNK2, and JNK3 are other components of the MAPK phosphorelay system and are classified as stress-activated protein kinases that bind and phosphorylate DNA binding protein c-Jun and increase its transcriptional activity (see Figure 9.2). c-Jun is a component of the AP-1 transcriptional complex, which is an important regulator of gene expression. It is activated in response to environmental stresses and growth factors and is important in controlling programmed cell death.

The most well characterized of the p38 kinases, the final MAPK phosphorelay pathway, is p38α, which is expressed by most cells and regulates the expression of many cytokines. Interleukin 1, IL-1, a product of inflammatory cells, has been implicated in modulating the response to mechanical loading in a number of tissues. IL-1 is a product of inflammatory cells thought to be involved in cartilage destruction in osteoarthritis and in bone resorption associated with total joint implant failure.

There are a number of biochemical components that are involved in mechanochemical transduction processes. We introduce some of the important molecules in an attempt to demonstrate how complex the mechanisms appear to be; however, the functions and interrelationships among these molecules are currently unclear.

A number of protein kinases have been implicated in mechanochemical transduction and activation of phosphorelay systems (see Figures 9.2 to 9.4 for some of these molecules). They include the following.

PKA, also known as cyclic adenosine monophosphate (cAMP)-dependent PK, because it uses cAMP as its secondary messenger.

PKC, a family of proteins some of which have activities that are regulated by intracellular concentrations of free Ca^{+2} and diacylglycerol.

PTK, a family of kinases that regulate phosphotyrosine (PT) levels including the Src family found in the cytoplasm and are associated with the cytoskeleton. Other members of this family include pp125^{FAK}, FAK, PTK, p120, a tyrosine-phosphorylated protein that is similar in sequence to cadherin-binding factors, an actin filament associated protein (AFAP-110), and contactin.

pp125^{FAK}, which functions as PTK that is localized at the focal adhesion plaque and is associated with the cytoskeleton through the protein paxillin.

As diagrammed in Figure 9.3, integrin clustering results in the formation of focal adhesion contacts (FCs) after integrins bind to extracellular matrix molecules including fibronectin, laminin, and collagen. FCs contain a multitude of anchor and cytoskeletal molecules including vincullin, paxillin, and talin. Association of the integrin legs with tensin, integrin-linked kinase (ILK), teraspan family transmembrane proteins (TM4), Shc, and FAK occurs and results in assembly and tyrosine phosphorylation of FCs (not all of these molecules are shown in Figure 9.4). FAK has docking functions and supports binding of paxillin and vinculin and leads to activation of MAP kinase through molecular interactions of FAK with cytoskeletal and signaling molecules (see Figure 9.4). A recent study suggests that local tyrosine phosphorylation is dependent on ECM rigidity regulating local tension at adhesion sites and recruiting a variety of plaque molecules to the sites.

Heterotrimeric G proteins transduce a variety of signals generated by interactions with hormones, growth factors, neurotransmitters, and other molecules with cell surface receptors. G proteins are activated in endothelial cells as a result of strain and increased strain-rate phosphorylation of ERK 1/2, which requires G proteins. G proteins are believed to be involved in mechanochemical transduction through adhesion coupled cell mechanosensing.

Although mechanosensing and mechanochemical transduction processes are likely to be quite complex from a biochemical point of view, we can begin to understand from an engineering point of view how changes in external loading of tissues leads to increased energy stored in ECMs, associated with structural changes. Structural changes are then converted into changes in energy states of cell membrane components that activate pathways within the cell that lead to changes in gene expression, cell division, and protein synthesis. The first step in this series of steps is the sensing of mechanical forces by cells found in the ECM.

9.3.1 How Does Mechanosensing Occur?

Classical mechanosensing is based on a five-component transmembrane system consisting of an ion channel, tethered via intracellular and extracellular links to cytoskeletal and external anchors (see Geiger and Bershadsky, 2002 for background reading). The ion specificity of the channel and the molecular nature of the links may vary from one system to another, but the general mode of action appears similar.

Much information is available about the sensing and response of cells to mechanical forces that appears to be related to cellular adhesion processes. Several lines of evidence suggest that integrin-mediated adhesion is a mechanosensory process. These include the dependence on both clustering and occupancy and the locality of the effect; namely, one receptor may not be influenced by the behavior of another receptor. The focal adhesion complex (see FAK in Figure 9.4), which is a multicomponent complex connecting the extracellular matrix with the actin cytoskeleton, is composed of integrin receptors that connect the ECM to a submembrane protein complex containing over 50 proteins. This complex is in turn connected to actin filament bodies. The exact relationship among ion channels, and integrin, growth factor, and hormone receptors is quite complex and is beyond the scope of this chapter.

9.3.2 Influence of Mechanical Forces on Protein Synthesis

As discussed in some detail in the following sections, mechanical loading appears to play a central role in regulation of gene expression, and protein synthesis and secretion of proteins (see Figures 9.2 through 9.4). Thus mechanical energy in the form of work (stretching force acting through a strain) is transduced within cells into chemical energy stored in the creation of macromolecular structures. Increased tensile loading of ECMs appears to result in increased energy storage in the form of increased amounts of high molecular weight components found in tissues. Just how this occurs is unknown, but physical forces can affect cells by influencing expression of so-called immediate early response genes including c-fos, c-Jun, c-myc, JE, and Egr-1, which encode proteins related to transcriptional factors and signal transduction (Komuro et al., 1991). Physical forces also increase mRNA levels of growth factors such as platelet-derived growth factor-B (PDGF-B) and ECM proteins such as collagen. The mechanism for energy storage and synthesis of high molecular weight materials is likely to vary from tissue to tissue; but the concept that mechanical energy is stored in tissues in the form of high molecular weight materials appears to occur in many different tissues. Below we look at several different factors that affect tissue formation.

9.3.3 Influence of Mechanical Forces on Intercellular Communication

Cell–cell and cell matrix interactions are essential to coordinate growth and repair of tissues and are affected by external loading. An example is the repair of skin defects, which first occurs by dermal tissue synthesis and then by epidermal proliferation over the repaired dermis. Mesenchymal–epithelial interactions are a controlling factor for development and maturation of a number of tissues including the lung. Cells commonly use factors that are mitogenic and cause cell division (such as growth factors), molecules released from inflammatory cells (termed cytokines), and other small soluble molecules. Cells also communicate through cell junctions via Ca^{+2} release through connexin channels in gap junctions and by force transmission involving cadherins in tight junctions. Stretch-activated ion channels, activated by membrane strain, have been identified as having stretch-inactivated channels that open at low force and close with increasing loads.

Before it is possible to analyze how physical forces affect signaling behavior in different types of tissue, it is first important to analyze what external and internal forces act in each tissue.

9.4 Stresses in Extracellular Matrices and Mechanochemical Transduction

Under normal physiological conditions, ECMs found in musculoskeletal, cardiovascular, pulmonary, and dermal tissue are all under tension even in the absence of external loading (see Figure 3.9). The tension is similar to the residual stress discussed by Fung (see, e.g., Fung, 1990). This tension sets up a state of dynamic mechanical loading at the collagen fibril–cell interface that stimulates mechanochemical transduction. Tension at the ECM–cell interface causes energy to be stored as an increase in free energy of the macromolecular components as a result of changes in both entropy and enthalpy (see, e.g., Silver and Christiansen, 1999). By definition, at equilibrium all external forces acting on tissues and organs and collagen fibrils must sum to zero. In addition, increases in external loading result in increases in internal stresses acting on collagen fibrils and at the collagen fibril–cell interface. Beyond this effect, the observation that cells grown in collagen lattices exert a contractile force (Grinnell, 2000) suggests that under normal physiological conditions, cells apply tension to the attached ECM. This tension not only leads to dynamic active stresses that are applied to the collagen network, but also to incorporation of passive tension into the collagen fibrils during development and maturation of tissue scaffolds. Because cells must pull back as a result of forces applied to them by their ECM, they must be able to convert chemical energy into mechanical energy reversibly.

For instance, the tensile forces that operate in skin have been studied extensively for over 100 years and serve as an example of the existence of passive stresses existing in the collagen fibrils found in skin ECM. It is well established that the collagen fibrils are laid down approximately parallel to Langer's lines, which characterize the direction of principal tension in skin. As noted earlier, in 1862 Langer found that circular holes punched in the skin of cadavers became elliptical, with the major axis of the ellipse along the direction of maximum tension. In the absence of external loading, cadaver skin, in which the cells are not viable, is under a biaxial tension; the magnitude of the tension varies from location to location, with the tension highest in the skin from the limbs (see Silver et al., 2003). Beyond the passive tension that exists in the skin of cadavers, in living skin there are both passive and active tensional forces. A circular punch defect made in living skin increases in size (a 2.0 mm biopsy site increases to at least 3.0 mm depending on the location) and remains approximately circular after the skin plug is removed. The circular nature of the skin defect in living skin, as opposed to the elliptical defect in cadaver skin, underscores the ability of skin fibroblasts to maintain approximately equal tension in both directions within the plane of the skin (see Figures 3.8 and 3.9). This suggests that the active tension exerted by skin cells occurs over and beyond the passive tension in the collagen network within skin, thus supporting the conclusion that mechanochemical transduction at the collagen fibril–cell interface must regulate tissue tension. A recent study suggests that skin fibroblasts can produce a tensional response to mechanical stresses applied to the surrounding ECM.

The concept that both active and passive stresses exist internally within ECM is important because any applied external stress affects the balance between active cellular-induced stresses and passive stresses transferred through the collagen fibrils of the matrix. At equilibrium any increase in external loading transmitted through the collagen fibrils found in the ECM must lead to an increase in the active stress produced by the attached cells. This increased active stress, or the resulting increase in cell strain, may cause conformational changes in membrane components or in cell–cell junctions that activate cellular pathways that modify gene expression and protein synthesis. Beyond this, changes in external stress and strain acting on cell membranes alters the amount of work done on the cell and this energy may be stored for use within the cell. Cells may balance the tension in skin across and along Langer's lines by transducing this energy into changes in the mass of the cell. Of particular interest is how mechanical loading may up-regulate expression of genes required for production of collagen and the resulting synthesis of new ECM, and also down-regulate genes that control production of collagenolytic enzymes that degrade the ECM.

Other examples of tissues exhibiting active and passive tension include cornea, cardiovascular tissue, and cartilage. When a corneal transplant is trephined from a cadaver eye, the corneal material to be transplanted shrinks from about 8.5 mm in diameter to about 8.0 mm as a result of unloading of the passive and active tensions that exist. In the cardiovascular system passive and active stresses along the longitudinal and transverse directions of the vessel wall provide in situ strains that are as high as 50% in the carotid artery. Thus the elastic arteries contract when they are removed from the cardiovascular system.

The basis of the active internal tension exerted by cells has been studied by growing a variety of cell types in collagen matrices. When isolated fibroblasts are grown in a reconstituted collagen matrix, they contract the substrate as a result of active cellular tension. In addition, these cells respond differently when the matrix is stressed as opposed to unstressed. Fibroblasts cultured in collagen matrices not only actively contract the matrix; they also remodel it. These examples underscore the importance of passive and active stresses in mechanobiology of a variety of extracellular matrices and suggest that mechanical loading is somehow intrinsically related to genetic expression of the resident cells.

The existence of tension in normal ECMs may be a means by which stored mechanical energy can be quickly converted into chemical energy as a result of environmental demand during development. However, this ability to convert mechanical energy into chemical energy is inhibited, possibly by down-regulation of mechanochemical transduction processes, during aging and in disease.

9.4.1 Internal Tensile Stresses in Skin

Of critical importance in attempting to understand mechanochemical transduction in a variety of tissues is to separate out active (cellular) and passive (noncellular) internal stresses. Results of a study of the biaxial mechanical properties of skin indicate that the prestress in cadaver skin is greatest for the arms, sternum, thigh, patella, and tibia, and it is lowest for the back; the prestress is about 10% of the maximum stress or about 1 MPa. This level of prestress varies from region to region and may be important in regulating ECM metabolism through mechanochemical transduction. Because of the passive biaxial prestress in skin, the actual stress–strain curve for skin in vivo is shifted to the left and condensed. Skin appears to operate in vivo in the upper linear region of the stress–strain curve. The maximum stress occurs in skin after a biaxial strain of about 20%, a value that decreases with increased age. The principal stresses occur at angle of +/− 10° with respect to Langer's lines and support the conclusion that the direction of Langer's lines is within an angle of about 10° to the direction of maximum prestress in skin.

The isotropy of skin is highest in areas that have the highest prestresses. However, the active contribution of fibroblast tension on the collagen network also needs to be taken into consideration. This would shift the curve even farther to the left suggesting that the in vivo extensibility of skin is limited by the presence of a biaxial strain on the collagen fibrils in the dermis. Because fibroblasts sit on the surface of collagen fibrils in dermis, external elastic tensile loading of skin would lead to direct stretching of fibroblast cell membrane–collagen fibril interfaces and fibroblast–fibroblast cell attachments. In skin, fibroblasts appear to form a dendritic pattern of intercellular connections that may be important in balancing active stress generation. These connections would be stressed by tensile loading. Active cellular tension also acts approximately along Langer's lines and is produced by fibroblast contraction of collagen fibrils in the extracellular matrix.

In the absence of external forces, internal tension acting on the collagen fibrils of the dermis causes tension to occur at keratinocyte–keratinocyte cell junctions. External forces applied to the skin surface at the air–epidermis interface increase the tension at keratinocyte–keratinocyte cell junctions as well as change the state of stress in the dermis. Transmission of external forces through the epidermis to the dermis occurs through a number of possible mechanisms including: (a) keratinocyte–keratinocyte interactions in the epidermis; (b) keratinocyte–ECM interactions that occur in basement membranes at the dermal–epidermal junction; (c) macromolecular–macromolecular interactions that occur in the dermis; (d) macromolecular–fibroblast interactions in the dermis; and (e) fibroblast–fibroblast interactions in the dermis. Therefore there are a number of possible mechanisms by which internal and external stresses are transmitted through skin. These stresses may also affect mechanochemical transduction. Conversely, active fibroblast tension applied to the surface of collagen fibrils could lead to increased tension in the dermis (see Figures 3.8 and 3.9).

9.4.2 Internal Tensile Stresses in Cartilage

Although the presence of internal stresses in skin was the first documented evidence that ECMs are prestretched during development and maturation, additional evidence suggests that tension exists in a variety of ECMs. Several reports suggest that the curling or distortion of intact articular cartilage after it is removed from the underlying bone is a result of the internal tensile stresses. Removal of the superficial layer of rib cartilage leads to

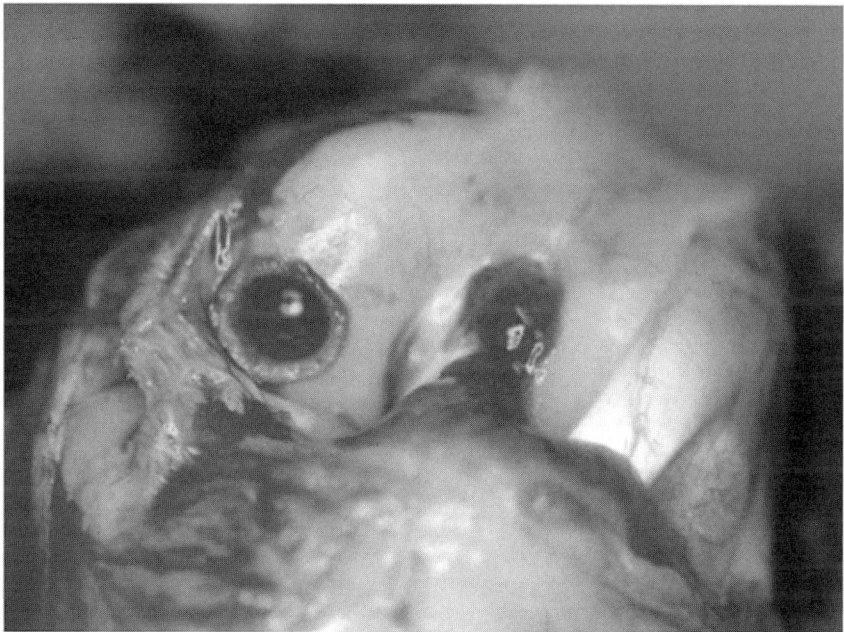

FIGURE 9.5. Photomicrograph showing contraction of the tissue around a circular defect made in the femoral condyle of a rabbit. The contraction is evidence that an active tension is present across the surface of the tissue.

straightening of the curled cartilage suggesting that it is the superficial zone that is under tension. In articular cartilage, removal of a circular defect results in differential retraction of the edges depending on the depth of the defect; the edges retract more in the superficial zone as compared to the deeper zones after a circular defect is removed with a punch (Figure 9.5). This observation underscores the presence of active and passive tensile forces in articular cartilage.

Normal human cartilage, with an intact superficial zone, curls similarly to rib cartilage when removed from the underlying bone. Osteoarthritic human cartilage lacks the superficial zone and does not curl when cartilage is removed from the underlying bone (see Figure 9.6), suggesting that active and passive tension is absent. This observation suggests that the changes in ECM structure and function associated with aging and disease processes may lead to changes in mechanochemical transduction mechanisms.

Interpretation of this curling phenomenon is complicated by observations of the swelling behavior of articular cartilage. Swelling is a potential mechanism for the existence of residual stresses and strains in cartilaginous tissues. Swelling-induced residual strain at physiological ionic strength is estimated to be tensile and to vary from 3 to 15%. The extent of curling

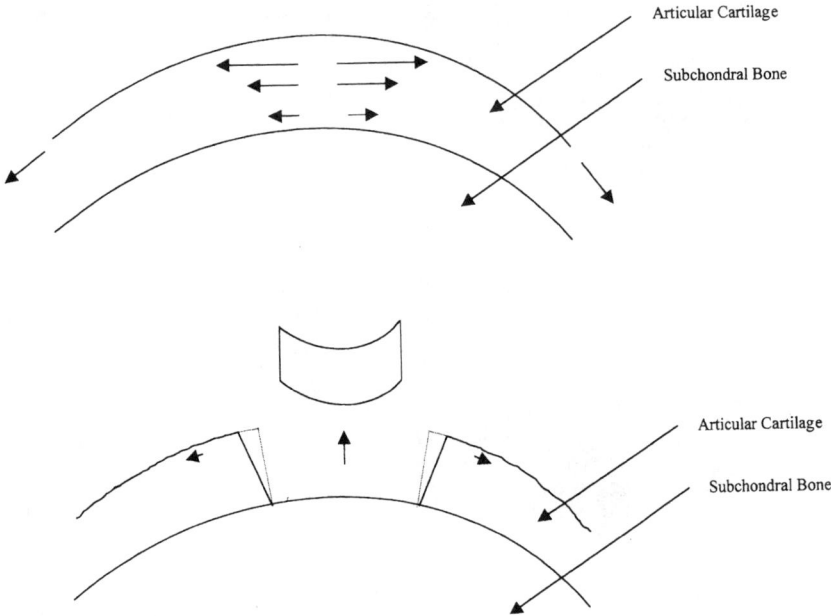

FIGURE 9.6. Diagram illustrating the pretension present in the superficial zone of articular cartilage. Normal articular cartilage shown at the top is loaded in tension across the surface like a drumhead that is pulled taut over a drum. When a piece of cartilage is cut from the surface, it curls as a result of release of this tension, as shown in the lower diagram. The presence of tension in the superficial zone makes articular cartilage behave like a drumhead, allowing compressive forces applied to the surface at specific points to be distributed across the surface to lower local stresses. The presence of tension on the chondrocytes in the superficial layer may be important to limit inflammation and support reparative processes by stimulating mechanochemical transduction.

decreases with increasing salt concentration from physiological levels to 2.0 m suggesting that breakage of electrostatic interactions decreases curling, leading to the conclusion that the surface zone greatly limits swelling of the entire cartilage. In contrast, another group reported that the swelling strain was tensile in the superficial and middle zones, and compressive in the deep zone of articular cartilage. They found a tensile modulus of 29 MPa for the middle zone suggesting that cartilage may be operating in the lower part of the linear region of the stress–strain curve. These results suggest that swelling may influence the tensile stresses existing in articular cartilage because the extent of swelling of cartilage is related to the amount of glycosaminoglycans. Thus collagen–proteoglycan (PG) interactions may not only affect swelling, but also affect the tensile internal stresses in articular cartilage.

9.4.3 Internal Stresses in Bone

Internal stresses in bone are more difficult to observe due to its high modulus and limited deformability. However, the existence of prestress in the osteon, the basic structural unit found in long bone has been point out. The existence of prestress in natural and synthetic materials impedes lesion formation at low stresses. In bone, prestress is hypothesized to protect circularly fibered lamellae from crack propagation in areas where collagen fibrils are transverse to the osteon axis. Preexisting shear strains in the circumferential–axial and normal axial strain directions reported for bone average 0.08 and 0.05, respectively. This corresponds to an average prestress of up to 0.11 GPa and occurs at the level of collagen bundles. Again, as in skin and cartilage, collagen fibrils and fibers are under internal tension in bone.

9.4.4 Internal Stresses in Vessel Wall

Fung reported many years ago that when a vessel wall is cut it does not open fully but can be characterized by an opening angle (see Fung, 1984). When a vessel is cut for excision from the body, the vessel contracts both radially and longitudinally as a result of internal stresses that are acting on the wall in vivo. These internal stresses arise from passive tension that is programmed into the ECM as well as active stresses that are generated by smooth muscle cell (SMC) contraction. The passive part of internal stress in arteries and veins has been termed residual stress and has been studied by several investigators (see Fung, 1984). These workers demonstrated that as the zero-stress state of a vessel wall is approached, making cuts in the vessel wall and analyzing the resulting opening angle could estimate the residual stress. Another group, using porcine and bovine aortas, reported a compressive stress of about 0.02 MPa for the intima (inner layer of vessel wall) and a tensile stress of about 0.2 MPa for the adventitia (outer wall layer) (see Fung, 1984). Compressive internal stresses existing on the intima due to residual stresses were believed to reduce the stress concentration that exists on the inner wall of a cylinder expanding under internal pressure.

Fung (1984, 1990, 1991) summarized that the opening angle, and therefore the residual stress, was highest in the aortic arch and then decreased to a minumum value just above the diaphragm; it increased again as the iliacs were approached. He also noted that when blood pressure was increased the vessel wall got thicker and the opening angle increased until it reached a peak, after which it decreased. He later concluded that increased stress in the vessel wall caused by hypertension causes remodeling that leads to increased wall thickness, which ends up lowering the circumferential wall stress (Fung, 1991). Wall thickening caused by hypertension is mainly due to hypertrophy of the lamella units of the media, especially in the subendothelial layer where stress is the highest. It is reported that the maximum circumferential and axial residual strains in the pulmonary arteries are 21.5% and 36.5%, respectively.

Active stresses exerted by smooth muscle cells appear to increase the internal stresses that exist in vessel wall. The effects of passive and active muscular contraction on the residual stress in the wall have been considered. Their results suggest that basal muscle tone, which exists under physiological conditions, reduces the strain gradient in the arterial wall and yields a near uniform stress distribution. Increased muscular tone that accompanies elevated blood pressure tends to restore the distribution of circumferential strain in the arterial wall, and to maintain the flow-induced wall shear stress at normal levels. It appears that the active stresses exerted by smooth muscle cells may balance the tension within the vessel wall in a similar manner to the way that active fibroblast tension balances the stress in the dermis.

9.4.5 Internal Stresses in Lung

Under normal physiological conditions the lungs are under a positive pressure that causes extension of lung tissue. In addition, other effects occur that increase the effective internal stress acting on lung tissue. Nonspecific airway or bronchial hyperresponsiveness (NSBH) is the exaggerated airway narrowing that occurs in response to pharmacological agents and nonspecific irritants including cold dry air and oxidant gases. These stimuli have the effect of causing airway smooth muscle cells to contract, thus increasing the internal forces acting on this tissue. Although airway remodeling may contribute to NSBH there is increasing evidence that in asthma, when the airways are narrowed, the bronchodilating response to cyclic and periodic stretch is impaired.

9.5 Influence of External Forces on the Behavior of Skin and Skin Cells

Beyond the effects of passive internal tension incorporated into the collagen fibers of the dermis and the active cellular contraction that is exerted also on the ECM, we must consider how external forces affect the behavior of skin. External forces acting on skin are transmitted through the epidermis to the dermis and underlying subcutaneous tissues (Figures 3.8 and 3.9).

The influence of mechanical forces on skin structure and remodeling has been studied extensively in an attempt to do the following: to understand wound healing and reduce hypertrophic scarring, to increase the skin surface area using balloon expanders, to study the reorganization and contraction of fibroblast-seeded collagen matrices in the absence and in the presence of external mechanical loading, and to understand the influence of gravity on the properties of skin.

Application of external forces to skin affects both cell–cell and cell–ECM interactions in vivo and in vitro. The use of compression dressings placed over areas of skin with hypertrophic scarring results in resorption of some of the underlying scar tissue, whereas the application of tension affects reorganization of fibroblast-seeded collagen matrices in vitro. Tissue expansion, performed using an expandable balloon placed in the dermis to apply tension to the epidermis and compression to the underlying dermal tissue, results in flattening of the basal cells of the epidermis and changes in keratinocyte morphology from columnar to cuboidal. Epidermal hyperplasia (increase in cell number) has been observed with increased cellularity in basal and suprabasal layers in balloon-expanded skin. The epidermis undergoes significant thickening whereas the dermis and subcutaneous tissue are significantly thinner after expansion. These studies underscore the response of both epidermal and dermal cells to the application of external loads. Both of these cell types appear to undergo synthesis, mitosis, and biosynthetic production of ECM as a result of tension, and catabolic tissue resorption is a result of compression.

The influence of the force of gravity on skin has also been proposed as playing a significant role in protecting against edema formation in the legs in the upright posture. Changes in skin elasticity and distensibility of the lower extremities in young individuals (associated with increased gravitational loading that occurs during standing) is higher compared to older individuals suggesting that a diurnal variation in skin elasticity occurs in younger individuals. However, in aged individuals, skin is less elastic and no diurnal variation in elasticity or distensibility is observed. Other reported effects of external loading on skin mechanics include gravity-induced facial edema and compression-dressing modification of skin mechanical properties of the leg.

Thus the balance between external and internal loading appears to play an important role in skin metabolism and healing. Tension applied to the epidermis appears to lead to cell division (hyperplasia) whereas compression of the dermis leads to scar tissue resorption. These observations appear to fit the general model that external tension promotes tissue augmentation and external compression leads to tissue destruction (catabolism).

9.5.1 Influence of External Forces on the Vessel Walls

All blood vessels experience internal tension under normal diastolic blood pressure. In addition to this static tension, chronic changes in wall stress that occur as a result of increases in blood pressure have been reported to lead to vascular remodeling, increased vascular wall diameter and thickness, in an attempt to restore normal values of vessel wall shear stress. Increased vascular wall thickness and diameter occur as a result of deposition of atherosclerotic materials in the vessel lumen and as a result of autoregulation in distributing arteries.

Distributing arteries regulate blood flow by a combination of myogenic, neuronal, and metabolic responses. Myogenic reactivity is the ability of vascular smooth muscle cells to contract in response to stretch or to an increase in transmural pressure and to relax in response to decreased wall pressure. Ischemia (decreased blood flow) has been shown to reduce vascular myogenic tone, which may lead to brain tissue damage during postischemic reperfusion, reportedly due to decreased vascular resistance to blood flow and impaired autoregulation of myogenic tone.

The increased vessel wall diameter and wall thickness seen in blood vessels from hypertensive individuals suggests that increased external tensile loading up-regulates tissue deposition. This observation suggests that mechanochemical transduction processes may play an important role in development of atherosclerosis as a result of elevated smooth muscle wall stresses.

9.5.2 Influence of External Forces on the Bone

Remodeling of bone and adaptation of skeletal tissues are examples of biological processes that are regulated by mechanical loading. Although metabolic intensity of physical activities is the most important component of cardiovascular fitness, the intensity of mechanical strain in terms of forces exerted by skeletal muscles and the reaction force at the interface with the ground are the most important criteria in developing and maintaining bone mass. Bone strain is thought also to act as a mechanostat, keeping the strain within certain thresholds that trigger formation of new bone at high strains and removal of old bone when the strains are decreased. Reduction of gravity leads to reduced muscle strength and bone loss; however, this is detected only in the lower limbs.

Although endogenous factors such as hormones are involved in remodeling and adaptation in skeletal tissues, exogenous factors including mechanical forces also play a role. Growth hormone (somatomedin) stimulates osteoblasts and has biomechanical implications in bone formation and resorption. During general healthy growth of the body, formation and resorption within tissue spaces cause bone surfaces to move during development of the growing bone. The degree of this progression determines outside bone diameter; however, external mechanical loading influences bone strength. The diameter and strength of a bone, in turn, influence the magnitude of strains a bone experiences when subjected to a given mechanical load.

The role of cells residing within skeletal tissues is to maintain a balance between the processes of resorption and formation. When these two processes are essentially equal, a conservation-mode of remodeling exists and an initial modeling threshold is established. If bone experiences a given mechanical load that produces strains in excess of the modeling threshold, increases in periosteal outside diameter result. When bone strength is increased by deposition of new bone, this same mechanical load now produces strains below the remodeling threshold (Frost, 1997). Where strains remain below this threshold, mechanically controlled remodeling is down-regulated. Under these conditions, postnatal osteoblastic activity also stops. Upon cessation of postnatal osteoblastic activity, a remodeling threshold lower than the initial threshold is established. Exceeding this new threshold, such as would occur with normal activity or with weight lifting, causes conservation-mode remodeling to turn on. If strains stay below the newly established threshold for extended periods, disuse-mode remodeling begins to remove bone that is directly adjacent to the marrow (see Frost, 1997 for background information). Additional considerations in mechanical loading effects on bone, which are discussed in detail by Frost (1998), involve issues relating to bone length and muscle forces especially as they relate to diseases. The effects of mechanical forces on bone structure and remodeling are affected by the magnitude and duration of loading as well as fluid shear stress created as a result of loading.

It is known that fatigue loading not only triggers bone resorption, but is also associated with stress fractures. A recent study quantified in vivo skeletal response to fatigue loading in adult rats by applying a load to the right forelimb. Osteogenic response to loading includes the formation of woven bone on the periosteal surface of the ulnar diaphysis and recovery time after loading correlates with increased bone cross-sectional area, moment of inertia, and mineral content. These results can be contrasted to tail-suspension studies involving rats. Tail-suspension is an accepted method for examining unloading of hindlimbs, because it removes the physiological resistance to hindlimb muscle contractions. Under unloading conditions, peak bone strains were sufficiently reduced to turn off mechanically controlled osteoblastic activity. In this instance, hormone supplements did not increase diaphyseal periosteal osteoblastic activity.

Studies involving fluid shear, hydrostatic compression, biaxial and uniaxial stretch, or a combination of two or more of these factors indicate that fluid shear is a major factor affecting bone cell metabolism and cells subjected to mechanical stress reshape and align themselves with their long axis perpendicular to the axis of force. Cells also exhibited remodeling of the actin cytoskeleton and increases in PKC levels, processes thought to be involved in the early phase of mechanochemical transduction.

At the physiological level, gravitational loading on skeletal tissues is known to be important. After space missions of varying periods, histomorphometric studies of bones from rats provided insight into the time course of cancellous bone cellular events: these events include a transient increase in resorption and sustained decrease in bone formation (Vico et al., 1998). Bone loss in rats occurred first in weight-bearing bones, and the time required to recover lost bone during postflight was greater than the mission length.

9.5.3 Influence of Internal and External Forces on Lung

The lung is subject to both external and internal loading due to breathing and pulmonary blood flow. The surface tension influences the magnitude of the forces acting on the air–alveolar surface interface. Pulmonary blood vessels are subject to shear stress due to fluid drag on the wall, blood pressure-induced radial expansion of the wall, and internal forces present in the ECM. Epithelial cells that line the airways and alveoli are exposed to compressive and tensile forces during the respiratory cycle, which affect both the function and phenotype of cells in the lung. Increased parenchymal tissue distension due to an increase in transpulmonary pressure results in increased surfactant secretion by type II epithelial cells and a reduction in airway smooth muscle shortening. Osmotic forces, which contribute to generating hydrostatic pressure, affect cells that transport ions and water such as pulmonary epithelium and endothelial cells. Mechanical forces introduced by a ventilator, induce injury as a result of repeated collapse and reopening of airways, leading to inflammation-induced injury.

Mechanical forces that occur during breathing have been reported to be capable of causing failure of the ECM at loci of stress and to contribute to the progression of emphysema. Chronic asthma is associated with a marked increase in the airway wall thickness, and an increased muscle cell mass. Hyperplasia of airway smooth muscle cells in vivo is probably accompanied by a change in contractile capacity and appears to be a consequence of the increased forces required for breathing through narrowed airways.

9.6 Summary of Effects of Internal and External Mechanical Forces

The presence of tension in ECMs from a variety of tissues suggests that humans have evolved a means of storing mechanical potential energy in their tissues possibly to convert mechanical energy into chemical energy. In macromolecules, increased mechanical energy causes conformational changes in matrix macromoleular components such as collagen. These conformational changes associated with mechanical energy storage may also play a role in activation of intracellular signal transduction.

To understand how mechanical energy is transduced into chemical energy within the cell requires that we understand the specific structure and composition of the ECM found in each tissue. Unfortunately, the structure and composition of each ECM is quite complex and therefore it is only possible to provide general relationships between structure and function to evaluate the effects of forces at the tissue level. In general, tensile forces appear to stimulate tissue deposition whereas compressive forces appear to result in tissue destruction. However, the application of tension is not always desirable inasmuch as this could lead to pathological tissue deposition. Some examples of pathological tissue deposition include vascular wall thickening and ectopic calcification of vascular tissue. Below we attempt to generalize the literature on ECM composition and structure in an effort to understand more completely how mechanical loading influences tissue remodeling responses to external forces.

9.7 The Influence of External Forces on Macromolecular Components of ECM

External forces applied to tissues lead to stretching of collagen, elastic fibers, and smooth muscle in the associated ECMs as well as proteoglycan deformation and fluid flow from within the matrix. The application of these forces ultimately leads to matrix remodeling and energy storage. The question arises as to how external mechanical events trigger cellular synthesis.

Up to a strain of about 2% in tendon, collagen molecular stretching predominates, whereas in comparison, beyond a strain of 2% increases in the collagen D-period reflect molecular slippage (Mosler et al., 1985; Sasakai and Odajima, 1996, 1996a). Mosler et al. (1985) reported that about 10% of the macroscopic strain in tendon reflects molecular deformation; and the remainder of the strain reflects molecular and fibrillar slippage. Models of molecular and fibrillar stretching suggest that deformation occurs in flexible domains of collagen that alternate with rigid domains (see Figures 2.23 and 3.25). The flexible domains coincide with the positive staining bands that characterize the collagen fibril. Two of the 12 positive staining bands in the collagen fibril, b2 and d, have been identified as the eukaryotic cell integrin binding sites. Recently it has been suggested that the application of external forces to ECMs results in stretching of the flexible domains leading to conformational changes and energy storage in collagen fibrils; these events could lead to stretching of integrin attachment sites on collagen resulting in conformational changes in integrin and reorganization of cell cytoskeletal elements (see Figure 9.1). Stretch-induced changes in cell membrane structure, gap junctions that occur between cells, and ion channels within the cell membrane are mechanisms by which external forces are transduced into activation of secondary messenger pathways within neighboring cells.

Although there is some information on the effects of mechanical forces on the structure of collagen and the possible mechanism by which this is transduced at the cell membrane, very little information is available on the possible role of other macromolecules. Ultimately, it will be important to understand how each component of the ECM acts to facilitate mechano-chemical transduction processes.

9.8 Effects of Physical Forces on Cell–ECM Interactions

The transduction of forces by cells leads to regulation of tissue synthesis and catabolism. Forces are transmitted to and from cells through the extra-cellular matrix as changes in mechanical forces and cell shape act as bio-logical regulators (Ingber, 1991). Ingber (1991) hypothesized that cells use a tension-dependent form of architecture, termed tensegrity, to organize and stabilize their cytoskeleton. Mechanical interactions between cells and their extracellular matrix appear to play a critical role in cell regulation by switching cells between different gene products (Chiquet et al., 1996; Chiquet, 1999). Understanding the relationship between external loading and changes in genetic expression of cells in ECMs is very important in elu-cidating the physical factors involved in the pathogenesis of disease.

Several mechanisms exist by which external loads affect gene expression of resident cells in ECM; these include cell–ECM interactions and cell–cell interactions. Integrin adhesion receptors that connect extracellular matrix components and cytoskeletal elements have been implicated in mediating signal transduction in both directions through the cell membrane (Liu et al., 2000). The mediation occurs by binding of ligands to integrins, which transmits signals into the cell and results in cytoskeletal reorganization, gene expression, and cellular differentiation (outside-in signaling). Media-tion also occurs when signals within the cell propagate through integrins and regulate integrin–ligand binding affinity and cell adhesion (inside-out signaling; Hynes, 1992; Giancotti and Ruoslahti, 1999). Integrin adhesion receptors are heterodimers of two different subunits termed α and β (Liu et al., 2000). They contain a large extracellular matrix domain responsible for binding to substrates, a single transmembrane domain, and a cytoplas-mic domain that in most cases consists of 20 to 70 amino acid residues (see Figures 9.1 and 9.3). Events that occur at the cell membrane are also affected by the systemic release of growth factors and hormones.

Eukaryotic cells directly attach to extracellular matrix collagen fibers via integrin subunits $\alpha1\beta1$ and $\alpha2\beta1$ (Xu et al., 2000) through a six-residue sequence (glycine-phenylalanine-hydroxyproline-glycine-glutamic acid-arginine; Knight et al., 2000) that is present in the b2 and d bands found in the collagen positive staining pattern (Figure 9.3). Integrins are transmem-brane molecules that associate via their cytoplasmic domains with a number of cytoplasmic proteins including vinculin, paxillin, tensin, and others (Yamada and Geiger, 1997; see Figure 9.2). These molecules are all involved in the dynamic association with actin filaments. In cultured cells, integrin-based molecular complexes form small (0.5 to 1μm) or point contacts known as focal adhesion complexes (Figure 9.3) and elongated streaklike structures (3 to 10 µm long). These elongated structures are associated with stress fibers containing actin and myosin, also known as focal contacts or focal adhesions (Figure 9.4). Integrin-containing focal complexes behave as mechanosensors exhibiting directional assembly in response to local force (Ingbar, 1991). It has been reported that collagen-binding integrins are involved in up- or down-regulating collagen $\alpha1(I)$ and collagenase (MMP1) mRNA depending on whether fibroblasts are unloaded or loaded.

During cell adhesion to collagen in the ECM, the initial binding of integrins to their ECM ligands leads to their activation and clustering, and to assembly of focal adhesion complexes that serve as "assembly lines" for signaling pathways. The signaling pathways include protein kinases, adaptor proteins, guanidine exchange factors, and small GTPases that are recruited to these sites and may directly trigger mitogen-activated protein kinase pathways or growth factor receptors.

The contacts between two adjoining cell membranes are stabilized by specific cell adhesion molecules (CAMs), which include the Ca^{+2}-dependent cadherins. These molecules appear to lead the way for cell–cell communications and are involved in mechanochemical transduction via cell–cell interactions. In some cell types, cadherins are concentrated within adherens junctions that are stretch-sensitive and their extracellular domains interact with cadherins on adjacent cells whereas their cytoplasmic domains provide attachment to the actin cytoskeleton via catenins and other cytoskeletal proteins. The Rho family is required for the establishment and maintenance of cadherin-based adherens junctions. The type of cadherin expressed in a cell can affect the specificity and the physiological properties of cell–cell interactions.

Major force-bearing elements within the cell cytoskeleton include actin microfilaments, microtubules, and intermediate filaments. Both microtubules and microfilaments have very high Young's moduli (1 GPa) with microtubules exhibiting a much greater bending stiffness compared to actin microfilaments. Actin microfilaments may provide the elastic response of the cell at strains less than 20%, whereas intermediate filaments may provide mechanical strength at higher strains, and are responsible for cell strain-hardening. Internal forces exerted by cytoskeletal components appear to balance external forces applied by collagen fibers found in the extracellular matrix. The net extracellular and intracellular forces may not only cause rearrangement of the ECM but also lead to changes in cell metabolic behavior.

9.8.1 Influence of Mechanical Forces on Skin Cells

In skin, both epidermal (keratinocytes) and dermal cells (fibroblasts) respond to mechanical loading by undergoing structural and biochemical changes. Wang and Stamenovic (2000) have shown that fibroblasts grown on a silicone membrane coated with fibronectin exhibit reorganization of the actin cytoskeleton to an angle of about 60° upon stretching. Upon fibronectin engagement, it has been shown that the short cytoplasmid domain of the $\beta 1$ subunit binds to proteins that in turn associate with and reorganize actin filaments to form focal adhesions (see Figure 9.4). Upon further activation of secondary messenger pathways in cells of the ECM through Rho GTPases, changes in the actin cytoskeleton lead to integrin clustering (see Figure 9.3), which facilitates the polymerization and assembly of ECM on the cell surface and enables stable substratum attachment.

Fibroblast–fibroblast interactions in the dermis may also contribute to generation of the internal tension in skin (see Figure 3.9). Ragsdale et al. (1997) showed that fibroblasts were stretched in tension after spontaneous contraction of neighboring cells. They postulated that mechanical transmission of tensile forces occurs through adherens junctions in fibroblasts possibly causing changes in secondary messenger activation through changes in cyclic AMP. Mechanical perturbation leads to a transient increase in intracellular calcium that propagates from cell to cell. Mechanical forces applied to fibroblast adherens junctions activate N-cadherin associated stretch-sensitive calcium permeable channels increasing actin polymerization.

Fibroblast–collagen interactions have been studied by culturing fibroblasts in a collagen matrix, scaffold, or lattice. Forces exerted by cells on the collagen matrix cause collagen matrix contraction in the absence of external mechanical loading. If the collagen resists deformation, forces exerted cause isometric tension; the state of cellular mechanical loading appears to regulate how fibroblasts respond. The ability of cells to contract a collagen matrix has been shown to depend on the actin cytoskeleton because agents that disrupt the cytoskeleton prevent matrix contraction.

The transcriptional profile of induced genes of fibroblasts grown in collagen lattices suggests that mechanical stimulation leads to a "synthetic " fibroblast phenotype characterized by induction of connective tissue synthesis while simultaneously inhibiting matrix degradation. Mechanically loaded cells grown on laminin or elastin or other substrates expressed higher levels of procollagen mRNA and incorporated more labeled proline into protein than do unstressed cells. Fibroblasts grown in a three-dimensional collagen lattice have been shown to align themselves with the direction of principal strain (Eastwood et al., 1998) and adopt a synthetic fibroblast phenotype characterized by induction of connective tissue synthesis and inhibition of matrix degradation. Under these conditions they show de novo transcription of the COL1A1 gene and pro-$\alpha 1$(I) collagen mRNA.

Fibroblasts grown in collagen lattices can generate a force of approximately $10 \times 10^{-10} N$ as a result of a change in cell shape and attachment; they maintain a tensional homeostasis of approximately 40–$60 \times 10^{-5} N$ per million cells. Cell contraction of 3-D collagen matrices is opposite to the direction of applied loads and increased external loading is followed immediately by a reduction in cell-mediated contraction. Stress relaxation in a collagen matrix results in activation of a Ca^{2+}-dependent adenyl cyclase signaling pathway that leads to an increase in cyclic adenosine monophosphate (cAMP) and free arachidonic acid. Fibroblasts in cell culture that are not aligned with the force direction show a severalfold increase in matrix metalloproteinase activity (MMP1, MMP2, and MMP3), suggesting cells unable to align with the direction of the applied load remodel their matrix more rapidly than oriented cells.

In collagen matrices that are restrained from contraction, cells develop isometric tension, whereas matrices floating in a cell culture medium remain mechanically unloaded. Fibroblasts that are mechanically loaded develop prominent actin stress fibers and organize a fibronectin matrix. Treatment of mechanically loaded cells with transforming growth factor beta (TGF-β) results in cellular differentiation into myofibroblasts, expression of α-smooth muscle actin and formation of stress fibers. Fibroblasts grown in collagen matrices in the absence of mechanical loading down-regulate the extracellular-signal-regulated kinases pathways leading to quiescence. Some of the cells undergo apoptosis as a result of loss of mechanical loading. Primary human fibroblasts are reported to display a marked reduction of apoptosis in mechanically relaxed collagen matrices in the presence of adhesion-blocking antibodies to integrins α1β1 and α2β1. Cells that lack α2 integrin, or that undergo depolymerization of F-actin, display no apoptosis in mechanically relaxed matrices.

Tenacin-C and type XII collagen are associated with collagen fibrils in ECM. In matrices bearing high mechanical loads, expression of mRNA and synthesis of tenacin-C and collagen XII are up-regulated. Stretch response promoter regions have been identified in tenascin-C and collagen type XII genes. These regions have the same motif as has been implicated in the response of endothelial cells to platelet-derived growth factor (PDGF-B) under the influence of shear stress.

There are numerous studies in the literature suggesting that keratinocytes as well as fibroblasts are capable of transducing mechanical forces in skin (see Silver et al., 2003). Application of a cyclic compressive force has been shown to increase keratin synthesis and decrease cell division of epidermal cells leading to the conclusion that cyclic loading promotes differentiation of epidermal cells. Strain-induced changes in keratinocyte function are modeled to be modulated through increased expression of mRNA of interleukin-1 (Il-1) (see Figures 9.1 and 9.2). Expression of IL-1 by keratinocytes as a result of mechanical strain may activate vascular endothelium and promote local inflammation. Cells from mechanically active environments appear to be able to couple signals from forces applied through β-integrins to up-regulate the production of cytoprotective cytoskeletal proteins including filamin A.

Isolated keratinocytes subjected to cyclic strain exhibit a significant increase in cell proliferation, DNA synthesis, and protein synthesis compared to stationary or constantly loaded cells, which appear to involve changes in cyclic AMP. Takei et al. (1997) reported a strain-induced reduction in the levels of cyclic adenosine monophosphate, protein kinase A (PKA), and prostaglandin E2 (PGE2) as compared to stationary controls. Takei et al. (1997) also studied the effects of cyclic strain on protein kinase C (PKC) activation and translocation in cultured keratinocytes.

In another study, mechanical stretching of skin was shown to alter cell shape and trigger biochemical signaling in keratinocytes through MAPKs. Cell stretch, which activates ERK1/2, was reversed by treatment with monoclonal antibodies to β1 integrins (Kippenberger et al., 2000). ERK 1/2 activation was decreased in aged skin suggesting that mechanochemical transduction may be affected by age.

Growth factors such as platelet-derived growth factor and angiotension II have been implicated in strain-induced cellular growth in vascular tissues. Two growth factors in skin, epidermal growth factor (EGF) and TGF-β, appear to act via cytoskeletal molecules or protein kinase families and may play a role in mechanical strain-induced changes in skin cells. The EGF receptor has been shown to bind to actin-binding proteins and cause reorganization of the actin microfilament system. Addition of EGF to its receptor leads to phosphorylation of tyrosine residues on the EGF receptor (EGFR), and to colocalization of F-actin and the EGF receptor. Cytoskeletal-linked EGFR has been postulated to induce cell proliferation through the microfilament system. Membrane ruffling occurs at sites that are rich in actin, EGFR, phospholipase C, and tyrosine phosphorylated proteins, all of which are essential for strain-induced signal transduction. EGF-induced effects on keratinocytes involve high-affinity receptors and the tyrosine kinase pathway.

The relationship between external loading and the effects of growth factors appears to occur through membrane bound growth factor receptors (see Figure 9.1). This synergy between these two factors appears important in promoting tissue growth as well as in adapting to demanding growth requirements during maturation. This occurs in a variety of tissues over the same time intervals so that tissue proportions are maintained.

9.8.2 *Influence of Mechanical Forces on Lung Cells*

The pulmonary alveolar epithelium is comprised of two morphologically distinct cells, type I and type II cells. Type I cells are extremely large, squamous cells that make up 95% of the alveolar surface. Type II cells are smaller cuboidal cells that secrete and recycle surfactant and cover the remaining 5% of the alveolar surface. Mechanical distention of fetal lung tissue has been shown to stimulate expression of the type I cell phenotype and inhibit expression of the type II phenotype. Lumenal mechanical stimulation of airway epithelial cells produces a propagating wave of elevated intracellular Ca^{+2} coordinating an integrated epithelial stress response that is initiated by the release of ATP and UTP across both apical and basolateral membranes. These nucleotides, after release into the extracellular compartment, interact with purinoceptors at both membranes triggering Ca^{+2} mobilization to coordinate local and distal airway defenses to mechanical stimuli.

Mechanical stretch has also been shown to promote differentiation of type II cells and to affect lung fibroblasts. Normal stresses in the range of those produced in collapsed airways exhibit gene up-regulation of TGF-β, endothelin, and early growth response-1; these genes are also associated with inflammation and remodeling of the lung in asthma. Mechanical stress communicated between stressed lung epithelium and unstressed lung fibroblasts elicits a remodeling response mimicking airway remodeling that is characterized by production of fibronectin, collagen types III and V, and matrix metalloproteinase type 9. Both lung epithelial cells and fibroblasts exhibit strain-induced increases in DNA synthesis either alone or in combination with each other.

Mechanical strain of rat fetal lung cells is reported to increase gene expression and protein synthesis of the platelet-derived growth factor (PDGF) β-receptor. Blockage of the receptor with a protein tyrosine kinase (PTK) inhibitor or antisense oligionucleotides abolished mechanical strain-induced cell proliferation. Mechanical stretch of human alveolar epithelial cells also is reported to result in secretion of interleukins, whereas continuously stretched fetal lung cells demonstrate increased expression of proinflammatory molecules including cytokines and chemokines. Increasing the magnitude of the cyclic biaxial strain significantly injures primary cultures of alveolar epithelial cells. Mechanical strain of human pulmonary epithelial cell line rapidly activates ERK 1/2. PDGF-BB production from fetal lung cells is stretch induced and stimulates MAPK activation, suggesting that stretch-induced MAPK activation is indirectly mediated via PDGF-BB production.

Boitano et al. (1994) reported that Ca^{+2}-conducting channels in airway epithelial cells were opened when mechanical stimulation was applied, causing a rapid depolarization of the stimulated cell. A mechanical strain-induced increase in cAMP content has been reported in fetal rabbit type II epithelial cells after continuous high amplitude strain for 24 h. An intermittent strain of 18% resulted in an increase in cAMP production in cultured rat fetal lung cells. Strain-induced fetal lung cell growth occurs via activation of PTK and PLC.

Airway smooth muscle cells isolated from canine tracheae and bronchi subjected to cyclic strain exhibit increased cell number and DNA synthesis in cell culture. The content of total cellular protein, especially contractile proteins including myosin, myosin light chain kinase, and desmin, was increased compared to cells cultured under static conditions.

Stretch-induced mechanical loading also appears to effect secondary messenger activation in airway smooth muscle cells. A 20% single static stretch of rat pulmonary smooth muscle cells increases both Ca^{+2} influx and efflux. Mechanical strain rapidly increases tyrosine phosphorylation of pp125[FAK] and paxillin in airway smooth muscle cells cultured in type I collagen matrices. Tyrosine kinase inhibitors hindered strain-induced reorientation and elongation of airway smooth muscle cells.

Agents that change contractile activity of smooth muscle modulate cytoskeletal stiffness of human airway smooth muscle cells. It was demonstrated that the orientation of airway smooth muscle cells and actin stress fibers within these cells was reorganized after mechanical strain, and was coupled with an increase in number and length of focal adhesions. This suggests that cytoskeletal components can be reorganized to increase the efficiency of transmitting signals from the ECM into the cell interior. Cyclic strain causes a rapid increase in tyrosine phosphorylation pp125FAK, paxillin, and focal adhesion-related proteins. A single 20% static stretch of rat pulmonary arterial smooth muscle cells increases both Ca^{+2} influx and efflux.

9.8.3 Influence of Mechanical Forces on Vessel Wall Cells

Vascular endothelial cells (ECs) provide an interface with the flowing blood, assist in maintaining blood fluidity, and act as a primary sensor of wall shear stress for the transduction of mechanical forces that occur at the blood–lumenal surface interface (see Davies, 1995 for a review). Shear stress appears to be associated with regulation of the expression of many genes and their products (including calcium in endothelial cells) by acting through several signaling pathways. These include the mitogen-activated protein kinases (MAPKs), ERKs and c-Jun terminal kinase (JNK), G-protein, ion channels, and the TGF-β receptor. Dynamic formation of new connections between integrins and their specific extracellular matrix ligands is critical in relaying the signals induced by shear stress to intracellular pathways. Receptor tyrosine kinases and integrins serve as mechanosensors to transduce mechanical stimuli (Chen et al., 1999). Specifically, Jalali et al. (1998) have reported that fluid shear stress activates endothelial cell p60src, which in turn activates RAS-MAPK signaling pathways. They concluded that p60src plays a critical role in the shear stress activation of MAPK pathways and induction of Activating Protein-1 (AP-1) mediated transcription in ECs.

The manner by which shear stress-induced cellular changes occur in endothelial cells involves cell membrane and cytoskeletal molecules that lead to a shape change. The cytoskeleton contains actin filaments, intermediate filaments, and microtubules, all of which are restructured upon exposure to external force. Under stress conditions, actin filaments coalesce to form stress fibers that anchor at the focal contacts, which are adhesion sites at the cell substrate interface.

Flow-dependent changes in vessel diameter contribute to the optimization of circulatory function and are mediated via shear stress-induced release of NO, vasodilator prostanoids, and a putative endothelium-derived hyperpolarization factor or EDHF (Griffith, 2002). There is growing evidence that NO/prostanoid-independent relaxations involve direct heterocellular signaling between endothelium and smooth muscle cells via gap junctions.

Shear stress induces release of bFGF from endothelium and is dependent on specific cell–matrix interactions via $\alpha v\beta 3$ integrins. The release may be the first step in adaptive remodeling in the vessel wall induced by shear stress. Shear stress also induces a rapid induction as well as nuclear translocation of the vascular endothelial growth factor (VEGF) receptor 2, and promotes binding of VEGF receptor 2 and adherence junction molecules (VE-cadherin and beta-catenin) to the endothelial cell cytoskeleton.

9.8.4 Mechanochemical Transduction by Vascular Smooth Muscle Cells (VSMCs)

Smooth muscle cells grown under conditions of dynamic stress exhibit increased orientation and contractile apparatus including microfilaments, dense bodies, and basement membranes. The growth-promoting effect of mechanical stress on vascular smooth muscle cells has been shown to involve mechanical stretch-activated extracellular signal-regulatcd kinases (ERK1/2) requiring intact actin filaments and may also involve stretch-activated ion channels. Specifically, steady intralumenal pressure in blood vessels activates the extracellular signal-regulated kinases, the ERK1/2 pathway. Dynamic cycling of smooth muscle cells stimulates the synthesis of collagen and total protein as well as induces an elevated secretion of TGF-β. Applied tensile forces are reported to increase microtubule mass whereas compressive forces decrease microtubule mass in smooth muscle cells.

Thrombin and angiotensin II act as mediators of vascular remodeling through their effects on stimulating bundling of actin filaments to form stress fibers; this effect appears to require c-Src kinase activity. Cyclic stresses increase fibronectin, collagen, and metalloproteinase-2 synthesis, as well as expression of TGF-β production. Cyclic strain significantly reduces DNA synthesis in smooth muscle cells suggesting that cyclic stretching may keep the proliferation of smooth muscle cells at a low level. Dynamic cycling increases smooth muscle myosin expression and decreases non-smooth muscle myosin expression; it also up-regulates the sodium pump by expression of both the $\alpha 1$ and $\alpha 2$ subunits of N^+, K^+-ATPase. Cyclic stress is believed to increase proliferation and orientation of smooth muscle cells through changes in tyrosine phosphorylation of adhesion-related proteins.

The mitogenic response in vascular smooth muscle cells caused by cyclic mechanical stimulation is due to the production of platelet-derived growth factor; this effect is blocked by fibronectin and the integrin binding peptide GRGDTP. Integrin specific antibodies, anti-β1 and anti-α2, inhibit smooth muscle cell adhesion to collagen and compaction of a tissue-engineered blood vessel. Proliferation and αVβ3-dependent migration of vascular smooth muscle cells as well as gene expression of a number of proteins are suppressed on polymerized type I collagen.

Recently, it has been demonstrated that mechanical stress rapidly induced phosphorylation of PDGF receptor, activation of integrin receptor, stretch-activation of cation channels, and production of G proteins. Once mechanical stress was sensed, protein kinase C and MAPKs were activated, leading to increased c-fos and c-Jun gene expression and enhanced transcription factor AP-1 DNA binding activity. The application of physical forces also rapidly resulted in expression of MAPK phosphatase-1 (MKP-1), which inactivates MAPKs.

Cyclic forces are required for activating extracellular signal-regulated MAPK in vascular smooth muscle cells; the frequency–response curve increases approximately twofold between 0.5 and 2.0 Hz and is reversed by blocking with antibodies to α2 and β1 integrin subunits. Mechanical activation of the ERK1/2–MAPK system requires a cyclic force of about 2.5 pN per cell. Elevated blood pressure imposes increased mechanical stress on the vascular wall, and mechanical strain is a mitogenic stimulus for vascular smooth muscle cells. Chronic cyclic mechanical strain increases fibronectin, collagen, and MMP-2 activity in human VSM cells in part through mechanical strain-induced TGF-β(1) production. Rhythmical stretch has also been reported to decrease DNA synthesis and presumably proliferation of VSM. Cyclic mechanical strain (1 Hz) causes a mitogenic reponse in neonatal rat vascular muscle smooth cells due to production and secretion of PDGF. Fibronectin and the integrin-binding peptide both block the mitogenic response to mechanical strain in cells grown on immobilized collagen. Cyclic loading potentiates proliferation and an increased expression of h-caldeson, a marker of differentiation of vascular smooth muscle cells.

Proliferation of αvβ3 integrin-dependent migration of VSMCs on polymerized collagen is suppressed and has been postulated to regulate gene expression and pericellular accumulation of ECM. Cyclic mechanical loading of VSMs increases collagen synthesis and appears to be promoted through release of angiotensin II and TGF-β (Li et al., 1998). TGF-β1 stimulates L-arginine transport and metabolism to growth stimulatory polyamines in VSMCs. Cyclic stretch stimulates TGF-β1 mRNA expression suggesting that autocrine/paracrine production of TGF-β may play a critical role in the progression of vascular remodeling associated with high blood pressure. Mechanical stretch activates ERKs in VSMCs, with a peak response observed after 20 minutes, followed by a significant decrease in

DNA synthesis. Mechanochemical transduction in VSMCs is dependent on intact actin filaments. Stretch responses involve Rho activation and Rho/p160ROCK and mediate stretch-induced ERK activation and cellular vascular hyperplasia in VSMCs.

Elastin peptides derived from the degradation of vessel wall are present in serum and are associated with vessel wall aging. Elastin peptides activate phospholipase C, inducing the production of inositol triphosphate (IP3) leading to an increase in intracellular free calcium and of diacylglycerol. This leads to phosphorylation of members of the MAPK family. A progressive age-dependent uncoupling of the elastin receptor occurs impairing transduction and altering calcium signaling, resulting in loss in calcium homeostasis in VSMCs.

9.8.5 Influence of Soluble Mediators and Mechanical Forces on Articular Cartilage Cells

Molecular signaling via soluble mediators as well as cyclic mechanical loading has been shown to be crucial to cartilage homeostasis. Therefore, it is important to consider the possibility of interactions between these mediators and mechanical forces that influence the metabolism of articular cartilage. A number of vitamins, hormones, growth/differentiation factors, and cytokines have been implicated in chondrocyte differentiation and cartilage metabolism. Retinoic acid (RA) is a well-characterized soluble mediator of cartilage development. RA treatment of differentiated chondrocytes in culture results in the reversion to a fibroblastic phenotype where types I and III collagen are synthesized. However, experimental evidence suggests that at later stages of differentiation, RA promotes hypertrophy and type X collagen production as well as expression of mineralization-related genes.

Vitamin D may regulate calcification of cartilage by regulating chondrocyte synthesis of proteoglycans. $1,25(OH)_2D_3$, an active metabolite of vitamin D has been shown to produce a concentration-dependent reduction in an immortalized rat chondrocyte cell line. Both vitamin D and $1,25(OH)_2D_3$ are considered secosteroids with vitamin D receptors, which includes cytosolic/nuclear proteins that bind specifically to the secosteroid as it transits the plasma membrane.

Ascorbic acid stimulates proliferation of rabbit chondrocytes at high cell density, chick embryo and bovine articular chondrocytes, and rabbit cartilage explants. Although evidence suggests that ascorbic acid treatment does not directly stimulate transcription of ECM products, cultured chondrocytes undergo changes in gene expression characteristic of hypertrophy.

Implicated in induction and differentiation of cartilaginous tissues during development are a number of growth factors and cytokines, including the TGF-β superfamily. Those implicated include bone morphogenetic protein subfamilies, fibroblast growth factor family, insulinlike growth factor family, growth hormone, chondromodulin I and II, interleukin-1b, indian hedge-hog, and parathyroid hormone-related peptide. In addition to biochemical influences on cartilage metabolism, mechanical loading is necessary for the homeostasis of articular cartilage and may involve cell deformation and associated distortion of intracellular organelles. It has been reported that under conditions of constant cell deformation due to static compression, nuclear deformation was reduced significantly over a 25 min period particularly in the axis perpendicular to the applied compression. These authors estimated a Young's modulus for the chondrocyte in compression to be about 3 kPa.

In many cell types mechanical stimulation induces increased Ca^{+2} concentration in the cytosol that propagates from cell to cell as an intercellular wave. Cell-to-cell communication underlies tissue coordination of metabolism and sensitivity through gap junctions and could coordinate metabolism and sensitivity to external mechanical stimuli. Intracellular Ca^{+2} levels may provide a mechanism for coordinating tissue responses in cartilage. Fluid flow-induced signals activate the MAP kinase signaling pathway in articular cartilage, leading to down-regulation of the aggrecan gene. Both CD 44 and integrins have the capability to influence signal transduction as well as modulate interactions with the cell cytoskeleton in chondrocytes. Increased levels of integrins $\alpha1\beta1$ and $\alpha2\beta2$ could be an indication of an attempt by chondrocytes to repair damaged ECM if integrin-mediated signaling is disrupted as a result of the disease process.

Chondrocytes from osteoarthritic cartilage do not show pressure-induced increases in aggrecan mRNA levels or decreases in MMP-3 mRNA levels that are observed in the presence of normal chondrocytes. The increase in aggrecan and decrease in MMP-3 mRNA in normal chondrocytes is blocked by RGD-containing peptides, suggesting that the response is integrin-mediated and that for unknown reasons, OA chondrocytes are deficient in this response. There is some evidence that gene expression may be altered in OA cartilage.

Hydrostatic pressure as well as other types of mechanical loading appear to stimulate extracellular matrix synthesis by chondrocytes. Pressure-induced strains at a frequency of 0.33 Hz result in chondrocyte membrane hyperpolarization by activation of Ca^{+2}-dependent K^+ small conductance potassium activated Ca^{+2} channels. Mechanochemical transduction involves $\alpha5\beta1$ integrin, stretch-activated ion channels, actin cytoskeletal elements, and protein kinase with a previously unrecognized role for an integrin-β-catenin involvement in the signaling pathway. Low magnitude (5 kPa) cyclic tensile force (CTF), whether high or low frequency, does not result in gene expression of cartilage degradation factors suggesting that this loading causes only small changes in cartilage ECM. In contrast, high magnitude (15 kPa) and frequency CTF has been shown to promote expression of IL-1 and MMP-9, resulting in cartilage degradation. Oscillating fluid flow causes higher levels of chondrocyte cytosolic Ca^{+2} that increases with peak flow rate and decreases at increasing frequencies. Increased static compression leads to down-regulation of chondrocyte aggrecan and type II collagen gene expression.

Stretch-induced matrix deformation regulates chondrocyte proliferation and differentiation by two different signal transduction pathways. Stretch-activated channels are involved in transducing the proliferative signals and calcium channels are involved in transducing the signals for both proliferation and differentiation. Nitric oxide has been implicated in inhibition of chondrocyte proliferation after cytokine stimulation. Dynamic compressive strain has been shown to produce a significant reduction in NO production by chondrocytes. Fluid flow-induced signals have been shown to cause activation of the MEK/ERK signaling pathway in articular chondocytes, leading to down-regulation of expression of the aggrecan gene. This effect does not require calcium mobilization.

Chondrocyte apoptosis occurs at peak compressive stresses as low as 4.5 MPa and is increased by raising the peak stress. Injurious compression appears to stimulate cell death at lower stresses than the stresses at which matrix degradation occurs, suggesting that cellular responses may be the initial event that leads to matrix degradation and changes in the mechanical properties of cartilage. Continuous cyclic compressive loading significantly reduces PG release from articular cartilage as much as 50% by decreasing the interstitial porosity through which PG can be removed.

9.8.6 Influence of Mechanical Forces on Bone-Forming Cells

Osteocytes are formed by the incorporation of osteoblasts into bone matrix. They remain in contact with other osteocytes via gap junction-coupled cell processes that form small channels through the matrix, the canaliculi. These small channels connect the cell body containing lacunae with each other and ECM external to the bone and play an important role in mechano-chemical transduction.

Osteocytes are by far the most abundant cell type in bone. Recent models suggest that a 3-D network of osteoblasts and bone-lining cells provides the cellular basis for mechanosensing in bone, and adaptive remodeling. Osteocytes appear to modulate signals arising from mechanical loading and as a result direct the appearance and disappearance of bone tissue, which allows bone to grow and to adapt to external and internal mechanical stresses. They possess receptors for parathyroid hormone/parathyroid hormone-related peptide and both α and β estrogen receptors.

Parathyroid hormone (PTH) is an 84-amino acid polypeptide hormone that mediates bone remodeling and is an essential regulator of calcium homeostasis. Prolonged exposure to PTH changes the phenotype of the osteoblast from a cell involved in bone formation to one directing bone resorption. Prolonged exposure to PTH leads to increased bone resorption; however, short-term use of PTH stimulates bone formation through activation of MAPKs, which is PKC-dependent in osteoblasts.

The role of osteocytes in mechanical adaptation, whereby bone maintains its strength with a minimum mass, has been postulated to involve transduction of mechanical loads into biochemical signals. Mechanical force applied to bone produces two localized mechanical signals on the cell: deformation of the ECM and extracellular fluid flow. Osteocytes and other bone-forming cells, including osteoblasts, regulate mechanochemical transduction via the flow of extracellular fluid through the lacunar–canalicular system. In fact, results of other studies suggest that the forces generated by fluid flow that occur as a result of mechanical loading result in increased osteopontin (OPN) expression in osteoblasts, an important protein in bone metabolism. In addition, nitric oxide and prostaglandin production associated with mechanical stretching may lead to nitric oxide release. Oscillatory flow induces Ca^{+2} mobilization via the L-type voltage-operated channel and the inositol 1,4,5-triphosphate pathway. Cyclic pressure-induced strain causes a rapid change in membrane potential of human bone cells because of opening of membrane channels; this response is mediated via integrins and requires tyrosine kinase activity and an intact cytoskeleton. Results of studies using 0.33 Hz pressure-induced strain indicates that a rapid, integrin-mediated release of IL-1β is potentially involved in bone remodeling.

Osteoblasts subjected to fluid shear increase expression of the early response gene, c-fos, and the inducible isoform of cyclooxygenase, COX-2, two proteins linked to anabolic response of bone to mechanical stimulation. Flow-induced responses in osteoblasts are mediated by inositol triphosphate intracellular calcium release. Flow-mediated stress is reported to induce both PGE2 and NO production. Fluid shear stress stimulates NO release by two distinct pathways: a G-protein and calcium-dependent phase sensitive to flow gradients, and a G-protein and calcium-independent pathway stimulated by sustained flow.

Only periosteal (surface) osteoblasts and not cells from the haversian systems are sensitive to strains within the physiological range. Strain-induced activation of phospholipase C leads to stimulation of cell division and increases in collagen and collagenase production. TGF-β, IGFI, IGFII, and PTH increase for collagen mRNA. Release of TGF-β from osteoblasts may be responsible for regulating bone metabolism as a result of mechanical stress.

The application of tensile stresses to osteoblasts appears to cause induction of the BMP-4 gene in preosteoblastic cells and adjacent spindle-shaped fibroblasts suggesting that BMP-4 may play a pivotal role by acting as an autocrine and a paracrine factor for recruiting osteoblasts in tensile stress-induced osteogenesis. Intermittent hydrostatic pressure (IHC) up-regulates osteopontin (OPN) mRNA expression in osteoblast cell cultures. In both proliferating and differentiating osteoblastlike cells, OPN expression and synthesis were enhanced by IHC suggesting that the loss of OPN expression by primary cells may be linked to mechanical disuse and down-regulation of the osteoblast phenotype. The presence of OPN is associated with reduction in bone formation by osteoblasts in unloaded mice tissues. Mechanical strain stimulated a redistribution of the $\alpha v \beta 5$ integrin to irregular plaquelike areas at the cell–ECM surface, increased the number and size of these plaques. and increased mineralization of the ECM that developed in these plaques. In a mouse tooth movement model, applied mechanical stress induced deposition of bone matrix primarily by stimulating differentiation of osteoblasts. Alkaline phosphatase is reported to be an early marker of mechanically induced differentiation of osteoblasts. Cyclic strain at physiological magnitude leads to an increase of osteoblast activities related to matrix production.

The structural integrity of microfilaments has been shown to be necessary for signal transduction within osteoblasts. Examination of the effects of mechanical strain on the expression of major structural elements of the cytoskeleton indicated that the amount of tubulin decreased by 75% and the amount of vinculin, a major component of focal adhesion complexes, increased by about 250%. Immunofluorescence microscopy demonstrated that mechanical strain led to increased formation and thickening of actin stress fibers, with dissociation of the microtubules and a clear increase in levels of vinculin at the peripheral edges of the cell. This suggests that mechanical strain leads to a coordinated change in both the cytoskeleton and in ECM proteins that facilitate tighter adhesion of osteoblasts and their ECM.

Mechanical strain, testosterone, and estrogen all stimulate proliferation of primary cultures of male rat long bone-derived osteoblastlike cells. Rat osteoblastlike cells from males or females exhibit strain-related proliferation mediated through the estrogen receptor in a manner that does not compete with estrogen but can be blocked with receptor modulators. Increases in parathyroid hormone and hypotonicity were observed to induce a rapid and progressive increase of cytosolic calcium primarily in osteocyte processes. In cyclically stretched osteocytes, parathyroid hormone also synergistically elevated IGF-1 mRNA levels, suggesting that PTH may influence volume-sensitive calcium-dependent pathways through elevation of IGF-1 levels.

Bone resorption in response to continuous mechanical deformation appears to be regulated by cells of osteoblast lineage such as preosteoblasts, osteoblasts, bone lining cells and osteocytes, and stretch-enhanced, osteoclastlike cell formation involves prostaglandins, but not PGE2 (Soma et al., 1996). Stretch-induced increases in osteopontin and osteonectin were inhibited by addition of the calcium channel antagonist nifedipine suggesting an important role for L-type calcium channels in early mechanical strain transduction pathways in osteoblasts.

9.9 Summary

From an engineering perspective, external tensile loads applied to tissues lead to bigger muscles, more skin, and stronger tendons and bones. The conversion of mechanical energy into chemical energy is an important transduction phenomenon that is key to adaptation of the musculoskeleton to a gravitational field. What this means is that mechanical energy applied to tissues is somehow converted into increased chemical energy in the form of new tissue. This conversion is inhibited during aging and in disease processes including the wasting diseases such as cancer. It is clear that the mechanisms by which cells in the ECM transduce external tensile mechanical signals into changes in cell behavior are complex. We are able to identify some of the molecules that are involved in mechanosensing and mechanochemical transduction. Tissue formation is not just controlled by the application of tensile forces on cells, but must be a result of perturbation of a balance between external tension and internal tension within the cell cytoskeleton.

As diagrammed in Figures 9.1 through 9.4, mechanical forces are likely to be transduced through conformational changes in ECM structural molecules such as collagen, because it is the primary structural protein in vertebrate tissues and bears external loads applied to tissues. Conformational changes in ECM macromolecules could putatively lead to alteration of integrin-mediated and nonintegrin-mediated changes in cell structure and behavior thus triggering a change in mechanosensing within the cell (Figures 9.1 and 9.2). However, other macromolecules besides collagen play important roles in transferring stress to the interfibrillar matrix and to cells through fluid shear stresses.

Integrin-mediated changes are likely affected through alterations in the structure of the cell cytoskeleton that lead to clustering of molecules found in focal adhesion complexes, G proteins, and attachment of proteins with docking functions that provide sites to which soluble secondary messengers can bind (Figure 9.4). Once bound, these secondary messengers can activate molecules involved in the phosphorelay pathways leading to changes in cell division, gene expression, and protein synthesis (Figures 9.3 and 9.4). Recent evidence suggests that these pathways are down-regulated during aging and in disease (Silver et al., 2003) suggesting that their up-regulation could have therapeutic value.

Nonintegrin-mediated activation of the phosphorelay and other secondary messenger pathways can occur by direct stretching of growth factor and hormone receptors, cell membranes, cell membrane ion channels, and stretch-sensitive ion channels (Figure 9.1). It is likely that these receptors are stretched passively in areas of tissue that not directly affected by external loading. These processes appear to be modified by the binding of growth factors and hormones to their cell surface receptor sites. Understanding the interrelationships between the molecules and pathways that are affected by mechanical loading is required to make advances in tissue engineering or in elucidating the pathogenesis of disease. It is likely that it will take much additional research to understand the complexity of the pathways by which external stimuli alter cellular behavior in a gravitational field considering the millions of years it has taken for these processes to evolve. Long-term habitation in a microgravitational environment may lead to evolutionary changes that down-regulate mechanosensing processes and literally cause "down-sizing" of the musculoskeletal system.

References

Aarden EM, Burger EH, Nijweide PJ. Function of osteocytes in bone. *J Cell Biochem*. 1994;55:287–299.

Alexander RM. Elastic energy stores in running vertebrates. *A Zool*. 1984;24:85–94.

Ascenci M-G. A first estimation of prestress in co-called circularly fibered osteonic lamellae. *J Biomech*. 1999;32:935–942.

Bakker AD, Soejima K, Klein-Nulend J, Burger EH. The production of nitric oxide and prostaglandin E(2) by primary bone cells is shear stress dependent. *Biomech*. 2001;34:671–677.

Birukov KG, Shirinsky VP, Stepanova OV, Tkachuk VA, Hahn AW, Resink TJ, Smirov VN. Stretch affects phenotype and proliferation of vascular smooth muscle cells. *Mol Cell Biochem*. 1995;144:131–139.

Boitano S, Sanderson MJ, Dirksen ER. A role for Ca+2 conducting ion channels in mechanically-induced signal transduction of airway epithelial cells. *J Cell Sci*. 1994;107:3037–3044.

Boudreau NJ, Jones PL. Extracellular matrix and integrin signaling: The shape of things to come. *Biochem*. 1999;J 339:481–488.

Breen EC. Mechanical strain increases type I collagen expression in pulmonary fibroblasts in vitro. *J Appl Physiol*. 2000;88:203–209.

Brown RA, Prajapati R, McGrowther DA, Yannas IV, Eastwood M. Tensional homeostasis in dermal fibroblasts: Mechanical responses to mechanical loading in three-dimensional substrates. *J Cell Physiol*. 1998;175:323–332.

Burger EH, Klein-Nulend J. Microgravity and bone cell mechanosensitivity. *Bone*. 1998;22:127S–130S.

Burger EH, Klein-Nulend J. Responses of bone cells to biomechanical forces *in vitro*. *Adv Dent Res*. 1999;13:93–98.

Burger EH, Klein-Nulend J. Mechanotransduction in bone—Role of the lacuno-canalicular network. *FASEB* 1999;13:S101–S112.

Burger EH, Klein-Nulend J, van der Plas A, Nijweide PJ. Function of osteocytes in bone—Their role in mechanotransduction. *J Nutr*. 1995;125:2020S–2023S.

Burger EH, Klein-Nulend J, Veldhuijzen JP. Mechanical stress and osteogenesis *in vitro*. *J Bone Miner Res*. 1992;7:S397–S401.

Chen K-D, Li Y-S, Kim M, Li S, Yuan S, Chien S, Shyy Y-J. Mechanotransduction in response to shear stress. *J Biol Chem*. 1999;274:18393–18400.

Chen NX, Ryder KD, Pavalko FM, Turner CH, Burr DB, Qiu J, Duncan RL. Ca^{2+} regulates fluid shear-induced cytoskeletal reorganization and gene expression in osteoblasts. *Am J Physiol Cell Physiol*. 2000;278:C989–C997.

Chess PR, Toia L, Finkelstein JN. Mechanical strain-induced proliferation and signaling in pulmonary ephthelial H441 cells. *Am J Physiol Lung Cell Mol Physiol.* 2000;279;L43–L151.

Chien S, Shyy JY. Effects of hemodynamic forces on gene expression and signal transduction in endothelial cells. *Biol Bull.* 1998;194:390–391.

Chiquet M, Matthison M, Koch M, Tannheimer M, Chiquet-Ehrismann R. Regulation of extracellular matrix synthesis by mechanical stress. *Biochem Cell Biol.* 1996;74:737–744.

Chiquet, M. Regulation of extracellular gene expression by mechanical stress. *Matrix Biol.* 1999;18:417–426.

Cobb MH, Goldsmith EJ. How MAP kinases are regulated. *J Biol Chem.* 1995; 270:14843–14846.

Collet P, Uebelhart D, Vico L, Moro L, Hartmann D, Roth M. Effects of 1- and 6-month spaceflight on bone mass area and biochemistry in two humans. *Bone.* 1997;20:547–551.

Damien E, Price JS, Lanyon LE. Mechanical strain stimulates osteoblast proliferation through the estrogen receptor in males as well as females. *J Bone Miner Res.* 2000;15:2169–2177.

D'Andrea P, Calabrese A, Capozzi I, Grandolfo M, Tonon R, Vittur F. Intercellular Ca^{+2} waves in mechanically stimulated articular chondrocytes. *Biorheol.* 2000; 37:75–83.

Dartsch PC, Hammerle H. Orientation response of arterial smooth muscle cells to mechanical stimulation. *Eur J Cell Biol.* 1986;41:339–346.

Davies PF. Flow mediated endothelial mechanotransduction. *Physiol Rev.* 1995; 75:519–560.

Davis MJ, Donovitz JA, Hood JD. Stretch-activated single-channel whole cell currents in vascular smooth muscle cells. *Am J Physiol.* 1992;262:C1083–C1088.

Diamond SL, Sachs F, Sigurdson WJ. Mechanically induced calcium mobilization in cultured endothelial cells is dependent on actin and phospholipase. *Arterioscler Thromb.* 2000;14:2000–2006.

Du W, Mills I, Sumpio BE. Cyclic strain causes heterogeneous induction of transcription factors. AP-1, CRE binding protein and NF-kB, in endothelial cells: Species and vascular bed diversity. *J Biomech.* 1995;28:1485–1491.

Durrant LA, Archer CW, Benjamin M, Ralphs JR. Organization of the chondrocyte cytoskeleton and its response to changing mechanical conditions in organ culture. *J Anat.* 1999;194:343–353.

Eastwood M, McGrouther DA, Brown RA. Fibroblast responses to mechanical forces. *Proc Inst Mech Eng.* 1998;212:85–92.

Eastwood M, Mudera VC, McGrouther DA, Brown RA. Effect of precise mechanical loading on fibroblast populated collagen lattices: Morphological changes. *Cell Motil Cytoskeleton.* 1998;40:13–21.

Edlich M, Yellowey CE, Jacobs CR, Donahue HJ. Oscillating fluid flow regulates cytosolic calcium concentration in bovine articular chondroctes. *J Biomech.* 2001;34:59–65.

Eisenberg SR, Grodzinsky AJ. Swelling of articular cartilage and other connective tissue: Electromechanical forces. *J Orthop Res.* 1985;3:148–159.

Frost HM. Perspectives: On increased fractures during the human adolescent growth spurt. Summary of a new vital-biomechanical explanation. *J Bone Miner Metab.* 1997;15:115–121.

Frost HM. Could some biomechanical effects of growth hormone help to explain its effects on bone formation and resorption? *Bone*. 1998;23:395–398.

Fry HJ. The interlocked stresses of articular cartilage. *Brit J Plastic Surg*. 1974; 27:363–364.

Fry HJ, Robertson WV. Interlocked stresses of articular cartilage. *Nature*. 1967; 215:53–54.

Fujisawa T, Hattori T, Takahashi K, Kuboki T, Yamashita A, Takigawa, M. Cyclic mechanical stress induces extracellular matrix degradation in cultured chondrocytes via gene expression of matrix metalloproteinases and interleukin-1. *J Biochem (Tokyo)*. 1999;125:966–975.

Fung, YC. *Biodynamics: Circulation*. New York: Springer-Verlag; 1984: Chapter 2.

Fung YC. *Biomechanics: Motion, Flow, Stress, and Growth*. New York: Springer-Verlag; 1990:388–393.

Fung YC, What are the residual stresses doing in our blood vessels? *Ann Biomed Eng*. 1991;19:237–249.

Geiger B, Bershadsky A. Exploring the neighborhood: Adhesion coupled cell mechanosensors. *Cell*. 2002;110:139–142.

Geng WD, Boskovic G, Fultz ME, Li C, Niles RM, Ohno S, Wright GI. Regulation of expression and activity of four PKC isozymes in confluent and mechanically stimulated UMR-108 osteoblastic cells. *J Cell Physiol*. 2001; 189:216–228.

Giancotti FG. Integrin signaling: Specificity and control of cell survival and cell cycle progression. *Curr Opin Cell Biol*. 1997;9:601–700.

Giancotti FG, Ruoslahti E. Integrin signaling. *Science*. 1999;285:1028–1032.

Gloe T, Sohn HY, Meininger GA, Pohl U. Shear stress-induced release of basic fibroblast growth factor from endothelial cells is mediated by matrix interaction via integrin $\alpha v \beta 3$. *J Biol Chem*. 2002;277:23453–23458.

Gniadecka M, Ginadecki R, Serup J, Sondergaard J. Skin mechanical properties present adaptation to man's upright position. *Acta Derm Venereol*. 1994;74: 188–190.

Goldschmidt ME, McLeod KJ, Taylor WR. Integrin-mediated mechanotransduction in vascular smooth muscle cells: Frequency and force response characteristics. *Cir Res*. 2001;88:674–680.

Gomar FE, Bernd A, Bereiter-Hahn J, Holzmann H. A new model of epidermal differentiation: induction by mechanical stimulation. *Arch Dermatol Res*. 1990;282:22–32.

Grandolfo M, Calabrese A, D'Andrea P. Mechanism of mechanically induced intercellular calcium waves in rabbit articular chondrocytes and in HIG-82 synovial cells. *J Bone Miner Res*. 1998;13:443–453.

Griffith TM. Endothelial control of vascular tone by nitric oxide and gap junctions: A haemodynamic perspective. 2002;39:307–318.

Grinnell F. Fibroblast-collagen-matrix contraction: Growth-factor signaling and mechanical loading. *Trends Cell Biol*. 2000;10:362–365.

Grinnell F, Ho C-H. Transforming growth factor β stimulates fibroblast-collagen matrix contraction by different mechanisms in mechanically loaded and unloaded matrices. *Exp Cell Res*. 2002;273:248–255.

Grinnell F, Zhu M, Carlson MA, Abrams JM. Release of mechanical tension triggers apoptosis of human fibroblasts in a model of regressing granulation tissue. *Exp Cell Res*. 1999;248:608–619.

Guignandon A, Usson Y, Laroche N, Lafrage-Proust MH, Sabido O, Alexandre C, Vico L. Effects of intermittent or continuous gravitational stresses on cell-matrix adhesion: Quantitative analysis of focal contacts in osteoblastic ROS 17/2.8 cells. *Exp Cell Res*. 1997;10:66–75.

Guilak F. Compression-induced changes in the shape and volume of the chondro-cyte nucleus. *J Biomech*. 1995;28:1529–1541.

Gumbiner BM. Cell adhesion: The molecular basis of tissue architecture and mor-phogenesis. *Cell*. 1996;84:345–357.

He Y, Grinnell F. Stress relaxation of fibroblasts activates a cyclic AMP signaling pathway. *J Cell Biol*. 1994;126:457–464.

Homolya L, Steinberg TH, Boucher, RC. Cell to cell communication in response to mechanical stress via bilateral release of ATP and UTP in polarized epithelia. *J Cell Biol*. 2000;150:1349–1360.

Honda K, Ohno S, Tanimoto K, Ijuin C, Tanaka N, Doi T, Kato Y, Tanne K. The effects of high magnitude cyclic tensile load on cartilage matrix metabolism in cultured chondrocytes. *Eur J Cell Biol*. 2000;79:601–609.

Hsieh Y-F, Silva MJ. *In vivo* fatigue loading of the rat ulna induces both bone for-mation and resorption and leads to time-related changes in bone mechanical properties and density. *J Orthop Res*. 2002;20:764–771.

Hung CH, Henshaw DR, Wang CC-B, Mauck RL, Raia F, Palmer G, Chao P-HG, Mow VC, Ratcliffe A, Valhmu WB. Mitogen-activated protein kinase signaling in bovine articular chondrocytes in response to fluid flow does not require calcium mobilization. *J Biomech*. 2001;33:73–80.

Hynes RO. Integrins: Versatility, modulation, and signaling in cell adhesion. *Cell*. 1992;69:11–25.

Iba T, Sumpio B. Morphological response of human endothelial cells subjected to cyclic strain in vitro. *Microvasc Res*. 1991;3:841–848.

Ikegame M, Ishibashi O, Yoshizawa T, Shimomura J, Komori T, Ozawa H, Kawashima H. Tensile stress induces bone morphogenetic protein 4 in pre-osteoblastic and fibroblastic cells, which later differentiate into osteoblasts leading to osteogenesis in the mouse calvariae in organ culture. *J Bone Miner Res*. 2001;16:24–32.

Ingber DE. Integrins as mechanochemical transducers. *Curr Opin Cell Biol*. 1991; 3:841–848.

Ingber D. How cells (might) sense microgravity. *FASEB J*. 1999;13: S3–S15.

Ishijima M, Rittling S, Yamashita T, Tsuji K, Kurosawa H, Nifuji A, Denhardt D, Noda M. Enhancement of osteoclastic bone resorption and suppression of osteoblastic formation in response to reduced mechanical stress do not occur in the absence of osteopontin. *J Exp Med*. 2001;193(3):399–404.

Iwamoto M, Shapiro IM, Yagami K, Boskey AL, Leboy PS, Adams SL, Pacifici M. Retinoic acid induces rapid mineralization and expression of mineralization-related genes in chondrocytes. *Exp Cell Res*. 1993;207:413–420.

Jalali S, del Pozo MA, Chen K-D, Maio H, Li Y-I, Schwartz MA, Shyy JY-J, Chien S. Integrin-mediated mechanotransduction requires its dynamic interaction with specific extracellular matrix (ECM) ligands. *Proc Natl Acad Sci USA*. 2001; 98:1042–1046.

Jalali S, Li Y-S, Soutoudeh M, Yuan S, Li S, Chien S, Shyy J. Shear stress activates p60src-ras-MAPK signaling pathways in vascular endothelial cells. *Arterioscl Thromb Vasc Biol*. 1998;18:227–234.

Johnson GL, Lapadat R. Mitogen-activated protein kinase pathways mediated by ERK, JNK, and p38 protein kinases. *Science.* 2002;298:1911–1912.

Joki N, Kaname S, Hirakata M, Hori Y, Yamaguchi T, Fujita, Katoh T, Kurokawa K. Tyrosine-kinase dependent TGF-β and extracellular matrix expression by mechanical stretch in vascular smooth muscle cells. *Hypertens Res.* 2000;23: 91–99.

Jones DB, Nolte H, Scholubbers JG, Turner E, Veltel D. Biochemical signal transduction of mechanical strain in osteoblasts-like cells. *Biomaterials.* 1991;12: 101–110.

Kanda K, Matsuda T, Oka T. Mechanical stress induced cellular orientation and phenotypic modulation of 3-D cultured smooth muscle cells. *Am Soc Artif Intern Organs J.* 1993;39:M686–M690.

Katz BZ, Yamada KM. Integrins in morphogenesis and signaling. *Biochimie.* 1997; 79:467–476.

Katz BZ, Zamir E, Bershadsky A, Kam Z, Yamada KM, Geiger B. Physical state of the extracellular matrix regulates the structure and molecular composition of cell-matrix adhesions. *Mol Biol Cell.* 2000;11:1047–1060.

Kessler D, Dethlefsen S, Haase I, Plomann M, Hirche F, Krieg T, Eckes B. Fibroblasts in mechanically stressed collagen lattices assume a "synthetic" phenotype. *J Biol Chem.* 2001;237:159–172.

Khachigian LM, Resnick N, Gimbrone MA Jr, Collins T. Nuclear factor-κB interacts functionally with the platelet-derived growth factor B-chain shear-stress response element in vascular endothelial cells exposed to fluid shear stress. *J Clin Invest.* 1995;96:1169–1175.

Kippenberger S, Bernd A, Loitsch S, Guschel M, Muller J, Bereiter-Hahn J, Kaufmann R. Signaling of mechanical stretch in human keratinocytes via MAP kinases. *J Invest Dermatol.* 2000;114:408–412.

Kirber MT, Walsh JV Jr, Singer JJ. Stretch-activated ion channels in smooth muscle: A mechanism for the initiation of stretch-induced contraction. *Pflugers Arch.* 1988;412:339–345.

Klein-Nulend J, Roclofsen J, Semeins CM, Bronckers AL, Burger EH. Mechanical stimulation of osteopontin mRNA expression and synthesis in bone cell cultures. *J Cell Physiol.* 1997;170:174–181.

Klein-Nulend J, Roelofsen J, Sterck JGH, Semeins CM, Burger EH. Mechanical loading stimulates the release of transforming growth factor-β activity by cultured mouse calvariae and periosteal cells. *J Cell Physiol.* 1995;163:115–119.

Kolodney MS, Wysolmerski RB. Isometric contraction by fibroblasts and endothelial cells in tissue culture: A quantitative study. *J Cell Biol.* 1992;117:73–82.

Komuro I, Katoh Y, Kaida T, Shibazaki Y, Kurabayashi M, Hoh E, Takaku F, Yazaki Y. Mechanical loading stimulates cell hypertrophy and specific gene expression in cultured rat cardiac myoctes. *J Biol Chem.* 1991;266:1265–1268.

Ko KS, Arora PD, McCulloch CA. Cadherins mediate intercellular mechanical signaling in fibroblasts by activation of stretch-sensitive calcium permeable channels. *J Biol Chem.* 2001;276:35967–35977.

Kononov S, Brewer K, Sakai H, Cavalcante FSA, Sabayanagam CR, Ingenito EP, Suki B. Role of mechanical forces and collagen failure in the development of elastase-induced injury. *Am J Respir Crit Care Med.* 2001:164:1920–1926.

Knight CG, Morton LF, Peachey AR, Tuckwell DS, Farndale RW, Barnes MJ. The collagen-binding A domains of integrins α1β1 and α2β1 recognize the same specific amino acid sequence, GFOGER, in native triple-helical collagens. *J Biol Chem.* 2000;275:35–40.

Knight MM, van de Breevarrt Bravenboer J, Lee DA, van Osch GJ, Weinans H, Bader DL. Cell and nucleus deformation in compressed chondrocyte-alginate constructs: Temporal changes and calculation of cell modulus. *Biochim Biophys Acta.* 2002;1570:1–8.

Laborador V, Chen KD, Li YS, Muller S, Stoltz JF, Chien S. Interactions of mechanotransduction pathways. *Biorheol.* 2003;40:47–52.

Landis WJ, Hodgens KJ, Block D, Toma CD, Gerstenfeld LC. Spaceflight effects on cultured embryonic chick bone cells. *J Bone Miner Res.* 2000;15:1099–1112.

Lee DA, Frean SP, Lees P, Bader DL. Dynamic mechanical compression influences nitric oxide production by articular chondrocytes seeded in agarose. *Biochem Biophys Res Commun.* 1998;251:580–585.

Lee HS, Millward-Sadler SJ, Wright MO, Nuki G, Salter DM. Integrin and mechanosensitive ion channel-dependent tyrosine phosphorylation of focal adhesion proteins and β-catenin in human articular chondrocytes after mechanical stimulation. *J Bone Miner Res.* 2000:15:1501–1509.

Lee RT, Briggs WH, Cheng GC, Rossiter HB, Libby P, Kupper T. Mechanical deformation promotes secretion of IL-1 alpha and IL-1 beta receptor antagonist. *J Immunol.* 1997;159:5084–5088.

Lehoux S, Esposito B, Merval R, Loufrani L, Tedgui A. Pulsatile stretch-induced extracellular signal-regulated kinase $\frac{1}{2}$ activation in organ culture of rabbit aorta involves reactive oxygen species. *Arterioscl Thromb Vasc Biol.* 2000;20: 2366–2372.

Lehoux S, Tronc F, Tedgui A. Mechanisms of blood flow-induced vascular enlargement. *Biorheol.* 2002;39:319–324.

Li C, Xu Q. Mechanical stress-induced signal transductions in vascular smooth muscle cells. *Cell Signal.* 2000;12:435–445.

Li Q, Muragaki Y, Hatamura I, Ueno H, Ooshima A. Stretch-induced collagen synthesis in cultured smooth muscle cells from rabbit aortic media and a possible involvement of angiotensin II and transforming growth factor-β. *J Vasc Res.* 1998;35:93–103.

Li S, Kim M, Hu YL, Jalai S, Schlaepfer DD, Hunter T, Chien S, Shyy JY. Fluid shear stress activation of focal adhesion kinase. Linking to mitogen-activated protein kinases. *J Biol Chem.* 1997;272:30455–30462.

Liu M, Tanswell K, Post M. Mechanical force-induced signal transduction in lung cells. *Am J Physiol Lung Cell Mol Physiol.* 1999;277:L677–L683.

Liu M, Xu J, Souza P, Tanswell B, Tanswell AK, Post M. The effect of mechanical strain on fetal rat lung cell proliferation: Comparison of two- and three-dimensional culture systems. *In Vitro Cell Dev Biol Anim.* 1995;31:858–866.

Liu S, Calderwood DA, Ginsberg MH. Integrin cytoplasmic domain-binding proteins. *J Cell Sci.* 2000;113:3563–3571.

Loening AM, James IE, Levenston ME, Badger AM, Frank EH, Kurz B, Nuttall ME, Hung H-H, Blake SM, Grodzinsky AJ, Lark MW. Injurious mechanical compression of bovine articular cartilage indices chondrocyte apoptosis. *Arch Biochem Biophys.* 2000;381:205–212.

Lorber M, Milobsky SA. A straining of the skin *in vivo*. A method of influencing cell division and migration in the rat epidermis. *J Invest Dermatol.* 1968;51: 395–402.

Lord JM, Bunc CM, Brown G. The role of protein phosphorylation in control of cell growth and differentiation. *Br J Cancer*. 1988;58:549–555.

Matsumoto T, Hayashi K. Mechanical and dimensional adaptation of rat aorta to hypertension. *J Biomech Eng*. 1994;116:278–283.

Matsumoto T, Hayashi K. Stress and strain distribution in hypertensive and normotensive rat aorta considering residual strain. *J Biomech Eng*. 1996;118: 62–73.

McAllister TN, Du T, Frangos JA. Fluid shear stress stimulates prostaglandin and nitric oxide release in bone marrow-derived preosteoclast-like cells. *Biochem Biophys Res Comm*. 2000;270:643–648.

McKnight NL, Frangos JA. Strain rate mechanotransduction in aligned human vascular smooth muscle cells. *Ann Biomed Eng*. 2003;31:239–249.

Meazzini MC, Toma CD, Schaffer JL, Gray ML, Gerstenfeld LC. (1998). Osteoblast cytoskeletal modulation in response to mechanical strain *in vitro*. *J Orthop Res*. 1998;16(2):170–180.

Miyauchi A, Notoya K, Mikuni-TakagakiY, Takagi Y, Goto M, Miki Y, Takano-Yamamoto T, Jinnai K, Takashashi K, Kumegawa M, Chihara K, Fujita T. Parathyroid hormone-activated volume-sensitive calcium influx pathways in mechanically loaded osteocytes. *J Biol Chem*. 2000;275(5):3335–3342.

Moore SW. Scrambled eggs: Mechanical forces as ecological factors in early development. *Evol Devel*. 2003;5:61–66.

Morris CE, Sigurdson WJ. Stretch-inactivated ion channels. *Science*. 1989; 243:807–809.

Mosler E, Folkhard W, Knorzer E, Nemetschek-Gansler H, Nemetschek T, Koch MH. Stress-induced molecular arrangement in tendon collagen. *J Mol Biol*. 1985; 182:589–596.

Mundlos S, Olsen BR. Heritable diseases of the skeleton. Part I. Molecular insights into skeletal development-transcription factors and signaling pathways. *FASEB J*. 1997;11:125–132.

Nishizuka Y. The family of protein kinase C for signal transduction. *JAMA*. 1989;262:1826–1833.

O'Callaghan CJ, Williams B. Mechanical strain-induced extracellular matrix production by human vascular smooth muscle cells: Role of TGF-β(1). *Hypertension*. 2000;36:319–324.

Owan I, Burr DB, Turner CH, Qiu J, Tu Y, Onyia JE, Duncan RL. Mechanotransduction in bone: Osteoblasts are more responsive to fluid forces than mechanical strain. *Am J Physiol*. 1997;273:C810–C815.

Pavlin D, Dove SB, Zadro R, Gluhak-Heinrich J. Mechanical loading stimulates differentiation of periodontal osteoblasts in a mouse osteoinduction model: Effect on type I collagen and alkaline phosphatase genes. *Calcif Tissue Int*. 2000; 67(2):163–172.

Putnam AJ, Schultz K, Mooney DJ. Control of microtubule assembly by extracellular matrix and externally applied strain. *Am J Physiol Cell Physiol*. 2001;280: C556–C564.

Rachev A, Hayashi K. Theoretical study of the effects of vascular smooth muscle contraction on strain and stress distributions in arteries. *Ann Biomed Eng*. 1999; 27:459–468.

Ragsdale GK, Phelps J, Luby-Phelps K. Viscoelastic response of fibroblasts to tension transmitted through adherens junctions. *Biophys J*. 1997;73:2798–2808.

Ray LB, Sturgill TW. Insulin-stimulated microtubule-associated protein kinase is phosphorylated on tyrosine and threonine in vivo. *Proc Natl Acad Sci USA*. 1988;85:3753–3757.

Resnick N, Collins T, Atkinson W, Bonthron DT, Dewey CF Jr, Gimbron MA. Platelet-derived growth factor B chain promoter contains cis-acting fluid shear-stress responsive element. *Proc Natl Acad Sci USA*. 1993;90:7908.

Reusch P, Wagdy H, Reusch R, Wilson E, Ives HE. Mechanical strain increases smooth muscle and decreases nonmuscle myosin expression in rat vascular smooth muscle cells. *Circ Res*. 1996;79:1046–1053.

Rittweger J, Gunga HC, Felensberg D, Kirsh KA. Muscle and bone-aging in space. *J Gravit Physiol*. 1999;6:P133–P136.

Riveline D, Zamir E, Balaban NO, Schwarz US, Ishizaki T, Narumiya S, Kam Z, Geiger B, Bershadsky AD. Focal contacts as mechanosensors: Externally applied local mechanical force induces growth of focal contacts by an mDia1-dependent and ROCK-independent mechanism. *J Cell Biol*. 2001;153:1175–1186.

Rolondncy MS, Wysolmerski RB. Isometric contraction by fibroblasts and endothelial cells in tissue culture. *J Cell Biol*. 1992;117:73–82.

Rosenfeldt H, Grinnell F. Fibroblast quiescence and the disruption of ERK signaling in mechanically unloaded collagen matrices. *J Biol Chem*. 2000;275:3088–3092.

Sachs F. Mechanical transduction in biological systems. *CRC Crit Rev Biomed Eng*. 1988;16:141–169.

Salter DM, Wallace WH, Robb JE, Caldwel H, Wright MO. Human bone cell hyperpolarization response to cyclical mechanical strain is mediated by an interleukin-1beta autocrine/paracrine loop. *J Bone Miner Res*. 2000;15:1746–1755.

Sanchez-Esteban J, Cicchiello LA, Wang Y, Tsai S-W, Williams LK, Torday JS, Rubin LP. Mechanical stretch promotes alveolar type II cell differentiation. *J Appl Physiol*. 2001;91:589–595.

Sasakai N, Odajima S. Elongation mechanism of collagen fibrils and force-strain relationships of tendon at each level of structural hierarchy. *J Biomech*. 1996; 9:1131–1136.

Sasaki N, Odajima S. Stress-strain curve and Young's modulus of a collagen molecule as determined by the x-ray diffraction technique. *J Biomech*. 1996a;29: 655–658.

Schwartz MA, Schaller MD, Ginsberg MH. Integrins: Emerging paradigms of signal transduction. *Ann Rev Cell Dev Biol*. 1995;11:549–599.

Seger R, Krebs EG. The MAPK signaling cascade. *FASEB J*. 1995;9:726–735.

Serup J. Gravitational or vertical ageing. In: Leveque JL, Agache PG, eds. *Aging Skin Properties and Functional Changes*. New York: Marcel Dekker; 1993;82–85.

Shadwick RE. Elastic energy storage in tendons: Mechanical differences related to function and age. *J Appl Physiol*. 1990;68:1033–1040.

Shay-Salit A, Shushy M, Wolfovitz E, Yahav H, Berviario F, Dejana E, Resnick N. VEGF receptor 2 and the adherens junction as a mechanical transducer in vascular endothelial cells. *Proc Natl Acad Sci*. 2002;99:9462–9467.

Sibonga JD, Zhang M, Evans GL, Westerlind KC, Cavolina J, Morey-Holton E, Turner RT. Effects of spaceflight and simulated weightlessness on longitudinal bone growth. *Bone*. 2000;27:535–540.

Silver FH, Christiansen DL. *Biomaterials Science and Biocompatibility*. New York: Springer; 1999;187–212.

Silver FH, DeVore D, Siperko LM. Role of mechanophysiology in aging of ECM: Effects of changes in mechanochemical transduction. *J Appl Physiol.* 2003;95: 2134–2141.

Silver FH, Freeman JW, Horvath I, Landis WJ. Molecular basis for elastic energy storage in mineralized tendon. *Biomacromol.* 2001;2:750–756.

Simon MI, Strathmann MP, Gautam N. Diversity of G proteins in signal transduction. *Science.* 1991;252:802–808.

Skinner SJM, Somervell CE, Olson DM. The effects of mechanical stretching on fetal rat lung cell prostacyclin production. *Prostaglandins.* 1992;43:413–433.

Smalt R, Mitchell FT, Howard RL, Chambers TJ. Induction of NO and prostaglandin E2 in osteoblasts by wall-shear stress but not mechanical strain. *Am J Physiol.* 1997;273:E751–E758.

Smith PG, Garcia R, Kogerman L. Strain reorganizes focal adhesions and cytoskeleton in cultured airway smooth muscle cells. *Exp Cell Res.* 1997;232:127–136.

Smith PG, Garcia R, Kogerman L. Mechanical strain increases protein tyrosine phosphorylation in airway smooth muscle cells. *Exp Cell Res.* 1998;239:353–360.

Smith PG, Janiga KE, Bruce MC. Strain increases airway smooth muscle proliferation. *Am J Respir Cell Mol Biol.* 1994;10:85–90.

Smith PG, Moreno R, Ikebe M. Strain increases airway smooth muscle contractile and cytoskeletal proteins in vitro. *Am J Physiol Lung Cell Mol Physiol.* 1997; 272(16):L20–L27.

Smith RL, Lin J, Trindale MC, Shida J, Kajiyama G, Vu T, Hoffman AR, van der Meulen MC, Goodman SB, Schurman DJ, Carter DR. Time-dependent effects of intermittent hydrostatic pressure on articular chondrocyte type II collagen and aggrecan mRNA expression. *J Rehabil Res Dev.* 2000;2:153–161.

Smith RL, Rusk SF, Ellison BE, Wessells P, Tsuchiya K, Carter DR, Caler WE, Sandell LJ, Schurman DJ. In vitro stimulation of articular chondrocyte mRNA and extracellular matrix synthesis by hydrostatic pressure. *J Orthop Res.* 1996; 14:53–60.

Soma S, Matsumoto S, Takano-Yamamoto T. Enhancement by conditioned medium of stretched calvarial bone cells of the osteoclast-like cell formation induced by parathyroid hormone in mouse bone marrow cultures. *Archs Oral Biol.* 1996;42(3):205–211.

Songu-Mize E, Liu X, Stones JE, Hymel LJ. Regulation of Na^+, K^+-ATPase alpha subunit expression by mechanical strain in aortic smooth muscle cells. *Hypertension.* 1996;27:827–832.

Stamenovic D, Wang N. Cellular responses to mechanical stress, invited review: Engineering approaches to cytoskeletal mechanics. *J Appl Physiol.* 2000;89: 2085–2090.

Stanford CM, Morcuende JA, Brand RA. Proliferative and phenotypic responses of bonelike cells to mechanical deformation. *J Orthop Res.* 1995;13:664–670.

Steinmeyer J, Ackermann B, Raiss RX. Intermittent cyclic loading of cartilage explants modulates fibronectin metabolism. *Osteoarth Cart.* 1997;5:331–341.

Sudhir K, Wilson E, Chatterjee K, Ives HE. Mechanical strain and collagen potentiate mitogenic activity of angiotensin II in rat vascular smooth muscle cells. *J Clin Invest.* 1993;92:3003–3030.

Swartz MA, Tschumperlin DJ, Kamm RD, Drazen M. Mechanical stress is communicated between different cell types to elicit matrix remodeling. *Proc Natl Acad Sci USA.* 2001;98:6180–6185.

Takahashi M, Berk BC. Mitogen activated protein kinase (ERK1/2) activation by shear stress and adhesion in endothelial cells. Essential role for a herbimycin-sensitive kinase. *J Clin Invest.* 1996;98:2623–2631.

Takei T, Han O, Ikeda M, Male P, Mills I, Sumpio BE. Cyclic strain stimulates isoform-specific PKC activation and translocation in cultured keratinocytes. *J Cell Biochem.* 1997a;67:327–337.

Takei T, Kito H, Du W, Mills I, Sumpio BE. Induction of interleukin-1 alpha and beta gene expression in human keratinocytes exposed to repetitive strain: Their role in strain-induced keratinocyte proliferation and morphological change. *J Cell Biochem.* 1998;69:95–103.

Takei T, Rivas-Gotz C, Delling CA, Koo JT, Mills I, McCarthy TL, Centrella M, Sumpio BE. Effect of strain on human keratinocytes in vitro. *J Cell Phys.* 1997:173:64–72.

Takeichi M. Cadherin cell adhesion receptors as a morphogenetic regulator. *Science.* 1991;251:1451–1455.

Takeichi M. Morphogenetic roles of classic cadherins. *Curr Opin Cell Biol.* 1995;7:619–627.

Tschumperlin DJ, Margulies SS. Equibiaxial deformation-induced injury of alveolar epithelial cells in vitro. *Am J Physiol Lung Cell Mol Physiol.* 1998; 275(19):L1173–L1183.

Vandenburgh H, Chromiak J, Shansky J, Del Tatto M, Lemaire J. Space travel directly induces skeletal muscle atrophy. *FASEB.* 1999;13:1031–1038.

Vico L, Lafage-Proust M-H, Alexandre C. Effects of gravitational changes on the bone system *in vitro* and *in vivo. Bone.* 1998;22:95S–100S.

Vlahakis NE, Schroeder MA, Limper AH, Hubmayr RD. Stretch induces cytokine release by alveolar epithelial cells in vitro. *Am J Physiol Lung Cell Mol Physiol.* 1999;277(21):L167–L173.

Walker LM, Publicover SJ, Preston MR, Said Ahmed MA, El Haj AJ. Calcium-channel activation and matrix protein upregulation in bone cells in response to mechanical strain. *J Cell Biochem.* 2000;79:648–661.

Wang M, Stamenovic D. Contribution of intermediate filaments to cell stiffness, stiffening and growth. *Am J Cell Physiol.* 2000;279:C188–C194.

Webb CM, Zaman G, Masley JR, Ticker RP, Lanyon LE, Mackie EJ. Expression of tenascin-C in bones responding to mechanical loads. *J Bone Miner Res.* 1997;12:52–58.

Westerlind KC, Turner RT. The skeletal effects of spaceflight in growing rats: Tissue specific alterations in mRNA levels for TGF-b. *J Bone Miner Res.* 1995; 10:843–848.

Whalen R. Musculoskeletal adaptation to mechanical forces on Earth and in space. *Physiologist.* 1993;36:S127–S130.

Wiersbitzky M, Mills I, Sumpio BE, Gewirtz H. Chronic cyclic strain reduces adenylate cyclase activity and stimulatory G protein subunit levels in coronary smooth muscle cells. *Exp Cell Res.* 1994;210:52–55.

Wilson E, Mai Q, Sudhir K, Weiss RH, Ives HE. Mechanical strain induces growth of vascular smooth muscle cells via autocrine action of PDGF. *J Cell Biol.* 1993;123:741–747.

Wilson E, Sidhir K, Ives HE. Mechanical strain of rat vascular smooth muscle cells is sensed by specific extracellular matrix/integrin interactions. *J Clin Invest.* 1995;96:2364–2372.

Wirtz HR, Dobbs LG. The effects of mechanical forces on lung functions. *Resp Physiol.* 2000;119:1–17.

Wong M, Siegrist M, Cao X. Cyclic compression of articular cartilage explants is associated with progressive consolidation and altered expression pattern extracellular matrix proteins. *Matrix Biol.* 1999;18:391–399.

Wozniak M, Fausto A, Carron CP, Meyer DM, Hruska KA. Mechanically strained cells of the osteoblast lineage organize their extracellular matrix through unique sites of alpha V beta 3-integrin expression. *J Bone Miner Res.* 2000;15:1731–1745.

Xia S-L, Ferrier J. Propagation of a calcium pulse between osteoblastic cells. *Biochem Biophys Res Commun.* 1992;186:1212–1219.

Xu Y, Gurusiddappa S, Rich RL, Owens RT, Keene DR, Mayne R, Hook A, Hook M. Multiple binding sites in collagen type I for the integrins $\alpha1\beta1$ and $\alpha2\beta1$. *J Biol Chem.* 2000;275:38981–38989.

Yamada KM, Even-Ram S. Integrin regulation of growth factor receptors. *Nat Cell Biol.* 2002;4:E75–E76.

Yamada KM, Geiger B. Molecular interactions in cell adhesion complexes. *Curr Opin Cell Biol.* 1997;9:76–85.

You J, Reilly GC, Zhen X, Yellowley CE, Chen Q, Donahue HJ, Jacobs CR. Osteopontin gene regulation by oscillatory fluid flow via intracellular calcium mobilization and activation of mitogen-activated protein kinase in MC3T3-E1 osteoblasts. *J Biol Chem.* 2001;276(16):13365–13671.

10
Mechanochemical Transduction and Its Role in Biology

10.1 Introduction

Mechanical loading and the conversion of mechanical energy into chemical energy in cells influences development, growth, and maturation by regulating the host's metabolism based on the balance between external and internal mechanical loading. External loading derives from the forces applied at the interface between the host and the environment to do work against gravity. Running involves the application of a force against the ground to create a displacement and is opposed by the force of gravity and the friction that results when the foot pushes against the ground. The external force of the ground against the foot in turn results in forces applied to a variety of ECMs including the skin, cartilages, tendons, and bones of the foot and leg. This causes the cells to be placed in tension due to the friction and muscular stretching of the tendons and compression due to the weight of the impact with the ground. The effect of this tension and compression applied to the foot is to increase the net tensional forces experienced by these tissues because the effect of compressing a stretched membrane is to cause the tension to increase. An example of this is a trampoline, which experiences internal tension due to stretching of the elastomeric sheet, which is attached to a rigid frame. The trampoline exhibits increased tension when it is loaded in compression by the weight of a person jumping up and down on it.

Increased loading causes the cells in ECMs to contract and generate their own tension to offset the increase in internal tension. In addition, they store some of this external energy in the form of chemical energy by synthesizing new tissue. We know this because as we walk more we get bigger muscles and thicker skin on our feet. This process of converting external mechanical energy into chemical energy in the form of new tissue defines the biological mechanochemical transduction process.

We know by deduction that mechanochemical transduction processes affect growth and development and their regulation is changed during maturation and aging. Teenagers are known to grow several inches during the summer months and this is likely to be associated with hormonal effects and increased activity. We know that when they reach an age of about 22 much of this rapid increase in height and weight falls off. This means there must be some change in the mechanochemical transduction regulatory processes that occurs as development ends and maturation begins. Beyond this, as we age, the ability of bodies to maintain musculature falls off, suggesting that some of these processes must be down-regulated possibly through the interactions between growth hormones and mechanical forces.

In this chapter we also suggest how mechanochemical transduction processes influence aging, implant design, and the pathobiology of implant–host interactions, connective tissue disorders, the design of tissue-engineered implants, and wound healing. Although molecular biologists hale the advent of genetic manipulation as the cure for most diseases, we still need to understand how mechanical forces affect the same pathways that are controlled by regulation of gene expression. Below we consider how mechanochemical transduction affects these processes.

10.2 Relationship Between Mechanotransduction and Aging

We have seen in Chapter 9 that mechanical stretching modifies mechanochemical transduction by up-regulating either integrin-dependent or integrin-independent processes that activate phosphorelay pathways including MAPK (phosphorelay pathways). Beyond this, reduction in mechanical loading down-regulates the same pathways and can lead to cellular apoptosis. In some tissues, such as skin, components of the phosphorelay pathways are decreased with increasing age suggesting that mechanochemical transduction may be down-regulated or impaired. Elastin-derived peptides have been implicated in activation of some of the members of the MAPK family; a progressive age-dependent uncoupling of the elastin receptor appears to impair mechanochemical transduction. These changes are accompanied by decreases in hormones and growth factors that also occur with increased age.

Hormones and growth factors and their receptors play an important role in mechanochemical transduction by altering the sensitivity of effector cells to mechanical loads as pointed out in Chapter 9. Reduced levels of mechano-growth hormone and insulinlike growth factor I (IGF-I) have been reported to occur during aging and to lead to reduced osteogenic responses to mechanical loads. IGF-I secretion and production by cultured rat chondrocytes have been reported to decrease with increasing age. Chondrocytes show a continuous age-dependent decline in their proliferative response in response to growth factors; young cells respond better to growth factors than older cells.

Mechanical strain, testosterone, and estrogen all stimulate proliferation of rat long-bone-derived osteoblastlike cells. Rat osteoblastlike cells from males and females exhibit strain-related proliferation mediated through the estrogen receptor in a manner that does not compete with estrogen. Decreases in testosterone and estrogen associated with aging may down-regulate or reduce the efficiency of mechanochemical transduction processes and lead to changes in tissue metabolism.

This information suggests that changes in activation and the efficiency of activation of mechanochemical transduction processes may negatively influence the balance between tissue formation and tissue destruction. If this occurs due to genetic errors or disease processes, then chemical energy is transformed into mechanical energy through catabolic processes. This may occur as a result of genetic defects, aging, abnormal wound healing, or the presence of an implant.

10.3 Design and Use of Medical Implants Including Engineered Tissues

Although the use of implants has been documented since the 1950s, little attention has been directed to how mechanical forces at the implant interface affect tissue metabolism and the fate of the implant. Some interest has been expressed concerning the use of balloons for tissue expansion, pressure-induced necrosis (cell death) of skin and fat cells, and bone resorption as a result of stress shielding by hip and knee implants. These effects may need closer examination in light of the recent findings that most tissues are normally stretched in tension and that any interruption of this tension adversely affects homeostasis via perturbations in normal mechanochemical transduction and could lead to implant failure.

It is well known from the results of experiments using balloon expanders placed below the surface of the skin that balloon inflation is associated with expansion of the epidermis (the epidermis gets thicker), which is under tension, and destruction of the dermis (the dermis gets thinner). Thus design of medical devices that interfere with the normal stresses in the body will affect the tissue metabolism. Vascular prostheses that are tethered to the surrounding vessel ends using tension lead to proliferation of the intimal cells and thickening of the vessel wall at these sites. This process is termed intimal hyperplasia and is also found with vascular stents that are expanded against the vascular wall to reverse narrowing (stenosis). Pressure due to stuffing an implant into a pocket above the fat in the breast, leads to fat necrosis, chronic inflammation, and tissue destruction. Use of pressure to close a femoral artery puncture site leads to periarterial fat necrosis and to deposition of scar tissue at the arterial puncture site. The use of metallic implants in the hip and knee leads to stress shielding and bone resorption that leads to the release of catabolic enzymes and interleukin-1 that contributes to implant loosening.

All of these processes involve an alteration of the mechanical forces present at the tissue interface leading to tissue resorption or deposition. For these reasons the consideration of mechanical forces at the implant–tissue interface is an important aspect of implant design. Design of tissue-engineered products therefore necessitates the consideration of mechanical loading requirements so that the implanted tissue isn't immediately resorbed and replaced with functional scar tissue. Another problem involves the growth of tissue-engineered implants and their implantation under conditions which maintain constant loading that simulates the normal loading found in the tissues to be replaced. The normal loading conditions in tissues likely depend on the age and location in the host.

10.4 Relationship Between Mechanochemical Transduction and Connective Tissue Diseases

The shape of our bodies is to a large extent a consequence of the evolution of our species in a gravitational field. Therefore our genetic programming is controlled in part by the need to produce a species that can efficiently locomote in a gravitational field. Without this need we would be able to hunt and replicate in a form like a jellyfish. But this is not possible for land dwellers because it would take us forever to hunt and find water if we could not run. Beyond this any genetic defect that impairs mechanochemical transduction would negatively influence our survival as a species.

Diseases such as osteoarthritis, Ehlers Danlos Syndrome, hardening of the arteries, mineralization of heart valves, and even metastatic cancer all appear to be influenced by changes in mechanochemical transduction. For instance, when we look at a picture of a patient with a form of Ehlers Danlos Syndrome (see Figure 10.1) we begin to realize that the skin of this patient is not only hyperextensible but the patient also has more skin on his face because of all the folds. What this means is that the skin cells of this patient must continue to synthesize skin proteins even after skeletal growth has stopped. Under the influence of gravity and other external forces applied to his facial skin, the skin cells appear to continue to synthesize skin components and make new skin. This suggests that underlying this disorder is a genetic defect that allows the skin to continue to replicate as a result of continued mechanical loading beyond the point where maturation normally limits skin growth. Even in "normal" patients the formation of wrinkles may reflect the ability of our bodies to continue to make new skin under the influence of the force of gravity, which continually acts to pull our skin down. The question arises as to whether our tissues sag with age due to disuse or genetic changes alone or do we sag because our skin cells continually try to make new skin proteins due to the pull of gravity?

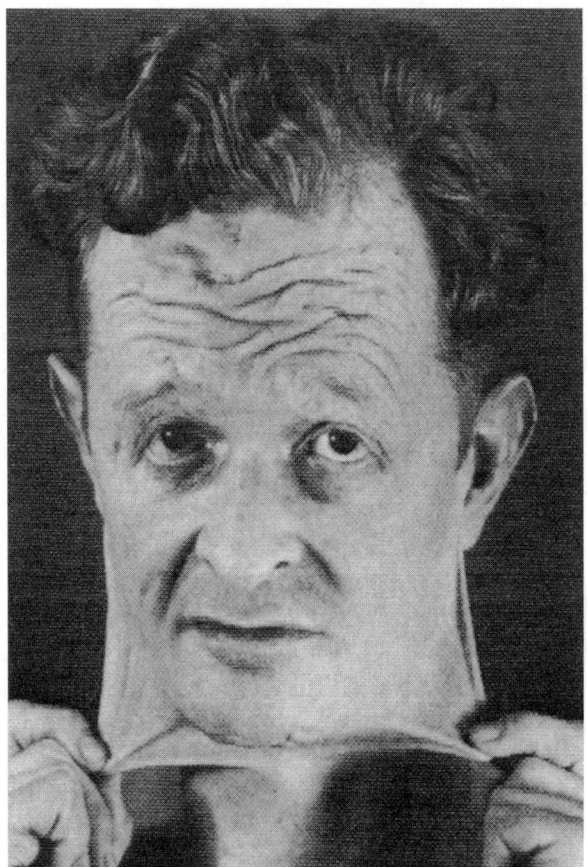

FIGURE 10.1. Dermal extensibility is a prominent feature of EEDS. On release the skin springs back to take up its former position.

In the case of osteoarthritis, it appears that abnormal mechanochemical transduction processes may be involved in triggering cartilage destruction. This may occur as a result of defects in the repair process at the cartilage surface. When defects arise that are not repaired, and tension is lost in the superficial zone, the ensuing catabolic destruction of the cartilage leads to tissue loss. Normally the cells in the superficial zone repair small defects to the cartilage surface. However, as we age, the superficial zone is gradually lost and cartilage destruction ensues. Because the superficial zone of young individuals is under tension, the loss of this tension through injury or aging may limit mechanochemical transduction and repair. This equivalent to tension-induced rapid healing in skin wounds discussed in Section 10.5 below. As shown in Figure 10.2 the mechanical properties of osteoarthritic

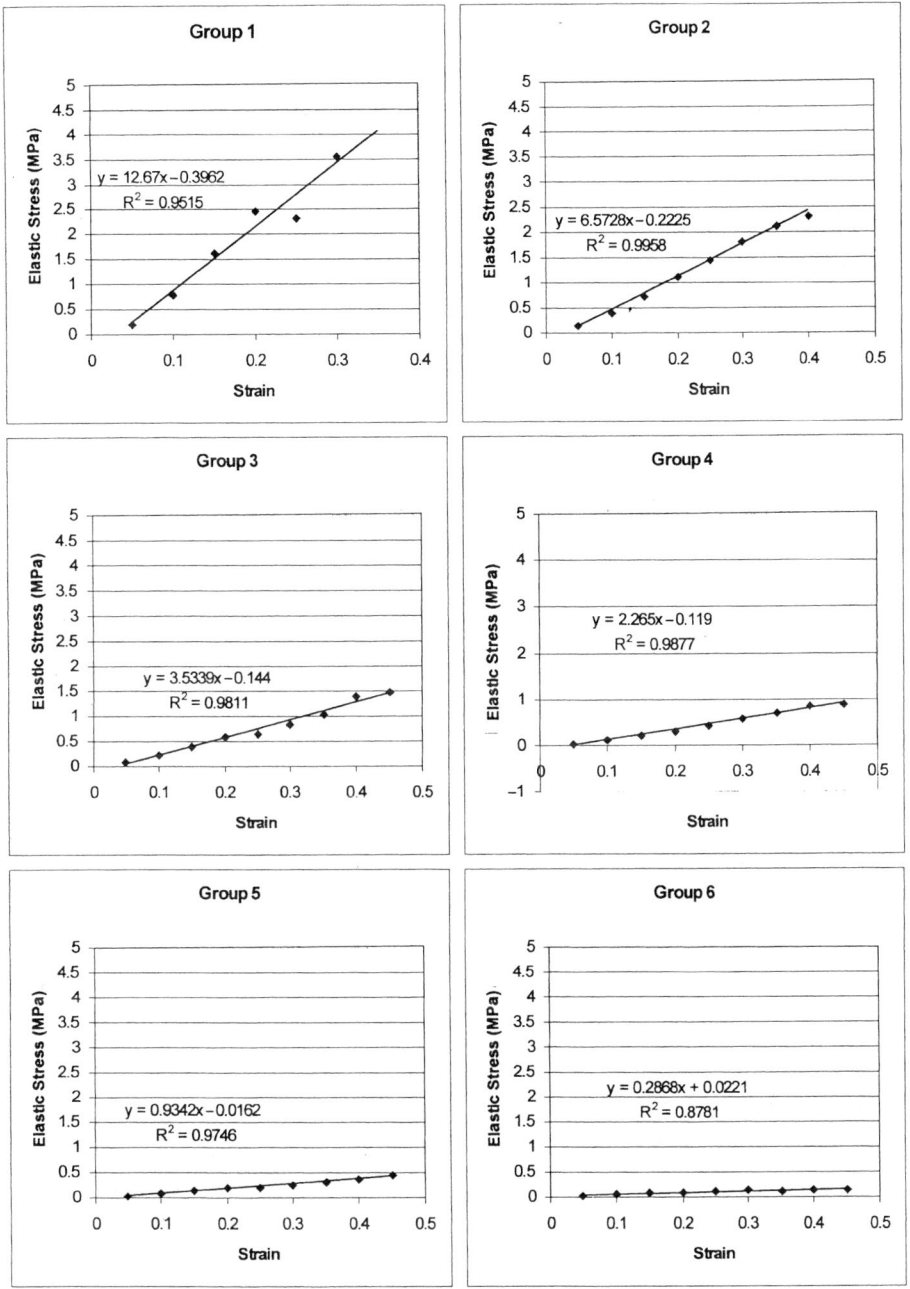

FIGURE 10.2. Mean elastic and viscous stress–strain curves for cartilage. Plot of elastic (**A**) and viscous (**B**) stress–strain curves for cartilage as a function of visual grade. The visual grade used was: 1, shiny and smooth; 2, slightly fibrillated; 3, mildly fibrillated; 4, fibrillated; 5, very fibrillated; and 6, fissured. The equation for the linear approximation for the stress-strain curve for each group is given, as well as the correlation coefficient. Note the decreased slope with increased severity of osteoarthritis. This data is consistent with down–regulation of mechanochemical transduction and tissue catabolism.

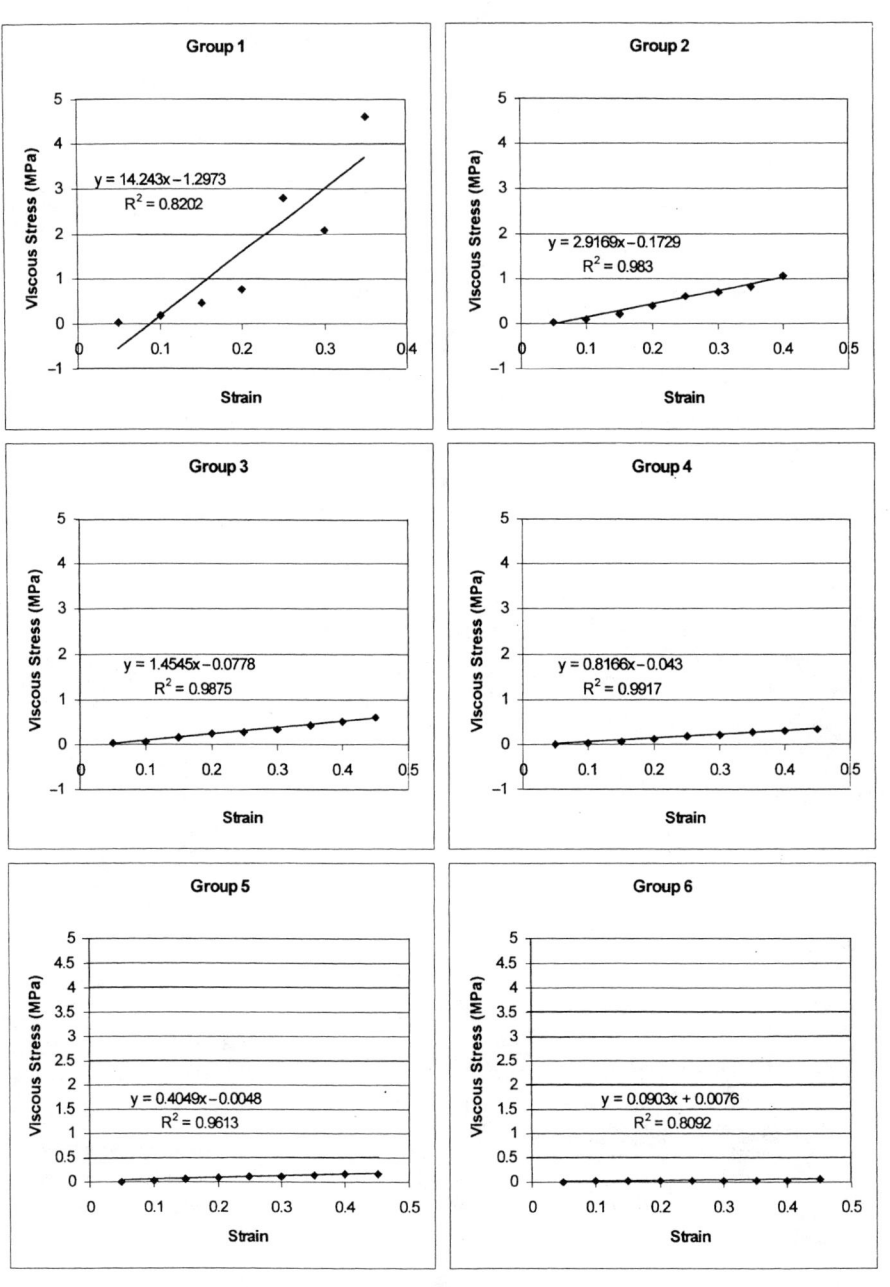

B

FIGURE 10.2. *Continued*

cartilage continue to decay as the disease progresses. This is consistent with the catabolic destruction of tissue as described in the biochemical literature describing the progression of osteoarthritis. However, it is not consistent with the disease being driven by inflammatory processes alone, because chondrocytes are seen proliferating near the tears that arise in the joint cartilage under destruction, but these chondrocytes do not appear to be able to make enough new collagen to support the compressive loads. These observations suggest that osteoarthritis involves changes in both inflammation and mechanotransduction. These types of changes also appear to occur in the circulatory system.

Pulsatile blood flow occurs in the arteries of the body as a result of the discontinuous pumping action of the heart. Although this sets up a time-dependent pressure throughout the arterial side of the vasculature, it also places a cyclic load on the endothelial and smooth muscle cells in the vessel wall. As we age, changes occur in mechanochemical transduction processes that lead to increased wall thickness and vessel diameter. This suggests that tissue deposition due to the constant mechanical cycling up-regulates mechanochemical transduction leading to more collagen in the vessel wall. Excessive mechanical tensile cycling leads to vessel wall and heart valve calcification probably due to the changes in collagen fibrillar structure brought about during mechanical loading. Abnormal changes at the cellular level may also be related to changes in cellular loading.

Because the signaling molecules involved in cell regulation are the same as those involved in mechanochemical transduction processes, it is difficult not to speculate that cell metastasis is in part due to loss of control of this process. What makes cells that normally do not break down the layer between the epidermis and dermis catabolize it and enter the bloodstream? These cells normally are under tension and therefore make skin components as opposed to catabolizing them. One can only speculate that these cells under genetic and environmental stresses lose their ability to recognize tensional barriers between tissues and under these conditions automatically revert to a catabolic state. If the maintenance of tissue structure requires continued tensional-based energy production through homeostatic mechanochemical transduction processes then loss of the controls for transduction processes is probably associated with cellular dedifferentiation and release of catabolic enzymes similar to those found in necrotic tissues. This would allow viable cells to migrate through the tissues into the blood vessels.

10.5 Wound Healing and Scarring

Perhaps the best example of how mechanical forces influence cell behavior is the story that has derived from the study of skin tension. Langer in the nineteenth century recognized that the direction of the principal tension in skin affected how skin wounds healed when sutured close. When Langer sutured in a manner that maintained the lines of tension in skin then the wounds healed with minimum scarring. However, when skin wounds were sutured closed across the lines of tension (Langer's lines) then skin scarring occurred. At that time it was known that connecting the edges of the wound with sutures so that the skin tension was preserved could minimize skin scarring. Today all plastic surgeons are taught to apply sutures across the edges of the skin to close wounds along the lines of maximum tension. Based on this analysis we can ask: why is it important to preserve the tension across the edges of a wound? Any medical student can tell you that skin wounds heal much more rapidly when they are closed by primary intension (the edges of the wound are pulled together and then sutured closed). This tells us that the presence of tension at the edges of a wound improves the tissue's ability to make new connective tissue components without causing the elaboration of catabolic enzymes. This means that mechanochemical transduction must play a role in wound healing and that application of tension to wound edges is required for homeostatic maintenance of mechanochemical transduction processes. We need to learn more about the relationship between mechanochemical transduction and optimization of wound healing so that we can "dial-in" the best tension to promote the most efficient healing.

10.6 Summary

The purpose of this text is to begin to educate the reader about how mechanical forces influence our lives. Although it is difficult to accept that much of our evolution as a species is dependent on living in a gravitational field, the sci-fi writers may have been correct in describing beings evolving on other planets as big bags of water. Our shape and form clearly derives from our need to locomote in a gravitational field. Because of the similarity between the signaling molecules and pathways that are used to respond to a variety of external stimuli we are stuck being mechanochemical machines. There has been recent interest in the processes that can be labeled mechanobiology, however, we need to expand this field until the interfaces with molecular and cell biology are continuous. Until the biologist and the physical scientist speak the same language we will continue to struggle with trial-and-error solutions to healthcare problems.

Index